FOOD SCIENCE AND TECHNOLOGY

OCHRATOXIN A

BIOSYNTHESIS, DETECTION AND BIOLOGICAL TOXICITY

FOOD SCIENCE AND TECHNOLOGY

Additional books in this series can be found on Nova's website under the Series tab.

Additional e-books in this series can be found on Nova's website under the eBooks tab.

FOOD SCIENCE AND TECHNOLOGY

OCHRATOXIN A

BIOSYNTHESIS, DETECTION AND BIOLOGICAL TOXICITY

WENTAO XU

nova publishers
New York

Library of Congress Cataloging-in-Publication Data

ISBN: 978-1-63484-450-5

Library of Congress Control Number: 2016933390

Published by Nova Science Publishers, Inc. † New York

CONTENTS

PREFACE

Ochratoxin is a kind of secondary metabolite produced by *Aspergillus and Penicillium*. It mainly contains Ochartoxin A (OTA), Ochratoxin B (OTB), Ochratoxin C (OTC), 4-hydroxyochratoxin A and Otα, and among which OTA has the strongest toxicity. Ochratoxin exists widely in a variety of food products, such as wheat, corn, soybean, oats, barley, coffee beans, meat and cheese. Ochratoxin carries potential health-associated risks, and it has caused wide public concern over the recent years. In 1993, OTA has been classified as a possible human carcinogen (group 2B) by the International Agency for Research on Cancer. OTA can induce various toxic effects, and its target organs are the kidney and liver. OTA has close relationship with many diseases. Therefore, a comprehensive introduction about OTA is extremely urgent. This book has been written with the aim of helping investigators in this field by providing information about the production, isolation, purification and the toxic mechanism of OTA.

This book consists of six chapters. Section I mainly introduces the discovery and the classification of ochratoxin. Section II describes the biosynthesis and metabolism of OTA and the ways of OTA control. Section III deals with some detection methods of OTA.

In the last three sections, toxicities of OTA are the emphasis of this book. The kidney and the liver are the main target organs of OTA. Nephrotoxicity, hepatotoxicity and other toxicities, such as intestinal metabolism toxicity and reproductive toxicity, are stated in section IV, V and VI, respectively. In these sections, methods of both traditional toxicology and molecular toxicology evaluation are applied to explore OTA-induced toxicity. Traditional toxicology describes the changes of OTA-induced toxicology indexes, containing blood biochemistry indexes and pathological parameters, and the molecular toxicology mainly focuses on the signaling pathways of its toxicity. A combinational approach encompassing miRNA, mRNA, and proteomic profiling and bioinformatics methods have been used to study the molecular mechanisms. The molecular toxicology is an important supplement and improvement of the traditional toxicology.

The book is fulfilled in the joint efforts of all contributors. I am very grateful for their suggestions and helps in preparing the manuscript. My special thanks go out to Beijing Advanced Innovation Center for Food Nutrition and Human Health, because some work in it could not be accomplished without their supports.

W. T. Xu

January 2016

Chapter 1

DISCOVERY AND DISTRIBUTION OF OCHRATOXIN A

ABSTRACT

Ochratoxin A (OTA) is a widespread mycotoxin that is produced mainly by the mold fungi *Aspergillus ochraceus* and *Penicillum verrucosum*. It is a pentaketide derived from a dihydrocoumarin family derivative coupled to β-phenylalanine. Before 1965, *Asp. ochraceus* was used as a food additive for its ability to induce a desirable change in flavor during the fermentation of coffee. However, in 1965, Vanderme et al. reported that this fungi can produce a toxic metabolite, ochratoxin A, and animal experiments showed that this mycotoxin induced liver damage in ducklings. In the following 50 years, ochratoxin A has attracted considerable attention for its toxicity. Risk assessments have been conducted and have shown that ochratoxin A is a nephrotoxic, hepatotoxic, teratogenic, immunotoxic, neurotoxic, genotoxic and carcinogenic toxin, and it has been classified by the IARC (International Agency for Research on Cancer) as a possible carcinogen to humans (Group 2B); however, evidence concerning humans was either inconclusive or limited. With the development of improved detection methods, this mycotoxin was gradually identified in many areas worldwide, particularly Europe. It was also widely distributed in many types of food materials and products. Detoxification methods to remove ochratoxin A from food based on its physical and chemical properties need to be explored.

In this chapter, we will present the basic information of ochratoxin A. We will start in 1965, when ochratoxin A as a toxic metabolite produced by *Aspergillus ochraceus* was discovered for the first time. Next, we will review the distribution of ochratoxin A in different food products and areas, including the contamination of ochratoxin A in plant products such as cereals, coffee, beer and wine, as well as in certain animal products. The distribution of ochratoxin A in different areas of the world, especially the Balkans, where ochratoxin A is suspected of being the main etiological agent responsible for human Balkan endemic nephropathy (BEN), is reviewed in this chapter. We also review other investigations, indicating that OTA is not only a European problem but an international one as well. Then, the risk assessment of ochratoxin A will be introduced, and we will provide the evidence supporting the OTA classification as a possible carcinogen to humans (Group 2B). As ochratoxin A is nephrotoxic, immunosuppressive, and teratogenic in many species and is a possible human carcinogen, detoxification methods are urgently needed to removal ochratoxin A from food; therefore, the physical and chemical properties of ochratoxin A, which can provide the theoretical foundation for detoxification of this mycotoxin, are also summarized in this chapter.

Keywords: Ochratoxin A, distribution, risk assessment, physicalco-chemical property

1. Occurrence of Ochratoxin A

1.1. Definition

Ochratoxin A (OTA) is a mycotoxin; it is a secondary metabolic product produced by many filamentous species belonging to the genera *Aspergillus* and *Penicillium* [1-5]. Its chemical structure is a pentaketide derived from a dihydrocoumarin derivative linked to β-phenylalanine. The chemical structure is presented in Figure 1 [6], and the full chemical name is: L-phenylalanine-N-[(5-chloro-3,4-dihydro-8-hydroxy-3-methyl-1-oxo-1H-2-benzopyrane-7 yl)carbonyl]-(R)-isocoumarin.

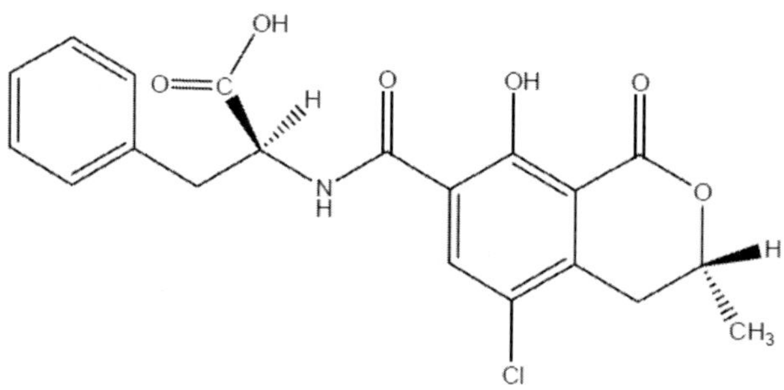

Figure 1. Chemical structure of ochratoxin A.

1.2. The Discovery of OTA

Since the discovery of aflatoxins, many toxicogenic fungi, as well as many toxic fungal metabolites, have attracted considerable attention in recent years. Among the latter, ochratoxins, a type of toxic fungal secondary metabolites produced by several species of *Aspergillus* and *Penicillium*, were discovered.

Of the many types of ochratoxins, OTA is the most famous. For a full understanding of OTA, we will discuss the fungi that produce ochratoxins. In 1965, an article reported that *Aspergillus ochraceus Wilh.* [7], which is widely distributed in the environment and is often found in soil and on decaying vegetation, can produce a toxic metabolite—ochratoxin A. The fungi are known to invade stored moist wheat. However, in some areas of the world, the fungi were not regarded as harmful microorganisms; on the contrary, they were added to food to improve the quality. For example, in Japan, *Aspergillus ochraceus* and allied species had been reported as constituents of the mycoflora characteristic of "katsuobushi" and other fermented fish preparations. In the United States, Patent No.1313209 covers the use of *Asp. ochraceus* for its ability to induce a desirable change in flavor during the fermentation of coffee.

The *Nature* article also reported the perniciousness of this mycotoxin. The authors used five strains of local origin grown in pure culture on sterile, moistened maize meal, and the moldy meals were tested for toxicity in feeding trials with day-old Peking ducklings and weaned white mice and rats. Three of the five strains caused acute death in all test animals.

Other agricultural products, including wheat, rye, sorghum, rice, buckwheat, soya beans and groundnuts, were also rendered toxic by inoculation with these strains. Then, one strain, designated K-804, was isolated from sorghum grain and was grown in bulk on sterilized maize meal, and the toxic characteristics were quantitatively extracted with methanol-chloroform. Fractionation of the toxic extract involved extraction with aqueous bicarbonate, chromatography of the acids on neutral silica and a final separation by ion-exchange chromatography using Dowex 1. Toxicity tests of all recovered fractions showed that the toxicity of the moldy meal was due to a new metabolite, called ochratoxin A, of which 500 mg was isolated from 20 kg moldy meal. Then, an acute toxicity test with the isolated metabolite was performed using ducklings as an animal model. Groups of eight ducklings were fed various doses in aqueous 0.1 M bicarbonate, with each bird receiving 0.5 mL by tube into the gizzard. The LD_{50} was ±25 μg, recorded six days after administration and calculated for a body weight of 50 g. The toxicity of ochratoxin A is thus of the same order as that of aflatoxin B_1 [8]. Thus, ochratoxin A as a toxic metabolite produced by *Aspergillus ochraceus* was identified.

2. Distribution of Ochratoxin A

OTA is widely distributed in food products around the world. We will describe the distribution of OTA in food and feed and the content of OTA in various food products worldwide.

2.1. The Distribution of OTA in Different Areas of the World

OTA is widely distributed throughout the world. One much-discussed case is the Balkans and surrounding areas, where a type of chronic renal disease with unknown etiology, Balkan endemic nephropathy (BEN), is found in geographically concentrated regions of Bulgaria, Romania, Serbia, Croatia, Bosnia and Herzegovina, Slovenia, and the former Yugoslav Republic of Macedonia. To clarify the etiology of the disease, a more than 20 years epidemiological investigation was carried out to search for the causative agent of this disease. The researchers searched for heavy metals, trace elements, infectious agents (bacteria, viruses) and genetic factors. Then, a preliminary survey of home-produced foodstuffs was carried out in an area of Yugoslavia, comparing OTA contamination in different areas, and this study revealed that OTA contamination is more frequent in the areas where BEN is prevalent (endemic areas) than in the areas where this disease is absent. These results indicated that OTA is related to BEN. Further evidence provided support for the hypothesis that OTA is the main determinant of BEN [9], and many reports found that OTA can be tested in human blood. In one study, blood samples were collected in 1984, 1986, 1989 and 1990 in the Balkan Peninsula. The results showed that OTA was found more frequently and at higher levels in blood from BEN patients and/or urinary tract tumors than in the blood of unaffected people from endemic and control areas. These findings also suggest that OTA is involved in the etiology of Balkan endemic nephropathy and the associated urinary tract tumors [10]. The OTA blood levels in healthy humans had a range of 1.0–10.0 ng/mL and were similar to the results obtained in a later investigation by Petkova-Bocharova et al. (2006) [11]. In the Vratza District, a high risk area for BEN in Bulgaria, the average OTA levels in the serum of volunteers from the two villages located in Vratza District were 1.59 ng/mL (standard deviation: sd = 0.44). The medians were 0.4 ± 0.22 (sd = 0.22) and 0.7 (sd = 0.15) in the two villages [11]. In Croatia, a ten-year investigation found that the incidence of OTA in human blood was up to 4.5% (range 2–50 ng/ml) in the hyperendemic village Kaniža and up to 2.4% (range 2–10 ng/ml) in non-endemic villages. Additionally, it was found that almost all samples of food and feed, collected randomly in the hyperendemic village, contained OTA [12].

In addition to the Balkan Peninsula, people in other regions of Europe were also exposed to OTA. Some reports found OTA in human serum/plasma of these areas. An investigation in 2005 found that the mean concentration of OTA in serum of the healthy group was 0.37 ng/ml in both men and women, while in patients subjected to nephrectomy it reached 1.06 ng/mL in men and 0.94 ng/mL in women, and the mean content of OTA in kidneys was 0.23 ng/g and 0.20 ng/g, respectively [13]. However, another study from 2007 obtained the blood samples of 61 grain workers and found that the OTA plasma levels ranged between 0.07 ng/ml and 0.75 ng/ml. The mean (0.28±0.13 ng/ml) and median (0.26 ng/ml) OTA values for this cohort were similar to average values previously reported for the German population [14]. OTA levels detected in serum samples were 0.63 ng/mL in the Czech Republic [15] and 0.56 ng/mL in Italy (range 0.12–2.84 ng/mL) [16, 17]. In Danish samples, higher concentrations were found, with mean values of 1.5, 2.3 and 1.6 ng/mL in blood samples from 1986, 1987, and 1988, respectively [17].

OTA was also found distributed in other areas of the world. In the Americas, an investigation in Canada by Scott et al. (1998) [18], demonstrated that OTA was found in all plasma samples with a mean concentration of 0.88 ng/mL (range 0.29–2.37 ng/mL). Another study also found OTA in Argentina; OTA concentration in human plasma in two different areas of the Buenos Aires province was assessed, and the results showed that 63.8% of human plasma samples from Mar del Plata and 62.3% from General Rodríguez were positive for OTA, with Winsorized means of 0.15 and 0.43 ng/mL, respectively [19]. In Asia, a study in Tokyo [20] suggested that the population was frequently exposed to OTA. During the 4 years from 1992 to 1996, it was found that 85% of the cases were positive for OTA (except 38% in 1994), and average levels were estimated at 68 pg/ml, although the levels in plasma were far less than those reported for Europe and Canada. A study in Pakistan found that OTA was also present in these areas by determining OTA blood levels in 96 bladder cancer patients and in 31 controls. The results showed that more than 92% of all blood samples contained quantifiable amounts of OTA with mean OTA concentrations of 0.33 ng/mL (sd=0.42; range: 0.03-3.41 ng/mL) in bladder cancer patients and 0.31 ng/mL (sd=0.29; range: 0.04-1.25 ng/mL) in healthy controls. These OTA concentrations are comparable to those reported for the general population in the European Union [21]. In Lebanon, OTA was detectible in 33% of human plasma samples with a mean of 0.17 ng/mL [22].

As described above, from the OTA contents in human blood, we can see that OTA is distributed widely throughout the world, and it has been linked to diseases in some areas. Therefore, OTA needs to be further studied.

2.2. The Distribution of OTA in Food

OTA are largely produced by *Aspergillus* species belonging to the sections *Circumdati* and *Nigri*, and *Aspergillus affinis* is a new OTA-producing species of the section *Circumdati* [23]. OTA has been found in many commodities and processed food, such as cereals, beans, raisins, groundnuts, coffee, wine and beer, and also in certain animal products [24].

2.2.1. The Presence of OTA in Cereal Products

OTA has been found in a myriad of foodstuffs, and cereal and its derivatives, such as flour, bread, breakfast cereals, and baby/infant foods, remain the major contributors to OTA exposure [25]. The OTA exposure contents are shown in Table 1. Many factors can simultaneously influence the OTA content of cereal and its derived products, such as the agricultural practices involved, harvest procedures and storage conditions, the type of grain, and the nature and extent of technological processing as well as the analytical quality of the sampling and analysis of the suspected ingredients or foods. Therefore, a direct relationship between any specific factor and the OTA content is very difficult to assess, and determining the correlation between each type of cereal and OTA is also very difficult [26].

Table 1. Occurrence of Ochratoxin A in unprocessed cereals and processed cereal products

Country	Samples	Incidence (%)	Contamination Range (mean) (μg/kg)	Reference
Spain	Cereal grain (Wheat, Barley and Corn)	58/115 (50.4%)	0.219	[27]
Spain	Wheat bread	93/93 (100%)	0.45	[28]
Spain	Bread	2/20 (10%)	2.55-1.82 (2.19)	[29]
	Organic wheat bread	4/20 (20%)	0.03-0.81	[30]
Croatia	Maize	19/49 (38.7%)	n.d.-2.54 (1.47)	[31]
Denmark	Conventional wheat kernels	217/405 (53.6%)	n.d.-32 (0.3)	[32]
	Organic wheat kernels	6/14 (42.9%)	n.d.-1.6 (0.3)	[32]
	Conventional rye kernels	257/405 (63.5%)	n.d.-63 (0.9)	[32]
	Organic rye kernels	14/17 (82.4%)	n.d.-45 (3.9)	[32]
Italy	Maize	19/70 (27.1%)	n.d.-5.2 (1.7)	[33]
Japan	Rye	9/10 (90%)	n.d.-1.59 (1.05)	[34]
	Wheat flour	28/50 (56%)	n.d.-0.48 (0.09)	[34]
Korea	Barley	5/22 (22.7%)	n.d.-0.9 (0.8)	[35]
China(six province)	rice	18/370(4.8%)	0.30-3.2(0.85)	[36]
Ivory Coast	Maize	16/16 (100%)	27-64 (44)	[37]
Morocco	Corn	40%	n.d.-7.22 (1.08)	[38]
	Wheat	40%	n.d.- 1.73 (0.42)	[38]
	Barley	55%	0.04-0.80 (0.17)	[38]
U.S.A.	Wheat bread	24/24 (100%)	0.41	[28]
Brazil	Wheat bread	15/15 (100%)	0.09	[28]

(n.d. - non-detected).

Van der Merwe et al. (1965) reported that OTA was originally described as a secondary metabolite of *Asp. ochraceus* [39] and later was reported as a secondary metabolite of several other *Penicillium spp.* and *Aspergillus* [40]. Ochratoxigenicity varies depending on species, but it is also affected by their geographic distribution relative to different niche products (substrate). For example, *P. verrucosum* grows well below 30^{0}C and down to a_w (water activity) of 0.80; therefore, it is often found in cereals in cool temperate regions of Canada and northern Europe, although occasionally in Mediterranean sea localities with temperate climates such as Spain, Italy, Portugal and France [41-43]). This species, under any condition, grows slowly and has been reported almost exclusively in grain and derived products [40]. *P. verrucosum* was also isolated more frequently in cool conditions [44]. *Asp. ochraceus* and its other mesophilic xerophile kindred species, *Aspergillus steynii* and *A. westerdijkiae*, belong to the *Aspergillus* section *Circundati*. They grow at warm temperatures of 24-31^{0}C, lower moisture contents and pH values between 3 and 10. Thus, these toxigenic species are often present in cereals [41-43]. *A. carbonarius,* the main representative of the *Aspergillus* section *Nigri*, and *Aspergillus niger* grow optimally at high temperatures and a_w values. Therefore, they often occur in maturing fruits, especially grapes, and less frequently in tropical cereals [41,42].

It must be stressed that only a small number of *Aspergillus niger* strains are thought to produce OTA in commercially important crops [43].

Therefore, because each product tends to carry specific OTA-producing molds, it is important to understand how environmental conditions and factors promote subsequent formation of OTA [45].

In addition, the composition of fungal crop groups, and the potential occurrence of mycotoxins, will be affected by the length and conditions of storage. During the harvest, fungi may decline or thrive with time, while during storage, fungi can increase rapidly [46]. Historically, mycotoxigenic fungi, belonging mainly to the *Fusarium*, *Alternaria*, *Aspergillus* and *Penicillium* genera, have been classified into two groups. The first, namely "field fungi," include *Fusarium* and *Alternaria,* which represent a high mycotoxicological risk in pre-harvested or freshly harvested drying plants. The second, called "storage fungi," are primarily species of *Aspergillus* and *Penicillium* (both can produce ochratoxins) because in stored grains, they are isolated and increased frequently. However, in both cases the fungal original source is the field, and the mycotoxicological aspects reviewed here are also related primarily to the field [47].

Therefore, it needs to be considered a problem that OTA can be produced in storage under certain environmental conditions and usually does not occur before harvest [48]. In a recent Danish survey, Elmholt and Rasmussen found that most of the combined harvested grain samples (82%) were contaminated with *P. verrucosum* prior to drying, which indicates contamination before multi-grain storage due to difficult harvest conditions [49]. Once established, this competitive fungal species can dominate at ambient conditions conductive to grain storage [41].

Bhattacharya and Raha found that stored seeds were easily invaded by fungi during the storage process, causing a gradual loss of carbohydrate (both soluble and insoluble) content and degradation of oil content during prolonged storage. Therefore, the seeds deteriorate, have difficulty germinating—causing a decrease in the value of planting—and are not suitable for food and feed. In addition, fungal infection constitutes a serious safety, quality and economic problem to producers [50]. Therefore, besides the risk of OTA, fungal contamination can reduce germination rate due to a quantum loss in carbohydrates, protein, and total oil content, which induces increased moisture and free fatty acid content, thereby enhancing other biochemical changes.

It is known that the best conditions for mold to produce toxin with regard to water activity and temperature are more stringent than the fungal growth [51]. As a general phenomenon, stress is frequently mentioned as a reason for mycotoxin synthesis [52]. The opposite is also true: because OTA generally remains stable, it may be detected shortly after the fungi have become extinct or a less dominant species [53]. In addition, the levels of ochratoxigenic fungi can also be considered as indicators of OTA formation. For example, Lund and Frisvad, found that more than 7% infestation of *P. verrucosum* can indicate OTA formation [54].

The fungal composition and quantity are important in potential interactions and competition between mycotoxigenic species and other spoilage bacteria in the cereal products. For example, the xerophilic spoilage fungus *A. ochraceus* plays an important role in maize and inevitably interacts and competes with other toxigenic species of *Fusarium*, *Aspergillus* and *Penicillium* for the maize grain niche. It has been recently shown that whether *A. ochraceus* becomes the dominant species depends on moisture content and temperature and that *in vitro* interactions and competition significantly influence the OTA-producing capability of this species. However, a study by Lee and Magan. (2000) [55] shows that, to a large extent, *A. ochraceus* cannot compete with other spoilage fungi in primary resource capture at a_w larger than 0.95 in maize grain, although *A. ochraceus* may change resource quality and influence secondary colonization by other species under the optimal conditions.

Grain harvest, the drying process and the quality of the storage facilities are considered to be the factors that are the most important for the grain production and incidence of OTA [56]. Specifically, in western and northern Europe, Canada and other temperate areas, cereals present a higher risk for incidence of OTA. Freshly harvested cereals contain a high moisture content (>20%), which is a requirement for ochratoxigenic species growth and OTA synthesis [57]. OTA formation in such grains is most likely due to insufficient drying time and the long grain storage times before drying [58]. If the above two conditions are met—freshly harvested cereals with inefficient drying and storage under humid conditions—high levels of OTA in cereals may be produced [32]. For example, rice has been shown to be a better substrate for the characterization of OTA-producing *A. ochraceus* strains in many studies. Thus, incidence of OTA in rice is often high (90% in Moroccan rice, 7.8% in non-organic rice and 30% in organic rice in Spain). This is attributable to the fact that rice is an aquatic plant, and at harvest time, there are usually very high moisture levels (35-50%) [59]. Mycotoxin-producing molds or spores which have already infected rice crops in the field and at harvest may later contaminate those products in storage, and OTA contamination may occur. If environmental conditions are suitable, OTA will be produced [60]. The level of OTA is often higher in wild rice than in conventional rice due to the differences in the environmental conditions and compositions of the two types of rice at harvest. In wild rice, the moisture level at harvest is higher, which is maintained for one or two weeks for fermentation. In addition, the amino acid and protein contents of wild rice are also twice as high as white rice, which can stimulate further OTA production [61]. This agrees with a previous study by Pena et al. (2005) [60] conducted in Portugal. This study showed that compared with the basmati, brown, and wild rice and aromatic samples, none of the white rice samples analyzed were found to be contaminated. Furthermore, proliferation of toxigenic molds might occur under improper storage conditions regardless of the season, which was confirmed by Nguyen et al. [62]. This study indicated that the OTA contamination in samples collected from Vietnamese local markets during the dry season was higher than that in the commercial samples during the rainy season. Although all sources of fungal infection cannot be completely eliminated, it is possible to reduce or even avoid mold growth under certain conditions. In sum, the cereals must be dried below 15% (a water activity of approximately 0.8) and kept at this level throughout storage. If there is a delay in drying grains, it will lead to subsequent storage problems [63]. Thus, because the weather conditions of the harvest vary, the sampling and analysis results will be different depending on the year, which was observed by Czerwiecki et al. (2002) [64] when analyzing two different consecutive harvests. Regardless of the type and variety of grain planting, the frequency and level of OTA contamination in cereals from 1998 were substantially higher than those from 1997. The authors in 1998 determined there was a higher mean degree of precipitation in Poland. In Denmark, Jørgensen et al. found similar results when analyzing the OTA contamination in rye and wheat [65]. The climatic conditions during harvest can also induce the production of OTA. Especially in years with wet weather during harvests, the daily intake of ochratoxin A for some individuals is much greater than the standard limits. If storage conditions are not suitable, after a long period OTA production can occur. Nevertheless, from Spain and Portugal, in organic and non-organic cereals and cereal products, the incidence of OTA was 33% and 14%, respectively [66]. The maximum value detected in whole meal rye organic samples was 27 ng/g. It is very important that the samples used are only whole grain or kernels because most fungi are located on the surface of the particles, thus this is where the highest contamination occurs. Therefore, use of whole grains will likely lead to higher levels, which is why the effect of sample treatment and methods is very important when elevated levels are detected. Depending on the samples used, the results may indicate a much higher level of contamination. Park et al. (2005) [35] reported that Korean polished rice samples presented levels higher than EU maximum limit (ML). Grain processing technology is an important factor, as described below. Cereal grain type also affects the incidence of OTA because of their nutritional composition For example, maize has a higher level of contamination. Araguás et al. (2003) [27] analyzed a total of 118 samples, and in corn, only one sample was higher than the maximum level permitted (5 pg/kg). Metabolism research reported that glutamate may be involved in the OTA production process. High levels of this amino acid in cereals, may be present as a cofactor for the OTA [30].

Other studies point to higher OTA concentration in rye grains [32,67]. Chemical properties affect seed moisture. For example, high levels of seed oil indicate low moisture absorption compared to those with high protein or starch levels. Water activity (a_w) plays a detrimental role during storage of the seeds. Seeds absorb moisture from the environment based on current needs and relative humidity in the storage atmosphere, and, especially during rainy months, a_w may increase [68]. The type of agricultural practices involved is considered the determining factor. In the study of Juan et al. [66], the organic cereal samples showed the highest incidence of contamination. Likewise, in 1997 and 1998, Czerwiecki et al. [64,67] studied 400 Polish cereal grain samples, and the presence of OTA was determined in samples from ecological and conventional farms. In the 1997 crop, in all types of grains, OTA frequency was substantially higher in grains originating from ecological farms than that from conventional farms. In particular, OTA contamination of the ecological rye samples was six times more frequent than conventional cultivation. Jørgensen and Jacobsen [32] found that while there are year to year changes, both in conventional and organic samples, OTA content in rye is higher than wheat. For rye, the multiyear mean concentration of OTA was higher for organic production than for conventional production for both the raw grain and flour, but for wheat flour, there was little difference in multiyear means between the different methods of production, and for wheat grain there was no difference at all. Gonzalez et al. [61] observed that in the case of rice samples, processed products possessed higher contamination than organic samples.

Another factor likely to affect the impact of contaminated cereal products is the quality of test and measurement results, including sampling, sample preparation, detection and interpretation [69]. Because there is a lack of uniform guidelines or instructions, reports showing variation in contamination levels do not necessarily reflect the actual situation. In Finland, Eskola et al. (2002) [56] did not detect OTA (n = 115) from the assorted samples, and in Spain, Blesa et al. (2004) [70] reported that most samples did not contain OTA (n = 43). The authors suggest that the methodology used had low sensitivity as an explanation.

Furthermore, the lack of regulation and/or control programs for OTA contamination at the national or regional level affects the grain harvest, storage and commercial conditions that lead to risks for human and animal populations. Especially in the developing countries, surveys and exposure assessments are needed.

While OTA is considered a natural contaminant mainly in northern temperate zones, a recent survey of grain in Africa reported a high level of average OTA concentration and incidence. According to Riba et al. (2008) [71], the fungi species isolated were *Aspergillus*, *Fusarium*, *Penicillium*, *Alternaria* and *Mucor*. Species of the *Aspergillus* genus were the most frequently detected in all of the samples. During harvesting, transportation, storage, and processing, microorganism propagules can contaminate grain in different ways, most often with dust from soil from the surface of plant remnants. Whenever there is a problem in the processing or in the storage of wheat and wheat-based feeds that allows fungal growth, the risk of OTA contamination should be taken into account. Ayalew et al. [53] indicated maximum levels that exceeded the ones reported in many European surveys [25].

As wheat is often processed before eating, and the OTA content will change during the processing, to estimate the dietary intake of ochratoxin A (OTA), information on its fate during wheat processing is very important. Wheat was contaminated with OTA at different levels (93.22 and 248.93 mg/kg) by artificial inoculation by *Aspergillus ochraceus*. Concentrations of OTA were 1.27-1.50 times higher in shorts than bran and 1.43-1.66 times higher in bran than whole wheat. This study also observed that when the wheat was processed into flour, the OTA concentration decreased by 43.29-55.16%, and the final OTA concentrations in flour were 41.80 and 138.33 mg/kg, respectively. Whereas, during the milling process, OTA was not transferred or decomposed, as the milling process cannot change the total amount of OTA. Further losses of OTA varied depending on different treatments. In Chinese traditional processed wheat products, different changes in OTA content were observed. When the flour was processed into Chinese steamed bread (CBS), the concentration of OTA increased by 11-16%, and the OTA concentrations were 49.2 and 161.35 mg/kg, respectively, in CBS. This revealed that fermentation and steaming did not decrease but increased the OTA levels in CBS. In Chinese fried bread sticks (CFBS), the loss of OTA was small, with a concentration reduction of 38.84 and 129.02 mg/kg and final concentrations of 38.84 and 129.02 mg/kg, respectively. In the noodle processing, a higher reduction of 23.27-23.92% for OTA was observed, and the final concentration of OTA was 32.98 and 108.24 mg/kg in noodles, respectively. The results revealed that the temperature, pH values, strains and the processes exerted different effects. High temperature (180^{0}C) and alkaline condition (pH > 7) facilitated reductions in the concentration of OTA in wheat products, while fermentation with dry yeast increased the OTA levels in these products [72].

2.2.2. The Presence of OTA in Coffee and Tea

Ochratoxin A (OTA) is produced by several fungal species. Due to factors such as climatic conditions, abnormally long storage and transportation and processing conditions such as wet processes and dry processes, OTA can be found in coffee beans [73]. OTA has been found in coffee beans and coffee beverages. Many studies have detected OTA in green coffee [74-78]. Eighty-four samples of green coffee beans from Africa, 60 from America, and 18 from Asia were collected to analyze OTA levels in green coffee samples from various countries. As a function of green coffee provenance, both the content and the variability of OTA levels were tested. In particular, African samples were more frequently contaminated compared with the frequency and level of OTA in samples from other origins; the highest levels were 18 and 48 μg/kg in two samples from the Congo. Samples from Africa were more highly contaminated by OTA than the samples from America. The crop processing and the climatic conditions may be responsible. For example, in a wet processing method, Arabica coffee is produced and in a dry processing method, Robusta and lower quality Arabica coffees are produced [73].

Some researchers also reported that OTA can be transferred to tea and coffee. A small survey assessing OTA content in black tea, fruit tea and ground roasted coffee was carried out. OTA content was detected by the HPLC-FLD method. The limit of quantification (LOQ) of this method is 0.35 ng/g. In black tea and fruit tea, the OTA levels were up to 250 ng/g and 104 ng/g, respectively. The tea infusions were prepared by using black tea and fruit tea, naturally contaminated. The transfer from black tea and fruit tea to the infusions was 34.8% ± 1.3% and 4.1% ± 0.2%, respectively. Seven types of coffee beverages were prepared using ground roasted coffee naturally contaminated at 0.92 ng/g. The transfer rate of OTA into coffee was from 22.3% to 66.1% depending on the type of process used. OTA intakes from black and fruit tea or coffee represent a non-negligible human source [79].

OTA content in raw black teas and fruit teas was also measured. The OTA contents in 12 samples of black tea and 12 samples of fruit tea were detected. The results showed that an OTA content above 0.35 ng/g was detected in 4 samples of black tea out of 12 (25%). Because no confirmation using artificially contaminated samples has been performed, only naturally contaminated samples were used to make the infusions. Infusions were made from six fractions of black and fruit teas, naturally contaminated.

This indicated that from a cup of black tea, which is the most contaminated, approximately 152.3 ng/250 mL OTA can be obtained. A similar percentage of OTA transfer (data not shown) was found in black tea made from the least contaminated sample (1.85 ng/g). For fruit teas, from the most contaminated sample (104 ng/g), the OTA transfer was much lower (7.5 ng/250 mL). It may be that the lower pH contributes to lower OTA transfer into fruit tea infusions [79].

2.2.3. The Presence of OTA in Alcoholic Drinks

As we have described above, mycotoxins are a group of naturally occurring toxic secondary metabolites that are contaminants of susceptible commodities. Mycotoxin contamination may occur if there are suitable environmental conditions for molds during crop growth, harvesting, storage, and processing operations. Many types of mycotoxins in foods and feeds, particularly in cereals, can be produced due to the complex ecology of mold growth and mycotoxin production. Beer and wine, as a type of drink made from must alcoholic fermentation, can contain OTA. In addition to alcoholic fermented drinks, red wine made from grapes can also contain OTA.

1) The Presence of OTA in Beer

As a drink, beer is made from must alcoholic fermentation, mainly using germ barley alone or mixed with other cereals, starchy substances, yeast and water. In the past twenty years, there have been several studies confirming the existence of OTA in beers around the world. The European Commission's report has stated that beer consumption contributes 5% of OTA intake [80]. The origin of OTA in beer comes from contaminated prime materials such as barley, malt and cereal derivatives to use in processing. Ochratoxigenic strains from *A. ochraceus* and *P. verrucosum* grow in barley grains, which are transmitted to beer. OTA produced by these strains on cereals is stable during the cooking process, but it will reduce during beer production because the yeast retains the original content of OTA (eliminating between 2 and 19% of OTA), and the rest is transferred to the final consumable beer during fermentation [81].

Because its presence in beer should be controlled in another more appropriate stage of the production chain, especially in the malt prime material, the EU has not set limits on OTA in beer (Directive (UE) No. 105/2010); however, some countries have established maximum permissible limits of OTA in beer, for example, Finland (0.5 ng/mL), Italy (0.2 ng/mL), and the Netherlands (0.3 ng/mL), and in research studies several authors have assumed a recommended limit of 0.2 ng/mL [82, 83].

2) The Incidence and Presence of OTA in Beer

Data related to OTA in beer or wine from southern Europe, specifically from the Mediterranean region, from the past decade have been collected. The incidence of OTA in beer and the range and OTA median concentrations found in beer in different studies are summarized in Table 2. Recent studies of OTA in beer show values ranging from 14% [84] to 100% [82], and the incidence in most of them was higher than values obtained by the European Commission [80] (set at 39% incidence in beer). The differences in the studies may occur as the result of the different detection methods and the malt quality used. When studies were examined per country, the highest presence (100%) of OTA in beer samples was found in a study carried out in Spain [82] compared with other published studies; however, only five samples presented concentrations above the assumed value of 0.2 ng/mL. The reports by the European Commission [80] for Spain showed similar levels. In Spain, OTA incidence was 97% of the analyzed samples, but this differed from results of other studies.

Another study reported that 77% of beer samples from Navarra markets in Spain contained OTA. However, no sample exceeded the proposed legal limit, and only one almost reached the proposed value (see Table 2). Several studies reported that approximately 70% of beer samples from Italy were contaminated by OTA [85,86], and the levels of OTA detected were lower than those obtained in beers from Spain [87] (see Table 2). The presence of OTA in beers from other countries such as Turkey decreases to 14%, with concentrations below the assumed reference [84]. To compare these results with studies carried out in other European countries, a study was carried out in Hungary, an eastern European country. The results of this study showed that all beer samples from Hungary contained small amounts of OTA (presence of 100%) but only one sample was detected at levels >0.2 ng/mL (see Table 2) [88]. A beer study carried out in the Czech Republic indicated that the incidence (39%) agrees with data obtained from the European Commission (2002), and only one pale beer sample had a higher concentration (0.2438 ng/mL) than the assumed reference. Other studies outside the region of interest of this review have been performed. Interestingly, in another study, analysis of beers of different origins (United States, Europe, Australia, Canada, Mexico, South America, and Asia) showed that no OTA levels were detected [89], correlating with results obtained in studies carried out in China [90].

Table 2. Incidence, Range, and Median Concentration of OTA in Beer [88]

Country (ref)		Product	no. of samples	Incidence (%)	Range (ng/mL)	median (ng/mL)
Spain [87]	beer		31	77	0.205a	0.044
		with alcohol	21	67	0.205a	0.038
		without alcohol	10	100	0.152	0.056
Spain [91]	beer		88	82.9	0.007–0.204	ni[b]
		from Spain	31	83.8	0.007–0.147	0.036
		imported	57	82.5	0.007–0.204	0.046
Spain [92]	beer		69	100	0.008–0.498	0.07
		from Spain	35	100	ni	0.085
		imported	34	100	ni	0.055
Italy [85]	beer		106	67.9	0.002–0.189	0.019
Italy [86]	beer		30	16.7 < 0.2	ni	
Turkey [84]	beer		35	14	0.012–0.045	0.02
		from Turkey	31	13	0.012–0.045	ni
		imported	4	25	0.012	ni
Czech Republic [81]	beer		115	39	0.001–0.244	ni
Hungary [93]	beer		25	100	0.030–0.250	0.13

[a]Maximum value. [b]ni, not indicated.

The variation in OTA mean levels obtained in beer is small, according to the published studies. The lowest OTA mean level was <0.2 ng/mL, and the maximum value for one sample was 0.5 ng/mL, as reported by Medina et al. Compared with those from the European Commission [50], the median OTA values obtained from samples of different studies are slightly higher than those from the European Commission (0.03 ng/g of 496 beer samples), although the levels were below 0.1 ng/mL.

Several studies have shown that in southern European countries (Italy, Spain, Turkey, Germany, and the Balkans), OTA levels in beer were below 0.04 ng/mL . It indicated that the levels from these countries were different with respect to all other analyzed beers; the highest concentration of OTA (0.189 ng/mL) was found in one beer sample from Denmark. One possible explanation for these low OTA values in beers from southern Europe is that *P. verrucosum*-produced OTA is important in northern European countries where environment conditions, such as temperature and humidity, are optimal for its growth [85] but not in the south. Studies on alcoholic and nonalcoholic beers have been carried out, and data showed that no differences were found in OTA levels between alcoholic and nonalcoholic beers [84, 87]; incidence and mean levels of OTA were higher in nonalcoholic beers than in alcoholic, likely due to the different procedures used in the two types of beer production. In nonalcoholic beer production, there is no fermentation step or it occurs under controlled conditions; as a consequence, it is expected to have higher contamination under the assumption that the fermentation step reduces OTA content. These results differ from the European Commission report [80], where OTA was detected at a proportion of 39%, and OTA levels (0.03 ng/g) were low. In short, average levels were never found to be higher than the assumed value (0.2 ng/ml), and this was detected from the difference between the different sources of beer.

3) The Estimation of OTA Intake in Beer

An evaluation of the OTA intake associated with beer consumption demonstrates that even considering the highest values, beer contributes only 3% OTA intake at its maximum 20 (<5% reported by the European Commission [83]. Nevertheless, these values are much lower than the tolerable daily intake (TDI) reported by the European Food Safety Authorities (EFSA).

Researchers from outside of Europe have carried out some studies of OTA and beer, such as one report from South Korea [35]. Beer consumption is a minor source of OTA intake, but it is recommended that a new assessment to measure actual consumption data be undertaken as the Asian population has experienced a significant increase in the consumption of beer in recent years [35].

In summary, beer does not appear to be an important factor, but it is an additional source to control and consider for OTA mycotoxin intake calculation.

4) The Presence of OTA in Wine

OTA is also found in wine. A survey of OTA in wine has been carried out around the world. In different countries, studies indicated that in most samples, OTA levels were not higher than 0.2 ng/ml. In red, rose red, and white wine, the concentration gradient of OTA decreases, but it also decreases with increasing latitude of the producing countries. OTA in wine is due to the presence of *Aspergillus niger* and mainly *A. carbonarius*, which can be grown in the vineyard grapes and produce toxin. Grape crushing and squeezing can transfer OTA, which is very stable, to contaminate the wine. Pre- and post-harvest treatment are needed to reduce possible contamination of wine [95].

Wine has been considered important in OTA intake assessment in the past decade. Wine, in addition to the grapes used, is one of the most important sub-products consumed in developed countries and developing countries. According to an estimation carried out by the Wine and Vineyard International Organization (OIV) world wine production in 2011 reached 265.8 million hectoliters (without counting grape juice or must), 600,000 hectoliters more than in 2010 [88], and Spain contributed 12.9% of this production with 34.3 million hectoliters [88].

During brewing, the presence of OTA in grapes can be transferred from grain grapes to grape wine. After cereals, wine daily intake is the main source of OTA in the European population and comprises 13% of the daily intake of OTA. OTA levels in wine depend on different factors, such as period of harvest, the wine brewing technique employed, weather (temperature, rain, relative humidity in the vineyards), pesticide treatments, and vineyard location (latitude). The European Union allows a maximum limit of 2 ng/g OTA in wine [88].

There is an enormous variety of wine typologies, which results in a very complicated classification. It was reported that levels of OTA in red wines are higher than rose wines, followed by white wine, is mainly due to the must impregnation with grape skins, which can improve the extraction of OTA. Sweet or special wine showed different concentrations of OTA compared with the dry wine due to the highly diverse enological practices. Under normal circumstances, sweet wines showed higher OTA concentrations than dry wines [95].

There is no standard related to OTA purity of wines except for liqueur wines and those with an alcoholic strength less than 15%. There are many studies of OTA in dry wines from different Mediterranean countries; however, only a few studies have showed the presence of OTA in sweet or special wines [96]. France, Italy, Spain, and Portugal produce the most wines in Europe. Consumption of wine is another aspect to be considered for OTA concerns.

Three countries with high wine production were investigated: (i) Spain, as the country with the third highest production of wine after France and Italy [88,97]; (ii) Portugal, as the country with the optimal environmental conditions for *Aspergillus* and *Penicillium* fungal growth and possible OTA secretion; and (iii) Italy, as the country with the second highest wine production according to OIV data [98] and the one with the highest index of wine consumption per inhabitant.

In the following sections, studies of OTA contamination in these countries are discussed. In Spain, 69 registered designations of origin (RDO) are accessible [88] (MAGRAMA, 2012). Recently, a few studies show that OTA presents in typical Spanish wines, especially chamomile wines and Rioja Alavesa wines in Jerez (Table 3).

The frequency of OTA present in muscatel sweet wines and Pedro Ximénez wines ranged from 75 to 100% [97]. It was reported that OTA was present in 97% of sweet wine samples in Spain. Furthermore, the percentage of OTA contamination in wines from southern Europe exceeds that in northern Europe, which is in accordance with the reported results in SCOOP [80].

A recent investigation of OTA contamination of wines in China indicated that dry red wine samples were OTA positive (57.1%), followed by dry rose and ice wines (25% each) and, finally, dry white wines (2.9%) [119]. OTA concentrations in red, rosé, ice, and white wines were 0.80, 0.22, 0.10, and 0.07 μg/L, respectively. The values of these OTA levels were higher than previous studies. In Chinese markets, OTA levels in red wines ranged from 0.0028 to 0.0437 μg/L [120]. Researchers evaluated the level of OTA intake through wine consumption in Spain. The results indicated that the values of OTA intake are below 0.1 ng/kg bw/day, which represents 1.31% of the maximum of TDI. The levels of OTA exposure in humans are always much lower than the TDI established by EFSA [121].

It was reported that OTA intake through wine consumption was 0.07 ng/kg bw/day in Spain, which was above the values reported in the EC [80]. Some studies assumed that the intake for starter wines is 50 mL/day [122], a value higher than that reported by the EU in the abovementioned report (2.93 g/person/day). The levels of OTA exposure could be higher in cases where the consumer is below 60 kg and the wine consumed has the highest OTA levels.

As a consequence, humans are easily exposed to OTA through wine consumption. It is necessary to assess OTA contamination in wine to avoid health hazards for humans.

Table 3. Percentage Detected and Median and Maximum Concentrations of OTA in Wines [88]

country	red wine			rosé wine			white wine			Ref
	% detected	median (ng/ mL)	max[a] (ng/ mL)	% detected	median (ng/ mL)	max[a] (ng/ mL)	% detected	median (ng/ mL)	max[a] (ng/ mL)	
Spain							75	1.81	3.42	[99]
							100	2.86	7.30	[100]
							97	0.5		
							99	0.65	4.63	
							50	1.49	27.79	[101]
							80	0.138	0.642	[102]
	57	0.035	0.179	33	0.007		67	0.014	0.015	[103, 104]
	53	0.037	0.179							
	88	0.051	0.237							
	88	0.085	0.138							
Italy	82[b]	0.29	1.4	71.4	0.32	0.82	62	0.16	0.42	[105]
	100[c]	0.33	0.87				66	0.14	0.17	
	84	ni[d]	4.0	56	ni	1.04	19	ni	0.21	[106]
	55	0.72	2.9	50	0.34	0.40	40	0.9	1.95	[107]
	ni	0.18[e]	0.75	ni	nd[f]	ni	ni	0.06	0.18	[108]
		0.16[g]	0.72		0.09			0.08	0.22	
	69	0.34	7.5	92	0.5	4.07	44.3	0.08	1.95	[109]
	69	0.12	2.63				61.3	0.08	1.36	[110]
	6.7	0.23	ni							[111]
Greece	66.7	0.17	1.5				81.5	0.57	2	[112]
Turkey	86	0.11	0.82	90	0.052	0.16	85	0.108	0.62	[113]
Hungary	0	nd								[114]
	97.7	0.01	0.53							[115]
Croatia	80	ni	0.02							[116]
	80		0.05							
China	57.1	0.8	5.65	25	0.22	0.22	2.9	0.07	0.07	[117]
	53	0.013	0.043							[118]

[a]OTA maximum concentration. [b]Bottle. [c]Box. [d]ni, not indicated. [e]Conventional. [f]nd, not detected. [g]Organic.

OTA was also studied in Portuguese wines. Portugal, based on its location, has several different climate conditions. It is affected by the Atlantic Ocean and Mediterranean Sea. As mentioned above, weather conditions and geographical locations are major factors that determine fungal growth and contamination of the grapes by OTA. Reports on wines from different regions in Portugal [123] (35 red wines, 25 white wines) indicated that 12% of white wine samples and 26% of red wine samples were OTA positive, and only 1 sample of white wine exceeded the EU's limitation (2 ng/g). In general, the levels of OTA in white wines were lower than those obtained in red or rosé wines; in fact, the highest levels found in white wine samples between studies were very different, which was possibly due to the low quality of the analyzed sample. The OTA tolerable weekly intake (TWI) was calculated by considering (i) the highest levels of OTA obtained (2.4% in white wine and 1.23 ng/mL in red wine), (ii) a wine consumption of 49.9 L/person/year (136.7 mL/person/day), and (iii) a body weight of 60 kg. OTA's TWI ranged from 20 to 38 ng/kg bw (from 2.8 to 5.5 ng/kg bw/day). These values represent between 17 and 32% of the TWI.

Italy has the second highest wine production according to OIV data [98] and the highest index of consumption per person. There are many publications in which the presence of OTA in wine has been studied. Table 3 reports the most recent studies of this mycotoxin in Italian wines

In general, OTA was detected in most of the Italian wines, although the levels detected were not high. The detection of OTA oscillates between 51 and 77% for its presence in Italian wines, but when comparing the different wine typologies, the highest frequency was 92% for rosé wines [124] and 100% for red wines [125]. In other publications which focused on developing new methods of detection, OTA was not detected in the analyzed samples [126, 127].

All studies carried out to detect the presence of OTA and its concentration in wines results showed that those values decrease in the following order: red wines, dessert wines, rosé wines, and white wines. This could be attributed to the different technological conditions used during their brewing. Nonetheless, an exception has been reported for rosé wines, where the highest value of OTA was detected [124].

The results of an OTA analysis revealed that 0.01% of the Italian wines are OTA positive, with levels exceeding the limits established by the EU [128]. The literature shows that red wine has the highest frequency and concentration of OTA, reaching its maximum at 2.92 ng/mL and exceeding the limits set by the EU [129].

There are studies in which all wine samples were below OTA levels established by the EU (2 ng/mL), and when organic and conventional wines were compared [130], there were no statistically significant differences for OTA levels.

The maximum OTA intake for Italian wines is 1.4 ng/kg bw/day, which represents 5.6–8% of the TDI established by EFSA [131]. In the worst evaluated case, when calculating OTA intake in regular wine consumers in southern Italy (including Italian islands) with consideration of the median of consumption, the exposure obtained was very low, representing 9.8% of the TDI (1.68 ng/kg bw/day) [124].

These values are similar to those reported by the EC [80]. OTA intake through wines is declared as 2.94 ng/kg bw/day for consumers and 0.86 ng/kg bw/day for the general population in the EC [80].

In summary, OTA intake through wine consumption is not likely to be a health risk in the Italian population [88].

Studies of OTA intake in wines have shown that for a person with a body weight of 60 kg and a consumption of 100 mL of wine per day, OTA intake was 0.18 ng, which is 3.6% of the daily intake established by the Scientific Committee on Food (SCF) (5 ng/kg bw/day) [93]. The OTA daily intake calculation for Chilean wines, using the formula of Breás et al. (2005) [128], was 0.24 ng/kg bw/day, which is 70 and 58 times lower than the current TDI established by the SCF of the EC [132]. It was concluded that commercial wines in Chile can be considered safe and are not associated with a toxicological risk. It is difficult to compare the percentage of positive samples and OTA levels in wines from different countries due to the great variety of wines analyzed and the differences between the limits of detection used in each analytical method; however, red and sweet wines appear to be the most contaminated wines, and wine samples from southern regions presented higher incidences and concentrations of this mycotoxin [88].

2.2.4. The Presence of OTA in Milk and Milk Products

(1) The Presence of OTA in Milk

1) The Presence of OTA in Breast Milk

Infants are considered to be more vulnerable than adults to the toxic effects of xenobiotics, in spite of the benefits of breastfeeding in early life. Thus, exposure of OTA in breast milk can be harmful to the infant, and attention should be given to this issue.

OTA exposure in nursing mothers and their infants over the first 6 months of breastfeeding was analyzed by HPLC. Analysis of OTA in maternal blood plasma, breast milk, and infant urine was carried out in a study of temporal and individual variations in the biomarker levels of a cohort of Chilean mother–child pairs. OTA in plasma of nursing mothers, who are exposed of dietary mycotoxin, will be transferred to breast milk through circulation. The ratios between OTA levels in breast milk and maternal plasma change due to individual differences and the study period. Colostrum (milk produced in the first days after delivery) has the highest OTA excretion. OTA intake of Chilean infants was found to exceed the chosen tolerable daily intake value of 5 ng/kg BW (body weight) at early stages of breastfeeding but not at later stages. The concentration of OTA in spot urine is considered a valid biomarker of exposure for infants because OTA levels in urine can indicate the OTA intake from breast milk. Biomonitoring studies for OTA exposure in infants is encouraged by the potential application of spot urine monitoring [133].

2) The Presence of OTA in Cow Milk

For ruminants' milk, several studies have demonstrated that OTA can be hydrolyzed to the less toxic metabolites, ochratoxin α, by rumen microbes and rumen pH value [134, 135]. In 2002, a report indicated that milk may not be a chief source of OTA, when researching the presence of OTA in milk [80]. At present, this matter has many researchers' attention.

A study analyzed 61 milk samples purchased in different stores and supermarkets. These samples included low fat (n = 20), fat-free (n = 12) and whole (n = 16) milk and also included substitute or imitation milk (n = 13). Only 3 out of 63 samples were positive for OTA, with the concentration between 0.07 and 0.11 ppb. None of the samples contained OTA above the detection level. This result was confirmed by the methyl ester formation step. No "non organic" sample were positive. However, although the OTA content in milk is low, the amount of people drinking milk is very large, and even low concentrations of OTA will also be a major source of toxins. The results with a report published in 2002 studying some European countries, such as Sweden, Norway and Germany, and the paper of González-Osnaya et al. (2008) [136].

Important comparative data in this research report and data in the published literature are complex because in many cases, the situations of the samples are unknown. For the data in this research report, the amount of OTA detected does not appear dangerous. It is difficult to guarantee consistency. If a man consumes milk at a rate of 1 L/week (data referred to Italy – CLAT, 2010) on average, the total intake of OTA in one week (TWI) would be approximately 1.52 ng/kg bw which is below the maximum tolerable (120 ng/kg bw), according to the European Food Safety Authority (EFSA) [137].

At present, when contaminated milk is processed to make cheese, there was no evaluation of the concentration factor of OTA. Aflatoxin M1, which is found in hard cheese, was calculated to have a concentration factor of 4.5 [138]; as for OTA, the factor is the same as Afatoxin M1, and CLAT (2010) calculated the consumption of the mycotoxin, which was double that calculated for milk but lower than that calculated by EFSA [137, 139].

(2) The Presence of OTA in Milk Products

Mycotoxins in milk and milk-based products are a public health concern and have been regularly monitored. In cow's milk, several *Penicillium* and *Aspergillus* species produce ochratoxins and aflatoxins, which are highly toxic secondary metabolites. The presence of mycotoxins in dairy products reflects the contamination of feedstuffs. Following ingestion of contaminated feedstuffs, OTA is largely transformed into the less toxic metabolite ochratoxin α by rumen microorganisms [140]. Then, ochratoxin A and OTα are mainly eliminated in the urine and feces, but they can also be found in milk and milk-based products. In 2003, a survey of raw bulk milk assessing ochratoxin A (OTA) was carried out in the main French milk-producing basin, located in the northwest of France. During the period of analyzing the contamination level of OTA in milk, at the farm level, the result was low, but it is possible that there is a relationship between mycotoxin contamination and feeding management practices [141]. In 2008, Rory Coffey developed a quantitative Monte Carlo exposure assessment model for mycotoxins in dairy milk to assess the potential human exposure levels. Mean concentrations of mycotoxins in milk were estimated using the simulation model (Aflatoxin M1 = 0.0161μg/kg, Ochratoxin A = 0.0002 μg/kg, Deoxynivalenol = 1 μg/kg) [142]. Recently, in milk-based infant formula and foods, a stable liquid chromatography tandem mass spectrometry (LC-MS/MS) method and isotope dilution assay were applied to detect the presence of 12 mycotoxins, including OTA. The average recoveries of ochratoxin A in fortified milk-based infant formula, baby yogurt, milk, and milk powder are 20 μg/kg [143].

In Italy, 133 (72%) infant formulas samples, purchased from a market, were analyzed for OTA the results showed that the contamination levels ranged from 35.1 to 689.5 ng/L [144]. In China, ochratoxin A (1.43 μg/L) was found in one of three milk samples [145]. In Egypt, milk and dairy products purchased at Egyptian markets and breast milk from lactating mothers in the Cairo and Giza governorates were analyzed for some mycotoxins, and 3 of 10 samples were positive for ochratoxin A (30 %) [146].

1) The Presence of OTA in Cheese

The presence of OTA in cheese has been only occasionally and recently reported. By using an improved HPLC-FLD method, Chiara Dall'Asta reported for the first time that ochratoxin A was occasionally found in blue cheeses, in particular in several typical manufactures from EU countries: Gorgonzola from Italy and Roquefort from France [147]. In 2009, a method was recommended that detected the distribution of OTA in a single cheese piece in different places by combined HPLC-MS/MS and solid-phase microextraction (SPME). The data indicated that the amount of OTA detected was directly proportional to storage time for moldy cheese [148]. This lack of knowledge depicts a lack of legislation because hygienic and microbiological conditions of hard cheese crust are not regulated by the EU. Furthermore, targeted and validated analytical methods do not exist so far. Because the OTA carry-over in animal products is not commonly considered a major health concern

(EFSA 2006), specific legal limits have not been set so far by the EU for OTA in meat and cheese [149].

2) The Presence of OTA in Milk Powder

OTA can be metabolized or transferred to raw milk because it is stable throughout processing of dairy products and therefore could exist in milk powder. To ensure the quality and safety of milk powder, 8 samples of milk powder were collected from supermarkets in Beijing in 2014, and all the samples were stored at −20℃ before analysis. The result showed that OTA in milk powder in Beijing was 27.0 ng/kg, which was very low [150]. A study detected the presence of ochratoxin A (OTA) in processed infant formula (milk powder) available in a Portuguese market, sold as conventional and organic origin, using the HPLC method. The results were below EU limits for OTA in baby and infant foods (0.50 μg/kg) [151]. Using the highly sensitive UHPLC–MS/MS method, powdered cereals infant formula, purchased in different supermarkets and drugstores of Castellon (Spain), was analyzed. Samples of cereals infant formula contained a mean of 0.050 $\mu g\ kg^{-1}$ OTA [152]. In another study, 63 infant formula and follow on formula and baby foods were randomly collected from pharmacies and supermarkets in Ankara, Turkey. OTA levels were assessed by commercially available enzyme-linked immunosorbent assay (ELISA) kits. OTA levels were found in 40% of the samples between 0.27-4.50 ppb [153].

3) The Presence of OTA in Yogurt

In yogurt, bifidobacteria and other yogurt bacteria can reduce OTA levels. When the OTA levels in milk samples do not exceed 0.1 mg/L, it can be reduced entirely due to *Lactobacillus bulgaricus* LB-51, *Bifidobacterium bifidum* L, *Lactobacillus delbrueckii* subsp, *salivarius* subsp and *Streptococcus thermophilus* T4. For milk samples with higher OTA concentrations levels, they can significantly reduce OTA. Furthermore, the yogurt bacteria can reduce the OTA content significantly during the milk fermentation period. In samples, including those with 0.05, 0.1 and 0.5 mg/L of toxins, OTA disappeared completely, while 36 and 26 percent decreases were found in the samples with OTA 1.0 and 1.5 mg/L, respectively [154, 155]. The conversion of OTA to isocoumarin can remove OTA (reduction of OTA). L-β phenylalanine and ochratoxin α (OTα) result from breakage of the amide bond that links those two moieties of OTA [156, 157].

2.2.5. The Presence of OTA in Meat and Meat Products

Many farm animal feedstuffs are based on cereals and cereal by-products, which are the preferred substrate for growth of *Penicillium* and *Aspergillus* spp [158], and therefore, the feed may become contaminated with OTA. Carry-over of OTA from feed to animal products has been demonstrated in pigs and poultry, as well as in other species [159]. Ruminants have a higher tolerance to OTA exposure than mono-gastric animals because protozoa in the rumen are able to degrade OTA into ochratoxin α and other less toxic metabolites [160]. Samples of fresh pork meat, cured pork meat, canned luncheon meat, poultry meat, lamb meat, sausages, eggs and cheese were analyzed by LC-MS/MS. OTA was found only in the samples of fresh pork meat (OTA concentrations 0.03 and 0.23 ng/g), cured pork meat (OTA concentrations 0.06 and 0.20 ng/g) and sausages (OTA levels 0.12 and 0.06 ng/g) [161]. And Pleadin et al. (2015) conducted a study to investigate OTA contamination of “Slavonski Kulen” during a six-month production process, results showed that no statistically significant differences ($p > 0.05$) were observed among the intact samples on any above-given production day, the maximal OTA value attributable to raw material contamination thereby being 3.18 mg/kg. As for the damaged samples, the mean OTA concentration established on the production day 180 was 7.92 ± 6.50 mg/kg, while the maximal observed OTA value equaled to 17.0 mg/kg [162].

Farmed animals exposed to OTA by plant source feed may cause contamination of fresh meat with OTA, an important source of exposure. Guillamont et al. (2005) detected 0.12μg/kg OTA in one sample of pig meat in Portugal [163]. A low frequency of contamination was reported in Romanian meat samples (17%), although matching samples of kidney and liver had higher frequencies of contamination or indirect transmission [164]. The other source is molds in fermented meat products, as molds of the *Penicillium* genus are used as typical starter cultures for dry sausage production in Europe and Croatia. These species are surface-inoculated because they are thought to improve sausage aroma, texture and appearance. Markov et al. (2013) detected the possible presence of ochratoxin A (OTA) in game sausages, semi-dry sausages and fermented dry-meat products, randomly taken from individual producers and the Croatian market. The analysis of different types of meat products resulted in OTA identification in 64.44% [165]. And the results of Pleadin et al. (2015) also demonstrated that casing damaging witnessed during a long-term dry-cured meat products' ripening can result in OTA entry from the surface into the product, causing a significant OTA contamination [162], it is also a source of OTA exposure for meat.

2.2.6. The Preventive Measures and the Limit Standard of OTA

The International Agency for Research on Cancer (IARC) classified OTA as a Group 2B human carcinogen in 1993 [166]. As OTA is widely distributed in many areas and many types of food, measures need to be taken to prevent exposure to OTA.

Due to the health hazards to humans and animals, the European Commission (2006) has set maximum permitted levels for OTA at 5 μg/kg for whole grains and 3 μg/kg for processed products. However, this has often been difficult to achieve in practice [167].

3. Safety Evaluation/Risk Assessment

There is great potential concern about the existing relationship of OTA with organ toxicity in pigs and rodent renal tumors, particularly among male rats. The International Agency for Research on Cancer concluded in 1993 that ochratoxin A is a “possible human carcinogen.” Food products, including rice, oats, wheat, beer, coffee, soy, corn, wine, raisins, and grapes, are susceptible to OTA contamination. People who frequently consume these foods may have higher exposures to OTA. Almost all consumers are potentially exposed to OTA. It is impossible to entirely eradicate OTA from our everyday diet. OTA is a mycotoxin with carcinogenic, immunosuppressive, teratogenic and nephrotoxic effects in many species, although there are differences relating to species and sex in the magnitude of these effects [168, 169]. The National Toxicology Program (NTP) found that OTA was one of the most potent renal carcinogens that has been identified so far [168]. Moreover, some studies indicated that OTA may be responsible for nephropathy in pigs and a chronic degenerative kidney disease with a high incidence of urinary tract tumors in humans [168, 170, 171],

Balkan endemic nephropathy (BEN). Thus, a good understanding of the foods that contribute to overall OTA exposure is important.

Risk assessment includes two processes, one is the hazard assessment and the other is the exposure assessment. The two processes are typically conducted independently, and an estimation of risk is provided through comparing the results of these assessments. The hazard assessment is the evaluation of toxicology data of a substance from human and/or animal studies. A quantitative estimate of dietary intake could be established by data from food intake surveys coupled with concentration data of the contaminants in foods to assess the exposures to contaminants from the diet. Based on the most sensitive adverse effects, which is found in humans, the dose-response curve can be determined. By comparing this dose to the results of the exposure assessment, the probability of adverse effects can be characterized.

3.1. Exposure Assessment

Extensive efforts have been made for exposure assessment in the past few years.

In 2003, Je Won Park conducted a survey for ochratoxin A in various domestic food commodities—14 unstrained rice wines (makkolli), 46 beers, 35 wheat flours, 22 barleys and 60 polished rice samples—to evaluate the exposure of Koreans to ochratoxin A [172]. They used immunoaffinity columns and high-performance liquid chromatography (HPLC)-fluorescence detection to analyze ochratoxin A, and using HPLC-tandem mass spectrometry, the positive samples were confirmed. Moreover, they obtained the Korean probable daily intakes (PDI) of OTA by combining the results of different surveys on the levels of OTA in selected foods. Staple rice is the major contributor (>90%) to the Korean dietary intake of OTA. The mean levels of OTA in the polished rice are the highest, which ranged from 0.2 (not detected, i.e., ND = 0) to 1.0 ng/g (ND = limit of detection, i.e., LOD). For all Koreans, the estimated PDI ranged from 0.8 to 4.1 ng/kg bw/day. While the estimate for heavy consumers did not exceed the PTDI value (14 ng/kg bw/day), it fell into the range of 1.7-9.1 ng/kg bw/day. They concluded that for the average Korean consumer, there is no considerable risk for OTA exposure at present.

In 2004, Wu and his team identified the nations that would be most heavily impacted by tighter mycotoxin regulations, examined costs and benefits as a function of regulatory stringency, and addressed risk-risk tradeoffs between health benefits and economic losses from compliance with those regulations. Their study informs that endeavor by a risk assessment and economic analysis of two important mycotoxins: aflatoxins and fumonisins. They found that the adoption of tighter standards for mycotoxins, from which there might be economic losses, would be unlikely to improve health significantly [173].

Later, T. Kuiper-Goodman and his team derived the negligible cancer risk intake (NCRI), re-evaluated the tolerable daily intake (TDI), and conducted a probabilistic risk assessment to manage risk of ochratoxin A (OTA) in foods [174]. It is of great significance for the development of a risk assessment associated with OTA. To derive 'usual' probabilistic exposure in the presence of variable occurrence data, a new approach was developed. Canadian occurrence data were used for various finished foods or raw food commodities and were combined with the United States Department of Agriculture (USDA) food consumption data, which included data on young children and infants. In the exposure estimates for various age/sex groups, not only the uncertainty but also the variability in input data were considered.

Wheat-based foods, such as raisins, rice, and oats, were the major contributors of OTA for children. In older individuals, coffee, wine, and beer contributed to total OTA exposure. Generally speaking, most people were exposed to OTA on a daily basis. For all age/sex strata, the mean adjusted exposures were generally below the NCRI, which is 4 ng/kg/bw, except for 1–4-year-olds because of their lower body weight. When European Commission maximum limits were applied to the occurrence data, predicted exposure to OTA decreased. As Kuiper-Goodman and his team stated in their experiment, mycotoxin concentration can vary by several orders of magnitude within a particular food commodity. Thus, the exposure estimates for the upper percentiles are based upon iterations which combined the highest contaminant estimates of the input ranges and the upper range of food consumption data. These data may be of great use for assessing the acute health risks but could not reflect the chronic intakes because the highest consumers do always eat the most contaminated foods. T. Kuiper-Goodman and his team examined the impact on risk for regular consumers of specific commodities at the same time. While the evidence to demonstrate adverse effects of OTA at current dietary exposures from human studies is inadequate in contrast to the animal data, it is reasonable to assume that OTA would cause similar toxicities in humans at high enough exposures based on the pig and rodent studies. Nevertheless, there are no obvious epidemiology studies to identify the conclusive link between OTA from the diet and the adverse health effects in humans. However, it has been implied in the study of Gluhovschi G in 2001 that OTA might be linked to BEN. Balkan endemic nephropathy (BEN) is a chronic tubulointerstitial nephritis observed primarily in countries in the Balkan Peninsula. The paper analyzes the main etiological factors suspected to play a role in BEN: mycotoxins, aristolochic acid, genetic factors, viruses and toxic substances from Pliocene lignite. Moreover, the OTA contamination in BEN-endemic areas was found in higher concentrations than those in non-endemic areas [175]. However, a link has not been proven so far. Recently, more studies have implicated aristolochic acid as the etiological agent. Aristolochic acids are human carcinogens that have recently been listed in the Twelfth Report on Carcinogens. In 2010, exposures to OTA were estimated by Kuiper-Goodman et al. They used various methods to analyze the input date and for treatment of the exposure models, specific to the present case of estimating exposure to OTA [174]. The approaches are outlined as follows:

1) Occurrence data: In the assessment, occurrence data of OTA were available for 37 different types of food commodities or categories known to have the potential to contain OTA .Canadian occurrence data for various finished foods and raw food commodities gathered over the past few years were used. Some of these commodities were peas, barley, oats, rice, raisins, breakfast cereals (including wheat-based multigrain, rice, corn, and oat cereals), pasta, infant cereal, and wheat (soft, hard, and durum). Beverages included grape juice, wine, beer, and coffee (decaffeinated and ground regular, decaffeinated and instant regular).
2) Processing factors: Processing factors were used to change the OTA concentration in the raw grain to the foods as processed to account for redistribution of OTA during the milling process.
3) Handling of censored data: To address occurrence values that fell below the limits of detection/quantification, two methods were used, including the use of ½ the LOQ or ½ the LOD as appropriate, and a distribution conducted by the authors.

4) Data on food consumption: The age/sex groups refer to representative population groupings based on the reported gender and ages of the dietary recall survey respondents. The United States Continuing Survey of Food Intakes by Individuals (CSFII) food intake survey of 1994–1996 and 1998 is the basis of the exposure estimates and includes food consumption data from two nonconsecutive survey days for >20,000 persons, of all ages. To ensure these data were appropriate for Canada, validation procedures were run.
5) The maximum levels (ML): In European countries, the maximum levels (ML) of OTA have been set for many food commodities, and those maximum levels were also used in the exposure assessment models. OTA occurrence values above these levels were assigned values equal to these levels ("ML-modified distribution of occurrence.") when the MLs were used in the models.

Recently, two studies assessing the concentrations of OTA in foods sold in Canada were published [176,177]. In the first study, approximately 200 rice samples from several different countries were analyzed for OTA in two different years [176]. The mean concentrations of ochratoxin A (OTA) were 0.05 and 0.005 ng/g in year 1 and year 2, respectively; incidences of samples containing ≥LOD were 43% and 1%, respectively, in the 2 years. The limit of detection (LOD) was 0.05 ng/g. In comparison, in Kuiper-Goodman's exposure assessment, the mean occurrence value of OTA in rice was reported to be 0.68 ng/g when positives exceeding the European ML were set to the ML and 0.80 ng/g with no ML [174]. On the other hand, Tam and his team developed a sensitive and selective analytical method for the quantitation of ochratoxin A (OTA) in the Canadian Total Diet Study [177]. A total of 140 diet composites were analyzed for OTA as part of the Canadian Total Diet Study. Samples were collected from two Canadian cities, Calgary and Quebec City, in 2008 and 2009, respectively. The results indicate that the levels of OTA in 73% (102/140) of the samples can be detected, and levels of OTA contamination in bread supply were found to be the highest in the Canadian study.

Lois A. Haighton reevaluated the data published in the Health Canada risk assessment and assessed risk on a lifetime basis rather than by individual age strata [178]. By using data from Canada to evaluate the results of exposure assessments for OTA in food and comparing these exposures to doses of OTA anticipated to be without appreciable risk, that review was accomplished.

3.2. Hazard Assessment

The toxicities of OTA that have been shown in animal studies include carcinogenicity, teratogenicity, immunosuppression and nephrotoxicity, and sensitivity differences relating to gender and species have been noted. In a follow-up study [179], exposure of six pigs to approximately 40 μg/kg bw/day for two years caused increased nephropathy but no further accumulation of ochratoxin A residue. By using concentrations of free fractions of OTA in plasma and glomerular filtration rates, Hagelberg calculated clearance of OTA by renal filtration. Then, he observed a significant difference in the kinetics of OTA in the pig, which suggests that the species is more sensitive to OTA [180]. The clearance of most species (such as humans, rat, mouse, quail, and monkey) is calculated to range from 1 to 14% and at least

10-fold lower than that in the pig by 110%. Other species differences in toxicokinetics of OTA have been recently reviewed [181]. OTA also caused an increased incidence of kidney tumors in rat studies at higher doses of 50 μg/kg/day, which was particularly evident in male rats. In a recent review, differences in the metabolism of OTA between various species is suggested to be responsible for the species differences in sensitivity to OTA toxicity [173]. In a 90-day subchronic animal study in groups of 6 to 11 pigs, renal enzyme dysfunction and renal function changes were obvious at all doses, and the lowest observed-effect level of 8 μg/kg bw/day was determined to be for renal function changes [182]. The tumorigenic response occurred at a higher doses than those related to effects in pigs. However, if associated with a non-threshold mechanism of action, cancer is considered to be a more sensitive endpoint.

Risk analysis is based mostly on natural occurrence data and published toxicological data in its widest sense. There are several considerations concerning ochratoxin A:

1) Published experimental toxicology mostly focuses on positive effects both with whole animals and with tissue preparations or tissue cultures, and in many cases, the doses of OTA used are much higher than the natural abundance. Due to the importance placed on publishing in late 20th century, there is little encouragement both in career development and funding to do well-replicated experiments determining undramatic nil effects. Nil effects data may be important for risk analysis.
2) It is difficult to extrapolate from experimental animals to the human. Humans are atypical experimental animals that are different from experimental animals in many ways. Human have longer lives, and they can obtain cleaner food in a mixed diet, particularly in some nations with sound food safety. It is very dangerous to extrapolate backwards from the more florid end of a dose–response curve because toxic effects are difficult to relate to any natural context and are generally unrealistic. What some researchers believe is needed is a large-scale low-dose study in the monkey over many years, but the cost may not justify the end.
3) The occurrence of OTA is increasingly widespread. It has been detected by more and more sophisticated analysis, especially involving selective but expensive immunoaffinity clean-up columns that are designed to find a few nanograms in kilograms of natural substrates. Data arising from such analysis might increase alarm, especially in analyses that are associated with human urine, milk or serum, and they are related to the use of the term ochratoxicosis. This term implies that there might be some easily associated diseases, but it is incorrect. In any case, it is necessary to confirm that the analytical data used for risk analysis come from laboratories with validated methodology.
4) Use of tissue cultures are increasingly being encouraged for ethical reasons. The regulations for animal project license applications have become stricter in the United Kingdom. However, there are some difficulties for risk analysis using tissue culture experiments with OTA, which has a wide spectrum of effects in the whole animal, does not enter into the cultured animal cells via a liver and is not greatly bound to serum proteins.

3.3. Other Risk Assessments

The PTWI of OTA was evaluated by The JECFA (Joint FAO/WHO Expert Committee on Food Additives) and was established to be 112 ng/kg bw at its 37th meeting. However, at the 44th meeting, the evaluation was essentially confirmed and the PTWI was established to be 100 ng/kg bw based on the pig data in a two year study [182]. Nevertheless, based on kidney function, this level was reported to be a LOEL in the 90-day study. The PTW was derived using an increased safety factor of 500. It has been claimed that the carcinogenicity of OTA was not considered by the JECFA in its evaluation [168]. The JECFA did review the carcinogenicity of OTA, and concluded that renal tumors were only observed at above the maximum tolerated dose (MTD). The data (including the available genotoxicity data) suggested that there was a secondary mechanism involving chronic toxicity to cells of the renal tubule, which could be considered a threshold by the chronic toxic NOEL. Therefore, using the most sensitive index (renal toxicity) in the most sensitive species (pig) might be the most conservative approach if tumors arise only above the MTD. This approach is more logical for relating food intake data to risk than the TDI (based on the human half-life, it would take several weeks to achieve steady state levels at a steady intake level).

An assessment used both the safety factor approaches and a quantitative risk assessment by the Canadian authorities [183]. This evaluation made particular note of the aggressive, invasive nature of the renal tumors in rodents, which it considered most probably reflected a genotoxic mode of action. The safety factor approach was applied to the NOEL, the NEL, the TD_{50} and the benchmark dose, and a safety factor of 5000 was used (50,000 for the TD_{50}) giving rise to TDI estimates of 1.5 to 5.7 ng/kg bw. The risk assessment that was used in this evaluation was made by a model-independent low dose extrapolation with an incremental risk of 10.5 and resulted in a TDI of 1.8 ng/kg bw. According to the principles of Environmental Health Criteria 170 [184], the reason why they used a factor of 5000 was not scientifically derived but appeared to be arbitrary due to "concerns about the toxic and pharmacokinetic properties." This has no real rationale for a number of reasons. Firstly, if it is the mechanism of carcinogenesis then the nature of the toxicity has no meaning below the threshold. Conversely, it is inappropriate to use the safety factor approach if it is assumed that there is no threshold for the carcinogenic effect. On the other hand, the relationship between the toxicity and pharmacokinetics is unclear because slower metabolism may lead to reduced risk if metabolism to a reactive electrophile is supposed. With regard to the toxicodynamics, it seems more appropriate to use data derived from the most sensitive species, such as the pigs, to calculate a data-derived safety factor. Several other metabolites of varying toxicity have been identified in urine or feces of rats, both of which are important routes of excretion as OTA is metabolized by the intestines, liver, and kidneys. Thus far, the rat is the only species in which a lactone-opened form of OTA with higher toxicity compared to OTA has been identified. Thus, further study is necessary to characterize the toxicity of the various metabolites and to clarify if differences in sensitivities of animals to OTA is related predominantly to different metabolic profiles.

The Scientific Committee for Food stated its opinion on OTA (European Commission, 1998) , cognizant of the earlier evaluations mentioned above. This committee concluded that OTA is a carcinogen, a potent nephrotoxic agent that has genotoxic properties. Hence, it would be prudent to reduce exposure as much as possible, ensuring that exposures are "towards the lower end of the range of tolerable daily intakes of 1.2 to 14 ng/kg bw estimated by other bodies e.g., below 5 ng/kg bw." Because it is not obvious what calculation was used to derive the TDI nor why this figure is optimal within the cited range or whether it is merely arbitrary, this conclusion is not transparent.

From the above, we can find that different jurisdictions are arriving at different interpretations of the toxicological data and hence at different TDIs or PTWIs [185]. In reviewing the epidemiological evidence, the IARC (International Agency for Research on Cancer) has concluded that OTA is a possible human carcinogen but that there is inadequate evidence of carcinogenicity in humans and, as indicated earlier, the evidence of a causal association with endemic nephropathy remains inconclusive. Therefore, the potential for conflicts concerning the standards for commodities in international trade is obvious. Indeed, nine countries have specific legislation for OTA with limits from 5-50 μg/kg applied either to particular cereals, some foods or all foods [189]. It is necessary to arrive at a consensus both for necessary risk management procedures and in relation to the risk assessment.

In February 2001, the Codex Alimentarius Committee on Food Additives and Contaminants placed OTA on its re-evaluation priority list and referred it to the JECFA at a meeting of the JECFA based on the divergent risk assessments and risk management measures. The most effective method for the protection of human health is to minimize the conditions that give rise to production of the toxin and the contamination with fungi producing OTA. Therefore, the introduction of maximum limits for cereals might be justifiable when these commodities are considered to be the major source of exposure. To control and identify the causes of contamination of coffee, a project is conducted by the FAO, which is likely to be the most useful in view of human health and economical issues [185]. And the summary of calculated tolerable human intakes of OTA by international organization is shown in Table 4 [187], and Table 5 shows the standards for ochratoxin A set by European Union and Canada.

Table 4. Summary of calculated tolerable human intakes of OTA by international organization [187]

Organization	Tolerable intake metric*	Limit	Reference
European Food Safety Authority (EFSA)	PTWI	120 ng/kg bw/week	[121]
Health Canada	PTDI	3 ng/kg bw/day	[188]
Health Canada	NCRI	4 ng/kg bw/day	[188]
Joint FAO/WHO Expert Committee on Food Additives (JECFA) 2007	PTWI	100 ng/kg bw/week	[189]
Nordic Expert Group on Food Safety	TDI	5 ng/kg bw/day	[190]
Scientific Committee of Food (SCF) of the European Union	PTDI	5 ng/kg bw/day	[121]

*TDI, tolerable daily intake; PTDI, provisional tolerable daily intake; PTWI, provisional tolerable weekly intake; NCRI, negligible cancer risk intake.

4. THE CLASSIFICATION AND STRUCTURES OF OTA

Ochratoxins are a family of mycotoxins, which are structurally related cyclic pentaketides: 3, 4-dihydromethylisocoumarin derivatives linked with an amide bond to the amino group of L-β-phenylalanine [192]. The chemical structures of OTA and OTA metabolites are shown in Figure 2 [6].

Table 5. Standards for ochratoxin A [191]

Food commodity	European Union[a] maximum levels (μg/kg)	Canada[b] (proposed) maximum limits (ppb)
Unprocessed cereals/raw cereal grains[c]	5	5
All products derived from unprocessed cereals, including processed cereal products and cereals intended for direct human consumption with exception of foods for infants and young children and foods for special dietary purposes	3	
Direct consumer grains (i.e., rice, oats, pearled barley)		3
Derived cereal products (flour[d])		3
Derived cereal products (wheat bran)		7
Breakfast cereals		3
Grape juice (and as ingredients in other beverages) and related products (concentrated grape juice, grape nectar, grape must intended for human consumption)	2	2
Dried vine fruit (currants, raisins, sultanas)	10	10
Wine, fruit wine, aromatized wine, aromatized wine-based drinks and aromatized wine-product cocktails	2	
Roasted coffee beans and ground roasted coffee excluding soluble coffee	5	
Soluble coffee (instant coffee)	10	
Baby foods and processed cereal-based foods for infants and young children	0.5	0.5
Dietary foods for special medicinal purposes intended for infants	0.5	0.5

[a] Commission Regulation (EC) No 1881/2006 of 19 December 2006 (also includes spices and liquorice; Commission of the European Communities, 2006);

[b] Health Canada's proposed Maximum Limits (Standards) for the Presence of the Mycotoxin Ochratoxin A in Foods in February 2009 (Health Canada, 2009)';

[c] takes into consideration the reducing effect of processing or redistribution;

[d] For bread, pastries and other flour-based foods, the guidelines to pertain to the flour portion. In the future, based on further monitoring data, Health Canada may consider modifying these maximum limits (MLs), or introduce MLs for products not yet cover.

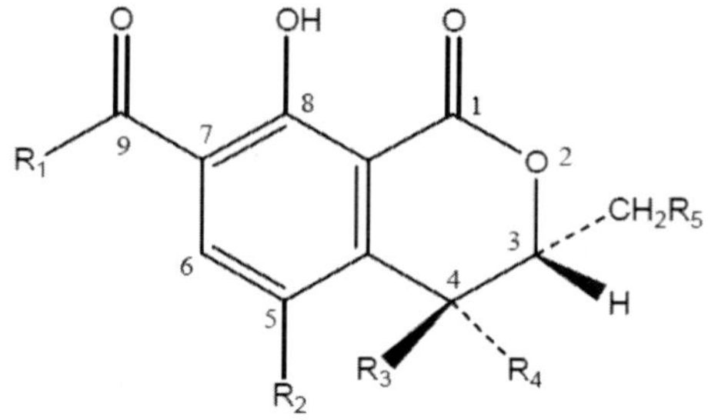

Name	R1	R2	R3	R4	R5
Natural ochratoxins					
Ochratoxin A	Phenylalanine	Cl	H	H	H
Ochratoxin B	Phenylalanine	H	H	H	H
Ochratoxin C	Ethyl-ester, phenylalanine	Cl	H	H	H
Ochratoxin A Methyl-ester	Methyl-ester, phenylalanine	Cl	H	H	H
Ochratoxin B Methyl-ester	Methyl-ester, phenylalanine	H	H	H	H
Ochratoxin B Ethyl-ester	Ethyl-ester, phenylalanine	H	H	H	H
Ochratoxin α	OH	Cl	H	H	H
Ochratoxin β	OH	H	H	H	H
4-R-Hydroxyochratixn A	Phenylalanine	Cl	H	OH	H
4-s-Hydroxyochratoxin A	Phenylalanine	Cl	OH	H	H
10-Hydroxyochratoxin A	Phenylalanine	Cl	H	H	OH
Tyrosine analog of OTA	Tyrosine	Cl	H	H	H
Serine analog of OTA	Serine	Cl	H	H	H
Hydroxyproline analog of OTA	Hydroxyproline	Cl	H	H	H
Lysine analog of OTA	Lysine	Cl	H	H	H
Synthetic ochratoxins					
d-Ochratoxin A	d-phenylalanine	Cl	H	H	H
Ochratoxin A Ethyl amid	Ethyl amid, phenylalanine	Cl	H	H	H
O-methyl Ochratoxin A	Phenylalanine, $OHCH_3$ on C-8	Cl	H	H	H

Figure 2. General structure of ochratoxin A metabolites.

There are several metabolites related to OTA. Four major metabolites have been identified in the past—ochratoxin B (OTB), ochratoxin C (OTC), ochratoxin α (OTα), and ochratoxin β (OTβ). OTB, $C_{20}H_{19}O_6$, is the dechloro analog of OTA, a colorless crystalline compound that differs from OTA by the absence of a halogen group. OTB has an H-group instead of a Cl-group, and it is less toxic than OTA [193-195]. Further structural alterations yield OTC, the amorphous ethyl ester form of OTA, which is generally perceived as having little or no toxic potential. OTα is a free carboxylic acid that represents the dihydroisocoumarin nucleus of OTA [193]. It was reported that when OTA is exposed to white and blue light, the carbon atom C-14 and the nitrogen atom will cleave, and the new compound OTα will be formed as a reaction product, which has been proposed based on mass spectrometry (MS) experiments, and its dechloro analog is ochratoxin β (OTβ). The four derivatives differ very slightly from each other in structure, and they are also different in elemental composition, as OTB and OTβ lack the halogen group. As a result, they have considerable differences with regard to toxic potentials, with ochratoxin A (OTA) being both the most commonly detected and the most toxic of the four major forms of OTA derivatives [8, 193-195]. Figure 2 presents the general structure common to these different metabolites, and also shows the characteristic composition of each one. Recently, some new OTA-derived metabolites have been identified, which include a dechlorinated OTA derivative identified by nano-ESI-IT-MS [196] and a quinone/hydroquinone metabolite with toxicological properties [196]. In addition, it has been predicted that OTA will form a benzoquinone electrophile following activation by cytochrome P450 enzymes and radical species following activation by enzymes with peroxidase activities [197, 198]. These electrophiles react preferentially with deoxyguanosine (dG) to form benzetheno adducts and C8-dG adducts, respectively [198].

4.1. The Physico-Chemical Properties of Ochratoxin A

In 1965, Vanderme et al. identified the properties of OTA. OTA: $C_{20}H_{18}O_6NCl$, white, odorless, crystalline solid, mol. wt. 403 (mass spec.), m.p. = 169 °C (from xylene), $[\alpha]_D^{2\nu}$-118° (c.1.1 in $CHCl_3$), and show λ_{max}^{Etofl} at 215 and 333 mμ (ε34,000 and 2,4000, respectively). Its infra-red spectrum showed the presence of a carboxyl (ν_{max} 1690 cm^{-1} and broad absorption between 2,700 and 2500 cm^{-1}), a secondary amide (ν_{max} 1,665, 1,535 and 3,430 cm^{-1}) and a lactone (ν_{max} 1,745 cm^{-1}) group, while a positive ferric chloride reaction indicated the presence of a phenolic hydroxyl group. Acid hydrolysis gave L-phenylalanine and an acid, $C_{11}H_9O_5Cl$. The methyl ester of this acid was reduced with lithium borohydride and gave a triacetate on acetylation. The nuclear magnetic resonance spectra of $C_{11}H_9O_5Cl$ acid and derived triacetate are only interpretable in terms of a dihydroisocoumarin system with a methyl group in position 3. The arrangement of the different substituents in this system was determined by ultra-violet spectroscopy and color tests [8].

4.1.1. Physical and Chemical Characteristics of OTA

OTA is a weak organic acid with a molar mass of 403.8 $g.mol^{-1}$. The crystalline structure varies from colorless to white, and this molecule possesses an intense green fluorescence under UV light in acid and blue fluorescence in alkaline conditions [4]. It has a melting point of approximately 90°C when crystallized with benzene as a solvate. However, non-solvated crystals with a melting point of 169°C have been obtained from xylene, which are suitable for X-ray structural analysis. The solubility of OTA is not dependent on the pH of the solution. In acid and at a neutral pH, OTA is slightly soluble in water, and it is insoluble in petroleum ethers and saturated hydrocarbons, but it is soluble in polar organic solvents (alcohols, ketones, chloroform). While in alkaline conditions, this molecule is soluble in aqueous sodium bicarbonate solution and in all alkaline solutions in general [199]. Extraction methods have been developed based on the solubility of OTA. One extraction step before determination of OTA was carried out with 50 mL methanol and 3% aqueous $NaHCO_3$ (1:1, v/v) [200]. OTA is optically active, and its spectral characteristics are shown in Table 6. Heat or radiation can lead to mycotoxin destruction. OTA exhibits UV adsorption: λ_{max}^{MeOH} (nm; ε) = 333 (6400) [199]. The fluorescence emission maximum is at 467 nm in 96% ethanol, while in absolute ethanol, the fluorescence emission maximum is at 428 nm. The infrared spectrum in chloroform includes peaks at 3380, 1723, 1678 and 1655 cm^{-1} [201].

Chemically, OTA is a weak organic acid. The pKa values are in the range 4.2-4.4 for the carboxyl group of the phenylalanine moiety and in the range 7.0-7.3 for the phenolic hydroxyl of the isocoumarin group, respectively [209].

Thermostability of OTA. The thermostability of OTA is closely related to the heating conditions. First, cooking style is one of the factors that affects the thermostability of OTA. OTA is generally stable at temperatures used during ordinary cooking. In 1982, Müller [210] showed that the OTA is only partially degraded under normal cooking conditions. Moreover, this molecule can resist three hours of high pressure steam sterilization at 121°C [211], and OTA is not completely destroyed even when the heating temperature is raised to 250°C [212]. At high temperatures during coffee roasting, researchers observed that OTA degraded into OTα, a thermal OTA product [213]. However, the degradation ratio was quite different. Initial studies indicated that OTA levels can be decreased by 77–87% [214], 80–90% [215], and even 90–100% [216] during the roasting process; however other reports also showed the reduction was only 0–12% [217]. This indicates that many factors may affect the degradation ratios. Temperature, time and moisture are three important factors that may influence the stability of OTA. Urbano et al. (2001) [218] found that temperature was an important factor for OTA degradation. OTA content was decreased by 22% when roasting at a temperature of 200°C for 10 minutes. However, the OTA content can be decreased by up to 94% by increasing the temperature to 220°C for 15 minutes. Boudra et al. (1995) [212] reported on the thermostability of OTA in wheat under two moisture conditions, as the moisture is an important factor for the stability of OTA. This work examined the decomposition of OTA under different temperatures and moisture conditions. The results showed that OTA is heat stable, and the calculated half-lives were different under different moisture conditions. For dry wheat heated at 100, 150, 200, and 250 ℃, the calculated half-lives were 707, 201, 12, and 6 min, respectively. When wheat was heated under wet conditions at temperatures of 100, 150, and 200 ℃, the half-lives of OTA were 145, 60, and 19 min, respectively. When heating at 100 and 150 ℃, the presence of water (50%) increased the degradation of OTA, but when the temperature increased to 200 ℃, the opposite case was observed. The half-life of OTA in the wet wheat is longer than the dry wheat, which indicates that OTA can be more easily degraded in dry conditions than wet conditions at this temperature. However, complete destruction of OTA within the limits of this study (100 to 250 ℃) was not obtained. In general, many reports indicated that initial contamination levels, spiking methods, selectivity and sensitivity values, inhomogeneous toxin distribution, and roasting and drying conditions are important factors for the thermostability of OTA [219].

OTA cannot be destroyed by freezing (-20^{0}C) or the defrosting (26^{0}C) process, but the conidia of mycotoxin-producing fungi can be destroyed under these conditions [220, 221].

OTA can be destroyed by irradiation, gamma-ray irradiation and electron beam (EB) irradiation. Degradation is related to the conditions—the solution that dissolves the OTA and the moisture level are two important factors for OTA degradation. Gamma irradiation (up to 7.5 mrad) of OTA in ethanol cannot induce degradation of OTA. However, degradation is observed at low moisture levels when OTA has been treated with an excess of sodium hypochlorite (NaOCl) [222]. Gamma irradiation from 2 to 5 kGy results in degradation of OTA [223]. Peng et al. also reported that electron beam irradiation (EB) can induce the degradation of OTA. OTA degradation efficiency increased in an irradiation dose-dependent manner and decreased with increasing substrate concentration, and the effect is more significant in water than in acetonitrile and methanol–water (60:40, v/v). This study also found that pH and the addition of other chemicals also affect the degradation efficiency, as alkaline conditions (pH 9.99) and the addition of H_2O_2 (<0.1% v/v) substantially improved the OTA degradation efficiency [224].

Table 6. Spectral characteristics of ochratoxin A [6]

Spectral	Solvents	Characteristics	References
UV-VIS	ETOH	λmax = 213nm (ε36.800) λmax = 332nm (ε 6.400)	[202]
Fluorescence	ETOH 96% ETOH/ABS.	λmax = 467nm λmax = 428nm	[202]
IR	$CHCl_3$	3380; 2988; 1723; 1674; 1528; 1425; 1381; 1304; 1260; 1170; 1140; 1107; 827 cm^{-1}	[203]
NMR ^{1}H250- MHZ	CDC13	δ 12,70; δ 10,80; δ 8,55 (3H); δ 7,23; δ 7,15 (H Aromatic); δ 4,71; δ 5,07 (CH); δ 2,78; δ 3,2 (CH_2); δ 1,55 (CH_3)	[199, 204-206]
MS	—	m/z 239/241 m/z 255/257 molecular ion m/z 403	[207, 208]

Moreover, it was also reported that exposure to fluorescent light can induce OTA degradation. Schmidt-Heydt et al. reported that OTA was sensitive to white and blue light [186], and, when exposed to white and blue light, the carbon atom C-14 and the nitrogen atom cleaved, with OTα as a reaction product [187]. Exposure to UVB for different periods of time is efficient in preventing OTA production in a liquid medium.

Multiple destruction methods have been explored to reduce the contamination of OTA in food based on its physico-chemical characteristics. Chapter 2 will introduce these methods specifically.

4.1.2. Chemical Agents Which Can Degrade OTA

Chemicals including various acids, bases, oxidizing agents or reducing agents, chlorinating agents, salts, and miscellaneous reagents, such as formaldehyde, are effective at destroying mycotoxins.

OTA could be eliminated by alkaline treatment, as many alkaline substance such as NH_3 and NaOH can interact with OTA at a high temperature. It was reported that OTA can be completely decomposed by ammoniation in corn, wheat, and barley [225]. In cocoa, more than 98% of the OTA could be eliminated by alkaline treatment [203]. Alkaline hydrogen peroxide, sodium hydroxide, and monomethylamine or ammonium with calcium hydroxide treatments can also destroy OTA [225]. Bazin et al. (2013) demonstrated that the increase in pH by adding polyethylene glycol (PEG) to wine led to an underestimation of OTA by conversion of OTA into open ring OTA OP-OA [226]. In coffee beans, it was found that treatments with methylene chloride, ethyl acetate, and dichloromethane supplemented with

2% formic acid were able to reduce OTA levels by up to 80% [227, 228]. Another method which can destroy OTA is ozonization; the application of ozone (O_3) for OTA removal up to undetectable levels in foodstuffs, such as vegetables, grains, nuts, is viable with the development of sophisticated electrochemical techniques [229, 230]. OTA was degraded using 10% wt O_3 in an aqueous solution in 15 sec [230]. Formic acid, sorbic acids, and propionic and sodium hypochlorite are effective chemicals that degraded OTA; when OTA was exposed to these chemicals at concentrations ranging from 0.25% to 1%, it was degraded after 3–24 hours [231]. It should be mentioned that chemical treatment is not allowed in the European Union for commodities destined for human consumption. These chemical properties only provide the theoretical basis for the degradation of OTA, and studies investigating the detoxification of OTA need to take into full consideration the toxicity of the breakdown products and the changes in the nutritional qualities of the foodstuffs. Chapter 2 will give an account of these studies.

REFERENCES

[1] Kuiper-Goodman, T., Scott, P.M., (1989). Risk assessment of the mycotoxin ochratoxin A. *Biomedical and environmental sciences : BES*. 2, 179-248.

[2] (1994) *Mycotoxins in grain: compounds other than aflatoxin.*

[3] Bredenkamp, M.W., Dillen, J.L.M., Vanrooyen, P.H., Steyn, P.S., (1989). Crystal-structures and conformational-analysis of ochratoxin A and ochratoxin B - probing the chemical-structure causing toxicity. *Journal of the Chemical Society-Perkin Transactions 2*, 1835-1839.

[4] Budavari, S.E., (1989). The merck index: an encyclopedia of chemical, drugs, and biologicals. *11th. ed.*

[5] Bredenkamp, M.W., Dillen, J.L., van Rooyen, P.H., Steyn, P.S., (1989). Crystal structures and conformational analysis of ochratoxin A and B: Probing the chemical structure causing toxicity. *Journal of the Chemical Society, Perkin Transactions 2*, 1835-1839.

[6] El Khoury, A., Atoui, A., (2010). Ochratoxin A: general overview and actual molecular status. *Toxins (Basel)*. 2, 461-493.

[7] Bräse, S., Gläser, F., Kramer, C., Lindner, S., Linsenmeier, A.M., Masters, K.-S., Meister, A.C., Ruff, B.M., Zhong, S. (2013) *The chemistry of mycotoxins*. Springer Science & Business Media.

[8] Vanderme.Kj, Steyn, P.S., Fourie, L., Scott, D.B., Theron, J.J., (1965). Ochratoxin A a toxic metabolite produced by aspergillus ochraceus wilh. *Nature*. 205, 1112-+.

[9] Krogh, P., Hald, B., Plestina, R., Ceovic, S., (1977). Balkan (endemic) nephropathy and food-borne ochratoxin-a - preliminary-results of a survey of foodstuffs. *Acta Pathologica Et Microbiologica Scandinavica Section B-Microbiology*. 85, 238-240.

[10] Petkovabocharova, T., Castegnaro, M. (1991) *Ochratoxin-A in human blood in relation to balkan endemic nephropathy and urinary-tract tumors in bulgaria.*

[11] Pfohl-Leszkowicz, A., Vrabcheva, T., Petkova-Bocharova, T., Garren, L., Grosso, F., Nikolov, I., Dragacci, S., Chernozemsky, I.N., Castegnaro, M. (2006) *Analysis of*

ochratoxin A in serum, urine and food consumed by inhabitants from an area with Balkan endemic nephropathy: A one month follow-up study.

[12] Radic, B., Fuchs, R., Peraica, M., Lucic, A., (1997). Ochratoxin A in human sera in the area with endemic nephropathy in Croatia. *Toxicology Letters*. 91, 105-109.

[13] Grajewski, J., Jarzemski, P., Twaruzek, M., Kuzminska, K., Trepala, M., (2007). The level of ochratoxin A in patients after nephrectomy. *Mycotoxin research*. 23, 22-26.

[14] Degen, G., Mayer, S., Blaszkewicz, M., (2007). Biomonitoring of ochratoxin A in grain workers. *Mycotoxin research*. 23, 88-93.

[15] Ruprich, J., Ostrý, V., (1993). Health risk assessment of the mycotoxin ochratoxin A to humans: Czech Republic--Brno--1991/92. *Central European journal of public health*. 1, 86-93.

[16] Palli, D., Miraglia, M., Saieva, C., Masala, G., Cava, E., Colatosti, M., Corsi, A.M., Russo, A., Brera, C., (1999). Serum levels of ochratoxin A in healthy adults in Tuscany: correlation with individual characteristics and between repeat measurements. *Cancer Epidemiology Biomarkers & Prevention*. 8, 265-269.

[17] Hald, B., (1990). Ochratoxin A in human blood in European countries. *IARC scientific publications*, 159-164.

[18] Scottt, P., Kanhere, S., Lau, B.Y., Levvis, D., Hayward, S., Ryan, J., Kuiper - Goodman, T., (1998). Survey of Canadian human blood plasma for ochratoxin A. *Food Additives & Contaminants*. 15, 555-562.

[19] Pacin, A., Ciancio Bovier, E., Motta, E., Resnik, S., Villa, D., Olsen, M., (2008). Survey of Argentinean human plasma for ochratoxin A. *Food Additives and Contaminants*. 25, 635-641.

[20] Ueno, Y., Maki, S., Lin, J., Furuya, M., Sugiura, Y., Kawamura, O., (1998). A 4-year study of plasma ochratoxin A in a selected population in Tokyo by immunoassay and immunoaffinity column-linked HPLC. *Food and Chemical Toxicology*. 36, 445-449.

[21] Aslam, M., Beg, A.E., Blaszkewicz, M., Degen, G.H., Golka, K., (2005). Ochratoxin A blood concentration in healthy subjects and bladder cancer cases from Pakistan. *Mycotoxin research*. 21, 164-167.

[22] Assaf, H., Betbeder, A.-M., Creppy, E.E., Pallardy, M., Azouri, H., (2004). Ochratoxin A levels in human plasma and foods in Lebanon. *Human & Experimental Toxicology*. 23, 495-501.

[23] Davolos, D., Pietrangeli, B., (2014). A molecular and bioinformatic study on the ochratoxin A (OTA)-producing Aspergillus affinis (section Circumdati). *Mycotoxin research*. 30, 113-122.

[24] Jorgensen, K., (2005). Occurrence of ochratoxin A in commodities and processed food - A review of EU occurrence data. *Food Additives and Contaminants*. 22, 26-30.

[25] Duarte, S.C., Pena, A., Lino, C.M., (2010). A review on ochratoxin A occurrence and effects of processing of cereal and cereal derived food products. *Food Microbiology*. 27, 187-198.

[26] Gonzalez-Osnaya, L., Soriano, J.M., Molto, J.C., Manes, J., (2007). Dietary intake of ochratoxin A from conventional and organic bread. *International Journal of Food Microbiology*. 118, 87-91.

[27] Araguas, C., González-Peñas, E., López de Cerain, A., Bello, J., (2003). Acerca de la posible contaminación por ocratoxina A en alimentos. I: Cereales cultivados en diversas zonas geográficas de la Comunidad Foral de Navarra. *Alimentaria*, 23-29.

[28] Legarda, T., Burdaspal, P., (2001). Occurrence of ochratoxin A in samples of bread marketed in Spain and twelve other countries. *Alimentaria*. 321, 89-96.

[29] Osnaya, L.G., del Castillo, J.M.S., Cortés, J.C.M., Vinuesa, J.M., (2006). Extraction and analysis of ochratoxin A in bread using pressurised liquid extraction and liquid chromatography. *Journal of Chromatography A*. 1113, 32-36.

[30] González-Osnaya, L., Soriano, J., Moltó, J., Mañes, J., (2007). Dietary intake of ochratoxin A from conventional and organic bread. *International journal of food microbiology*. 118, 87-91.

[31] Domijan, A.-M., Peraica, M., Jurjević, Ž., Ivić, D., Cvjetković, B., (2005). Fumonisin B1, fumonisin B2, zearalenone and ochratoxin A contamination of maize in Croatia. *Food additives and contaminants*. 22, 677-680.

[32] Jørgensen, K., Jacobsen, J., (2002). Occurrence of ochratoxin A in Danish wheat and rye, 1992-99. *Food Additives & Contaminants*. 19, 1184-1189.

[33] Palermo, D., Pietrobono, P., Palermo, C., Rotunno, T., (2002). Occurrence of ochratoxin A in cereals from Puglia (Italy). *Italian journal of food science*. 14, 447-453.

[34] Kumagai, S., Nakajima, M., Tabata, S., Ishikuro, E., Tanaka, T., Norizuki, H., Itoh, Y., Aoyama, K., Fujita, K., Kai, S., (2008). Aflatoxin and ochratoxin A contamination of retail foods and intake of these mycotoxins in Japan. *Food Additives and Contaminants*. 25, 1101-1106.

[35] Park, J.W., Chung, S.-H., Kim, Y.-B., (2005). Ochratoxin A in Korean food commodities: occurrence and safety evaluation. *Journal of agricultural and food chemistry*. 53, 4637-4642.

[36] Lai, X., Liu, R., Ruan, C., Zhang, H., Liu, C., (2015). Occurrence of aflatoxins and ochratoxin A in rice samples from six provinces in China. *Food Control*. 50, 401-404.

[37] Sangare-Tigori, B., Moukha, S., Kouadio, H.J., Betbeder, A.-M., Dano, D.S., Creppy, E.E., (2006). Co-occurrence of aflatoxin B1, fumonisin B1, ochratoxin A and zearalenone in cereals and peanuts from Côte d'Ivoire. *Food additives and contaminants*. 23, 1000-1007.

[38] Zinedine, A., Brera, C., Elakhdari, S., Catano, C., Debegnach, F., Angelini, S., De Santis, B., Faid, M., Benlemlih, M., Minardi, V., (2006). Natural occurrence of mycotoxins in cereals and spices commercialized in Morocco. *Food Control*. 17, 868-874.

[39] van der Merwe, K.J., Steyn, P.S., Fourie, L., Scott, D.B., Theron, J.J., (1965). Ochratoxin A, a toxic metabolite produced by Aspergillus ochraceus Wilh. *Nature*. 205, 1112-1113.

[40] Cabañes, F., Accensi, F., Bragulat, M., Abarca, M., Castellá, G., Minguez, S., Pons, A., (2002). What is the source of ochratoxin A in wine? *International journal of food microbiology*. 79, 213-215.

[41] Cabanas, R., Bragulat, M.R., Abarca, M.L., Castella, G., Cabanes, F.J., (2008). Occurrence of Penicillium verrucosum in retail wheat flours from the Spanish market. *Food Microbiol*. 25, 642-647.

[42] Logrieco, A., Bottalico, A., Mule, G., Moretti, A., Perrone, G., (2003). Epidemiology of toxigenic fungi and their associated mycotoxins for some Mediterranean crops. *Eur J Plant Pathol.* 109, 645-667.

[43] Meeting, J.F.W.E.C.o.F.A., Organization, W.H. (2001) *Safety evaluation of certain mycotoxins in food.* Food & Agriculture Org.

[44] Aldred, D., Cairns-Fuller, V., Magan, N., (2008). Environmental factors affect efficacy of some essential oils and resveratrol to control growth and ochratoxin A production by Penicillium verrucosum and Aspergillus westerdijkiae on wheat grain. *J Stored Prod Res.* 44, 341-346.

[45] Scudamore, K.A., (2005). Prevention of ochratoxin A in commodities and likely effects of processing fractionation and animal feeds. *Food additives and contaminants.* 22 Suppl 1, 17-25.

[46] Domijan, A.M., Peraica, M., Jurjevic, Z., Ivic, D., Cvjetkovic, B., (2005). Fumonisin B1, fumonisin B2, zearalenone and ochratoxin A contamination of maize in Croatia. *Food additives and contaminants.* 22, 677-680.

[47] Miller, J.D., (1995). Fungi and Mycotoxins in Grain - Implications for Stored-Product Research. *J Stored Prod Res.* 31, 1-16.

[48] Magnoli, C., Hallak, C., Astoreca, A., Ponsone, L., Chiacchiera, S., Dalcero, A.M., (2006). Occurrence of ochratoxin A-producing fungi in commercial corn kernels in Argentina. *Mycopathologia.* 161, 53-58.

[49] Elmholt, S., Rasmussen, P.H., (2005). Penicillium verrucosum occurrence and ochratoxin A contents in organically cultivated grain with special reference to ancient wheat types and drying practice. *Mycopathologia.* 159, 421-432.

[50] Bhattacharya, K., Raha, S., (2002). Deteriorative changes of maize, groundnut and soybean seeds by fungi in storage. *Mycopathologia.* 155, 135-141.

[51] Magnoli, C.E., Astoreca, A.L., Chiacchiera, S.M., Dalcero, A.M., (2007). Occurrence of ochratoxin A and ochratoxigenic mycoflora in corn and corn based foods and feeds in some South American countries. *Mycopathologia.* 163, 249-260.

[52] Birzele, B., Prange, A., Kramer, J., (2000). Deoxynivalenol and ochratoxin A in German wheat and changes of level in relation to storage parameters. *Food additives and contaminants.* 17, 1027-1035.

[53] Ayalew, A., Fehrmann, H., Lepschy, J., Beck, R., Abate, D., (2006). Natural occurrence of mycotoxins in staple cereals from Ethiopia. *Mycopathologia.* 162, 57-63.

[54] Lund, F., Frisvad, J.C., (2003). Penicillium verrucosum in wheat and barley indicates presence of ochratoxin A. *Journal of applied microbiology.* 95, 1117-1123.

[55] Lee, H., Magan, N., (2000). Impact of environment and interspecific interactions between spoilage fungi and Aspergillus ochraceus on growth and ochratoxin production in maize grain. *International journal of food microbiology.* 61, 11-16.

[56] Eskola, M. (2002). EELA Publications. Helsinki.

[57] Scudamore, K., Banks, J., MacDonald, S., (2003). Fate of ochratoxin A in the processing of whole wheat grains during milling and bread production. *Food Additives and Contaminants.* 20, 1153-1163.

[58] Uysal, Ü.D., Öncü, E.M., Berikten, D., Yilmaz, N., Tuncel, N.B., Kıvanc, M., Tuncel, M., (2009). Time and temperature dependent microbiological and mycotoxin (ochratoxin-A) levels in boza. *International journal of food microbiology.* 130, 43-48.

[59] Zinedine, A., Soriano, J.M., Juan, C., Mojemmi, B., Molto, J.C., Bouklouze, A., Cherrah, Y., Idrissi, L., Aouad, R.E., Manes, J., (2007). Incidence of ochratoxin A in rice and dried fruits from Rabat and Salé area, Morocco. *Food Additives and Contaminants*. 24, 285-291.

[60] Pena, A., Cerejo, F., Lino, C., Silveira, I., (2005). Determination of ochratoxin A in Portuguese rice samples by high performance liquid chromatography with fluorescence detection. *Analytical and bioanalytical chemistry*. 382, 1288-1293.

[61] Gonzalez, L., Juan, C., Soriano, J., Molto, J., Manes, J., (2006). Occurrence and daily intake of ochratoxin A of organic and non-organic rice and rice products. *International journal of food microbiology*. 107, 223-227.

[62] Nguyen, M.T., Tozlovanu, M., Tran, T.L., Pfohl-Leszkowicz, A., (2007). Occurrence of aflatoxin B1, citrinin and ochratoxin A in rice in five provinces of the central region of Vietnam. *Food Chemistry*. 105, 42-47.

[63] Scudamore, K.A., (2005). Prevention of ochratoxin A in commodities and likely effects of processing fractionation and animal feeds. *Food additives and contaminants*. 22, 17-25.

[64] Czerwiecki, L., Czajkowska, D., Witkowska-Gwiazdowska, A., (2002). On ochratoxin A and fungal flora in Polish cereals from conventional and ecological farms. Part 2: Occurrence of ochratoxin A and fungi in cereals in 1998. *Food Additives & Contaminants*. 19, 1051-1057.

[65] Jørgensen, K., Rasmussen, G., Thorup, I., (1996). Ochratoxin A in Danish cereals 1986-1992 and daily intake by the Danish population. *Food Additives and Contaminants*. 13, 95-104.

[66] Juan, C., Moltó, J.C., Lino, C.M., Mañes, J., (2008). Determination of ochratoxin A in organic and non-organic cereals and cereal products from Spain and Portugal. *Food Chemistry*. 107, 525-530.

[67] Czerwiecki, L., Czajkowska, D., Witkowska-Gwiazdowska, A., (2002). On ochratoxin A and fungal flora in Polish cereals from conventional and ecological farms-Part 1: Occurrence of ochratoxin A and fungi in cereals in 1997. *Food Additives & Contaminants*. 19, 470-477.

[68] Bhattacharya, K., Raha, S., (2002). Deteriorative changes of maize, groundnut and soybean seeds by fungi in storage. *Mycopathologia*. 155, 135-141.

[69] Duarte, S., Pena, A., Lino, C., (2010). A review on ochratoxin A occurrence and effects of processing of cereal and cereal derived food products. *Food Microbiology*. 27, 187-198.

[70] Blesa, J., Berrada, H., Soriano, J., Moltó, J., Mañes, J., (2004). Rapid determination of ochratoxin A in cereals and cereal products by liquid chromatography. *Journal of Chromatography A*. 1046, 127-131.

[71] Riba, A., Mokrane, S., Mathieu, F., Lebrihi, A., Sabaou, N., (2008). Mycoflora and ochratoxin A producing strains of Aspergillus in Algerian wheat. *International journal of food microbiology*. 122, 85-92.

[72] Peng, C., Wang, L., An, F., Zhang, L., Wang, Y., Li, S., Wang, C., Liu, H., (2015). Fate of ochratoxin A during wheat milling and some Chinese breakfast processing. *Food Control*. 57, 142-146.

[73] Romani, S., Sacchetti, G., Chaves López, C., Pinnavaia, G.G., Dalla Rosa, M., (2000). Screening on the occurrence of ochratoxin A in green coffee beans of different origins and types. *Journal of agricultural and food chemistry*. 48, 3616-3619.

[74] Levi, C.P., Trenk, H.L., Mohr, H.K., (1974). Study of Occurrence of Ochratoxin A in Green Coffee Beans. *Journal of the Association of Official Analytical Chemists*. 57, 866-870.

[75] Levi, C., (1980). Mycotoxins in coffee. *Journal-Association of Official Analytical Chemists*. 63, 1282-1285.

[76] Norton, D., Toule, G., Cooper, S., Partington, S., Chapman, W. (1982), *Proceedings, Fourth Meeting on Mycotoxins in Animal Disease*. Ministry of Agriculture, Fisheries and Food: Alnwick, Northumberland, pp. 77-81.

[77] Cantafora, A., Grossi, M., Miraglia, M., Benelli, L., (1983). Determination of ochratoxin A in coffee beans using reversed-phase high performance liquid chromatography. *La Rivista della Societa Italiana di Scienza dell'Alimentazione*. 12, 103-108.

[78] Tsubouchi, H., Yamamoto, K., Hisada, K., Sakabe, Y., Udagawa, S.-i., (1987). Effect of roasting on ochratoxin A level in green coffee beans inoculated with Aspergillus ochraceus. *Mycopathologia*. 97, 111-115.

[79] Malir, F., Ostry, V., Pfohl-Leszkowicz, A., Toman, J., Bazin, I., Roubal, T., (2014). Transfer of Ochratoxin A into Tea and Coffee Beverages. *Toxins*. 6, 3438-3453.

[80] Miraglia, M., Brera, C., (2002). Assessment of dietary intake of ochratoxin A by the population of EU member states. *Reports on tasks for scientific cooperation. Reports of experts participating in SCOOP Task*. 3.

[81] Běláková, S., Benešová, K., Mikulíková, R., Svoboda, Z., (2011). Determination of ochratoxin A in brewing materials and beer by ultra performance liquid chromatography with fluorescence detection. *Food Chemistry*. 126, 321-325.

[82] Medina, A., Valle-Algarra, F.M., Gimeno-Adelantado, J.V., Mateo, R., Mateo, F., Jimenez, M., (2006). New method for determination of ochratoxin A in beer using zinc acetate and solid-phase extraction silica cartridges. *Journal of Chromatography A*. 1121, 178-183.

[83] Medina, A., Jimenez, M., Gimeno-Adelantado, J.V., Valle-Algarra, F.M., Mateo, R., (2005). Determination of ochratoxin A in beer marketed in Spain by liquid chromatography with fluorescence detection using lead hydroxyacetate as a clean-up agent. *Journal of Chromatography A*. 1083, 7-13.

[84] Kabak, B., (2009). Ochratoxin A in cereal-derived products in Turkey: occurrence and exposure assessment. *Food and Chemical Toxicology*. 47, 348-352.

[85] Bertuzzi, T., Rastelli, S., Mulazzi, A., Donadini, G., Pietri, A., (2011). Mycotoxin occurrence in beer produced in several European countries. *Food Control*. 22, 2059-2064.

[86] Prelle, A., Spadaro, D., Denca, A., Garibaldi, A., Gullino, M.L., (2013). Comparison of Clean-Up Methods for Ochratoxin A on Wine, Beer, Roasted Coffee and Chili Commercialized in Italy. *Toxins*. 5, 1827-1844.

[87] Araguás, C., González-Peñas, E., De Cerain, A.L., (2005). Study on ochratoxin A in cereal-derived products from Spain. *Food chemistry*. 92, 459-464.

[88] Bellver Soto, J., Fernandez-Franzon, M., Ruiz, M.-J., Juan-Garcia, A., (2014). Presence of Ochratoxin A (OTA) Mycotoxin in Alcoholic Drinks from Southern European Countries: Wine and Beer. *Journal of Agricultural and Food Chemistry*. 62, 7643-7651.

[89] Al-Taher, F., Banaszewski, K., Jackson, L., Zweigenbaum, J., Ryu, D., Cappozzo, J., (2013). Rapid Method for the Determination of Multiple Mycotoxins in Wines and Beers by LC-MS/MS Using a Stable Isotope Dilution Assay. *Journal of agricultural and food chemistry*. 61, 2378-2384.

[90] Cao, J., Kong, W., Zhou, S., Yin, L., Wan, L., Yang, M., (2013). Molecularly imprinted polymer-based solid phase clean-up for analysis of ochratoxin A in beer, red wine, and grape juice. *Journal of Separation Science*. 36, 1291-1297.

[91] Medina, Á., Jiménez, M., Gimeno-Adelantado, J.V., Valle-Algarra, F.M., Mateo, R., (2005). Determination of ochratoxin A in beer marketed in Spain by liquid chromatography with fluorescence detection using lead hydroxyacetate as a clean-up agent. *Journal of Chromatography A*. 1083, 7-13.

[92] Medina, Á., Valle-Algarra, F.M., Gimeno-Adelantado, J.V., Mateo, R., Mateo, F., Jiménez, M., (2006). New method for determination of ochratoxin A in beer using zinc acetate and solid-phase extraction silica cartridges. *Journal of Chromatography A*. 1121, 178-183.

[93] Varga, J., Kiss, R., Mátrai, T., Mátrai, T., Téren, J., (2005). Detection of ochratoxin A in Hungarian wines and beers. *Acta Alimentaria*. 34, 381-392.

[94] Mateo, R., Medina, A., Mateo, E.M., Mateo, F., Jimenez, M., (2007). An overview of ochratoxin A in beer and wine. *International journal of food microbiology*. 119, 79-83.

[95] Covarelli, L., Beccari, G., Marini, A., Tosi, L., (2012). A review on the occurrence and control of ochratoxigenic fungal species and ochratoxin A in dehydrated grapes, non-fortified dessert wines and dried vine fruit in the Mediterranean area. *Food Control*. 26, 347-356.

[96] Valero, A., Marín, S., Ramos, A.J., Sanchis, V., (2008). Survey: Ochratoxin A in European special wines. *Food chemistry*. 108, 593-599.

[97] Hernández, M.J., García-Moreno, M.V., Durán, E., Guillén, D., Barroso, C.G., (2006). Validation of two analytical methods for the determination of ochratoxin A by reversed-phased high-performance liquid chromatography coupled to fluorescence detection in musts and sweet wines from Andalusia. *Anal. Chim. Acta*. 566, 117-121.

[98] Rothen, F., (2006). International Organisation of Vine and Wine. *Obst-und Weinbau. Schweizerische Zeitschrift fuer Obst-und Weinbau der Eidg. Forschungsanstalt Waedenswil (Switzerland).*

[99] Hernandez, M.J., Garcia-Moreno, M.V., Duran, E., Guillen, D., Barroso, C.G., (2006). Validation of two analytical methods for the determination of ochratoxin A by reversed-phased high-performance liquid chromatography coupled to fluorescence detection in musts and sweet wines from Andalusia. *Anal Chim Acta*. 566, 117-121.

[100] Burdaspal, P., Legarda, T., (2007). Occurrence of ochratoxin A in sweet wines produced in Spain and other countries. *Food additives and contaminants*. 24, 976-986.

[101] Valero, A., Marin, S., Ramos, A.J., Sanchis, V., (2008). Survey: Ochratoxin A in European special wines. *Food Chem*. 108, 593-599.

[102] Murillo-Arbizu, M.T., Amezqueta, S., Gonzalez-Penas, E., de Cerain, A.L., (2010). Occurrence of ochratoxin A in Southern Spanish generous wines under the

denomination of Origin "Jerez-Xeres-Sherry and 'manzanilla' Sanlucar de Barrameda". *Toxins*. 2, 1054-1064.
[103] Quintela, S., Villaran, M.C., de Armentia, I.L., Elejalde, E., (2012). Ochratoxin A in Spanish exportation wine market. *Food Control*. 25, 501-504.
[104] Quintela, S., Villaran, M.C., de Armentia, I.L., Elejalde, E., (2011). Occurrence of ochratoxin A in Rioja Alavesa wines. *Food Chem*. 126, 302-305.
[105] Bacaloni, A., Cavaliere, C., Faberi, A., Pastorini, E., Samperi, R., Lagana, A., (2005). Automated on-line solid-phase extraction-liquid chromatography-electrospray tandem mass spectrometry method for the determination of ochratoxin A in wine and beer. *Journal of agricultural and food chemistry*. 53, 5518-5525.
[106] Brera, C., Soriano, J.M., Debegnach, F., Miraglia, M., (2005). Exposure assessment to ochratoxin A from the consumption of Italian and Hungarian wines. *Microchem J*. 79, 109-113.
[107] Aresta, A., Vatinno, R., Palmisano, F., Zambonin, C.G., (2006). Determination of Ochratoxin A in wine at sub ng/mL levels by solid-phase microextraction coupled to liquid chromatography with fluorescence detection. *Journal of chromatography. A*. 1115, 196-201.
[108] Chiodini, A.M., Scherpenisse, P., Bergwerff, A.A., (2006). Ochratoxin a contents in wine: comparison of organically and conventionally produced products. *Journal of agricultural and food chemistry*. 54, 7399-7404.
[109] Brera, C., Debegnach, F., Minardi, V., Prantera, E., Pannunzi, E., Faleo, S., de Santis, B., Miraglia, M., (2008). Ochratoxin a contamination in italian wine samples and evaluation of the exposure in the italian population. *Journal of agricultural and food chemistry*. 56, 10611-10618.
[110] Spadaro, D., Lore, A., Garibaldi, A., Gullino, M.L., (2010). Occurrence of ochratoxin A before bottling in DOC and DOCG wines produced in Piedmont (Northern Italy). *Food Control*. 21, 1294-1297.
[111] Prelle, A., Spadaro, D., Denca, A., Garibaldi, A., Gullino, M.L., (2013). Comparison of clean-up methods for ochratoxin A on wine, beer, roasted coffee and chili commercialized in Italy. *Toxins*. 5, 1827-1844.
[112] Labrinea, E.P., Natskoulis, P.I., Spiropoulos, A.E., Magan, N., Tassou, C.C., (2011). A survey of ochratoxin A occurence in Greek wines. *Food additives & contaminants. Part B, Surveillance*. 4, 61-66.
[113] Var, I., Kabak, B., (2007). Occurrence of ochratoxin A in Turkish wines. *Microchem J*. 86, 241-247.
[114] Berente, B., Moricz, A., H-Otta, K., Zaray, G., Leko, L., Racz, L., (2005). Determination of ochratoxin A in Hungarian wines. *Microchem J*. 79, 103-107.
[115] Varga, J., Kiss, R., Matrai, T., Matrai, T., Teren, J., (2005). Detection of ochratoxin A in Hungarian wines and beers. *Acta Aliment Hung*. 34, 381-392.
[116] Flajs, D., Domijan, A.M., Ivic, D., Cvjetkovic, B., Peraica, M., (2009). ELISA and HPLC analysis of ochratoxin A in red wines of Croatia. *Food Control*. 20, 590-592.
[117] Zhang, X.X., Chen, L., Li, J.M., Zhu, B.Q., Ma, L.Y., (2013). Occurrence of Ochratoxin A in Chinese wines: influence of local meteorological parameters. *Eur Food Res Technol*. 236, 277-283.

[118] Mao, J.F., Lei, S.R., Yang, X.F., Xiao, D., (2013). Quantification of ochratoxin A in red wines by conventional HPLC-FLD using a column packed with core-shell particles. *Food Control*. 32, 505-511.

[119] Zhang, X., Chen, L., Li, J., Zhu, B., Ma, L., (2013). Occurrence of Ochratoxin A in Chinese wines: influence of local meteorological parameters. *European Food Research and Technology*. 236, 277-283.

[120] Remiro, R., Irigoyen, A., González-Peñas, E., Lizarraga, E., de Cerain, A.L., (2013). Levels of ochratoxins in Mediterranean red wines. *Food Control*. 32, 63-68.

[121] Alexander, J., Autrup, H., Bard, D., Benford, D., Carere, A., Costa, L., Cravedi, J., di Domenico, A., Fanelli, R., Fink-Gremmels, J., (2006). Opinion of the scientific panel on contaminants in the food chain on a request from the commission related to ochratoxin A in food. *EFSA J*. 365, 1-56.

[122] Murillo-Arbizu, M., Amézqueta, S., González-Peñas, E., De Cerain, A.L., (2010). Occurrence of Ochratoxin A in Southern Spanish Generous Wines under the Denomination of Origin "Jerez-Xérès-Sherry and 'Manzanilla'Sanlúcar de Barrameda". *Toxins*. 2, 1054-1064.

[123] Pena, A., Cerejo, F., Silva, L., Lino, C., (2010). Ochratoxin A survey in Portuguese wine by LC–FD with direct injection. *Talanta*. 82, 1556-1561.

[124] Brera, C., Debegnach, F., Minardi, V., Prantera, E., Pannunzi, E., Faleo, S., De Santis, B., Miraglia, M., (2008). Ochratoxin A contamination in Italian wine samples and evaluation of the exposure in the Italian population. *Journal of agricultural and food chemistry*. 56, 10611-10618.

[125] Bacaloni, A., Cavaliere, C., Faberi, A., Pastorini, E., Samperi, R., Laganà, A., (2005). Automated on-line solid-phase extraction-liquid chromatography-electrospray tandem mass spectrometry method for the determination of ochratoxin A in wine and beer. *Journal of agricultural and food chemistry*. 53, 5518-5525.

[126] Anfossi, L., Giovannoli, C., Giraudi, G., Biagioli, F., Passini, C., Baggiani, C., (2012). A Lateral Flow Immunoassay for the Rapid Detection of Ochratoxin A in Wine and Grape Must. *Journal of agricultural and food chemistry*. 60, 11491-11497.

[127] Longobardi, F., Iacovelli, V., Catucci, L., Panzarini, G., Pascale, M., Visconti, A., Agostiano, A., (2012). Determination of ochratoxin A in wine by means of immunoaffinity and aminopropyl solid-phase column cleanup and fluorometric detection. *Journal of agricultural and food chemistry*. 61, 1604-1608.

[128] Brera, C., Soriano, J., Debegnach, F., Miraglia, M., (2005). Exposure assessment to ochratoxin A from the consumption of Italian and Hungarian wines. *Microchemical Journal*. 79, 109-113.

[129] Aresta, A., Vatinno, R., Palmisano, F., Zambonin, C.G., (2006). Determination of Ochratoxin A in wine at sub ng/mL levels by solid-phase microextraction coupled to liquid chromatography with fluorescence detection. *Journal of Chromatography A*. 1115, 196-201.

[130] Chiodini, A.M., Scherpenisse, P., Bergwerff, A.A., (2006). Ochratoxin A contents in wine: comparison of organically and conventionally produced products. *Journal of agricultural and food chemistry*. 54, 7399-7404.

[131] Larsen, J.C. (2006). European Food Safety Authority.

[132] Alvarado, M., Galarce-Bustos, O., Vega, M., Aranda, M., (2013). Fast and Selective Determination of Ochratoxin A in Wines Using an Optimized and Validated Liquid Chromatographic Method. *Food Analytical Methods*. 6, 621-629.

[133] Munoz, K., Blaszkewicz, M., Campos, V., Vega, M., Degen, G.H., (2014). Exposure of infants to ochratoxin A with breast milk. *Archives of Toxicology*. 88, 837-846.

[134] Boudra, H., Morgavi, D., (2006). Development and validation of a HPLC method for the quantitation of ochratoxins in plasma and raw milk. *Journal of Chromatography B*. 843, 295-301.

[135] Sørensen, L., Elbaek, T., (2005). Determination of mycotoxins in bovine milk by liquid chromatography tandem mass spectrometry. *Journal of chromatography B*. 820, 183-196.

[136] González-Osnaya, L., Soriano, J., Moltó, J., Mañes, J., (2008). Simple liquid chromatography assay for analyzing ochratoxin A in bovine milk. *Food chemistry*. 108, 272-276.

[137] Nebbia, C. (2009) *Residui di farmaci e contaminanti ambientali nelle produzioni animali*. EdiSES.

[138] Manetta, A.C., Giammarco, M., Di Giuseppe, L., Fusaro, I., Gramenzi, A., Formigoni, A., Vignola, G., Lambertini, L., (2009). Distribution of aflatoxin M 1 during Grana Padano cheese production from naturally contaminated milk. *Food Chemistry*. 113, 595-599.

[139] Pattono, D., Gallo, P., Civera, T., (2011). Detection and quantification of ochratoxin A in milk produced in organic farms. *Food Chemistry*. 127, 374-377.

[140] Kiessling, K.H., Pettersson, H., Sandholm, K., Olsen, M., (1984). Metabolism of aflatoxin, ochratoxin, zearalenone, and three trichothecenes by intact rumen fluid, rumen protozoa, and rumen bacteria. *Applied and environmental microbiology*. 47, 1070-1073.

[141] Boudra, H., Barnouin, J., Dragacci, S., Morgavi, D.P., (2007). Aflatoxin M1 and ochratoxin A in raw bulk milk from French dairy herds. *Journal of dairy science*. 90, 3197-3201.

[142] Coffey, R., Cummins, E., Ward, S., (2009). Exposure assessment of mycotoxins in dairy milk. *Food Control*. 20, 239-249.

[143] Zhang, K., Wong, J.W., Hayward, D.G., Vaclavikova, M., Liao, C.D., Trucksess, M.W., (2013). Determination of mycotoxins in milk-based products and infant formula using stable isotope dilution assay and liquid chromatography tandem mass spectrometry. *Journal of agricultural and food chemistry*. 61, 6265-6273.

[144] Meucci, V., Razzuoli, E., Soldani, G., Massart, F., (2010). Mycotoxin detection in infant formula milks in Italy. *Food Additives and Contaminants Part a-Chemistry Analysis Control Exposure & Risk Assessment*. 27, 64-71.

[145] Chen, D., Cao, X., Tao, Y., Wu, Q., Pan, Y., Huang, L., Wang, X., Wang, Y., Peng, D., Liu, Z., Yuan, Z., (2012). Development of a sensitive and robust liquid chromatography coupled with tandem mass spectrometry and a pressurized liquid extraction for the determination of aflatoxins and ochratoxin A in animal derived foods. *Journal of chromatography. A*. 1253, 110-119.

[146] El-Sayed Abd Alla, A., Neamat-Allah, A., Aly, S.E., (2000). Situation of mycotoxins in milk, dairy products and human milk in Egypt. *Mycotoxin research*. 16, 91-100.

[147] Dall'Asta, C., Lindner, J.D., Galaverna, G., Dossena, A., Neviani, E., Marchelli, R., (2008). The occurrence of ochratoxin A in blue cheese. *Food Chem.* 106, 729-734.

[148] Zhang, X., Cudjoe, E., Vuckovic, D., Pawliszyn, J., (2009). Direct monitoring of ochratoxin A in cheese with solid-phase microextraction coupled to liquid chromatography-tandem mass spectrometry. *Journal of chromatography. A.* 1216, 7505-7509.

[149] Biancardi, A., Piro, R., Galaverna, G., Dall'Asta, C., (2013). A simple and reliable liquid chromatography-tandem mass spectrometry method for determination of ochratoxin A in hard cheese. *International journal of food sciences and nutrition.* 64, 632-640.

[150] Huang, L., Zheng, N., Zheng, B., Wen, F., Cheng, J., Han, R., Xu, X., Li, S., Wang, J., (2014). Simultaneous determination of aflatoxin M 1, ochratoxin A, zearalenone and α-zearalenol in milk by UHPLC–MS/MS. *Food chemistry.* 146, 242-249.

[151] Alvito, P.C., Sizoo, E.A., Almeida, C.M., van Egmond, H.P., (2010). Occurrence of aflatoxins and ochratoxin A in baby foods in Portugal. *Food Analytical Methods.* 3, 22-30.

[152] Beltrán, E., Ibáñez, M., Sancho, J.V., Cortés, M.Á., Yusà, V., Hernández, F., (2011). UHPLC–MS/MS highly sensitive determination of aflatoxins, the aflatoxin metabolite M1 and ochratoxin A in baby food and milk. *Food Chemistry.* 126, 737-744.

[153] Baydar, T., Erkekoglu, P., Sipahi, H., Sahin, G., (2007). Aflatoxin B~ 1, M~ 1 and Ochratoxin A Levels in Infant Formulae and Baby Foods Marketed in Ankara, Turkey. *Journal of Food and Drug Analysis.* 15, 89.

[154] Škrinjar, M., Rašić, J.L., Stojičić, V., (1996). Lowering of ochratoxin A level in milk by yoghurt bacteria and bifidobacteria. *Folia Microbiol. (Praha).* 41, 26-28.

[155] Juodeikiene, G., Bartkiene, E., Basinskiene, L., Matusevicius, P. (2012) *Mycotoxin decontamination aspects in food, feed and renewables using fermentation processes.* INTECH Open Access Publisher.

[156] Abrunhosa, L., Santos, L., Venâncio, A., (2006). Degradation of ochratoxin A by proteases and by a crude enzyme of Aspergillus niger. *Food Biotechnology.* 20, 231-242.

[157] BLAGOJEV, N., 2012 Control of mold growth and mycotoxin production by lactic acid bacteria metabolites MC MKT (301).

[158] Petzinger, E., Weidenbach, A., (2002). Mycotoxins in the food chain: the role of ochratoxins. *Livestock Production Science.* 76, 245-250.

[159] Monaci, L., Palmisano, F., Matrella, R., Tantillo, G., (2005). Determination of ochratoxin A at part-per-trillion level in Italian salami by immunoaffinity clean-up and high-performance liquid chromatography with fluorescence detection. *Journal of Chromatography a.* 1090, 184-187.

[160] Fink-Gremmels, J., (2008). Mycotoxins in cattle feeds and carry-over to dairy milk: A review. *Food Additives and Contaminants.* 25, 172-180.

[161] Tam, J., Pantazopoulos, P., Scott, P., Moisey, J., Dabeka, R., Richard, I., (2011). Application of isotope dilution mass spectrometry: determination of ochratoxin A in the Canadian Total Diet Study. *Food Additives and Contaminants.* 28, 754-761.

[162] Pleadin, J., Kovacevic, D., Persi, N., (2015). Ochratoxin A contamination of the autochthonous dry-cured meat product "Slavonski Kulen" during a six-month production process. Food Control. 57, 377-384.

[163] Guillamont, E.M., Lino, C., Baeta, M., Pena, A., Silveira, M., Vinuesa, J.M., (2005). A comparative study of extraction apparatus in HPLC analysis of ochratoxin A in muscle. *Analytical and bioanalytical chemistry*. 383, 570-575.

[164] Curtui, V.G., Gareis, M., (2001). A simple HPLC method for the determination of the mycotoxins ochratoxin A and B in blood serum of swine. *Food Additives & Contaminants*. 18, 635-643.

[165] Markov, K., Pleadin, J., Bevardi, M., Vahčić, N., Sokolić-Mihalak, D., Frece, J., (2013). Natural occurrence of aflatoxin B 1, ochratoxin A and citrinin in Croatian fermented meat products. *Food Control*. 34, 312-317.

[166] Castegnaro, M., Wild, C.P., (1995). IARC Activities in Mycotoxins Research. *Natural Toxins*. 3, 327-331.

[167] Commission, E., (2006). Commission Regulation (EC) No. 1881/2006 of 19 December 2006 setting maximum levels for certain contaminants in foodstuffs. *Off. J. Eur. Union*. 364, 5-24.

[168] Kuiper-Goodman, T., Scott, P.M., (1989). Risk assessment of the mycotoxin ochratoxin A. *Biomedical and environmental sciences : BES*. 2, 179-248.

[169] O'Brien, E., Heussner, A.H., Dietrich, D.R., (2001). Species-, sex-, and cell type-specific effects of ochratoxin A and B. *Toxicological sciences : an official journal of the Society of Toxicology*. 63, 256-264.

[170] Stefanovic, V., Polenakovic, M., (2009). Fifty years of research in Balkan endemic nephropathy: where are we now? *Nephron. Clinical practice*. 112, c51-56.

[171] Palma, N., Cinelli, S., Sapora, O., Wilson, S.H., Dogliotti, E., (2007). Ochratoxin A-induced mutagenesis in mammalian cells is consistent with the production of oxidative stress. *Chemical research in toxicology*. 20, 1031-1037.

[172] Park, J.W., Chung, S.H., Kim, Y.B., (2005). Ochratoxin A in Korean food commodities: occurrence and safety evaluation. *Journal of agricultural and food chemistry*. 53, 4637-4642.

[173] Wu, F., (2004). Mycotoxin risk assessment for the purpose of setting international regulatory standards. *Environmental science & technology*. 38, 4049-4055.

[174] Kuiper-Goodman, T., Hilts, C., Billiard, S.M., Kiparissis, Y., Richard, I.D., Hayward, S., (2010). Health risk assessment of ochratoxin A for all age-sex strata in a market economy. *Food additives & contaminants. Part A, Chemistry, analysis, control, exposure & risk assessment*. 27, 212-240.

[175] Gluhovschi, G., Margineanu, F., Velciov, S., Gluhovschi, C., Bob, F., Petrica, L., Bozdog, G., Trandafirescu, V., Modalca, M., (2011). Fifty years of Balkan endemic nephropathy in Romania: some aspects of the endemic focus in the Mehedinti county. *Clinical nephrology*. 75, 34-48.

[176] Bansal, J., Pantazopoulos, P., Tam, J., Cavlovic, P., Kwong, K., Turcotte, A.M., Lau, B.P., Scott, P.M., (2011). Surveys of rice sold in Canada for aflatoxins, ochratoxin A and fumonisins. *Food additives & contaminants. Part A, Chemistry, analysis, control, exposure & risk assessment*. 28, 767-774.

[177] Tam, J., Pantazopoulos, P., Scott, P.M., Moisey, J., Dabeka, R.W., Richard, I.D.K., (2011). Application of isotope dilution mass spectrometry: determination of ochratoxin A in the Canadian Total Diet Study. *Food Addit Contam A*. 28, 754-761.

[178] Haighton, L.A., Lynch, B.S., Magnuson, B.A., Nestmann, E.R., (2012). A reassessment of risk associated with dietary intake of ochratoxin A based on a lifetime exposure model. *Critical reviews in toxicology*. 42, 147-168.

[179] Krogh, P., Elling, F., Friis, C., Hald, B., Larsen, A.E., Lillehoj, E.B., Madsen, A., Mortensen, H.P., Rasmussen, F., Ravnskov, U., (1979). Porcine nephropathy induced by long-term ingestion of ochratoxin A. *Veterinary pathology*. 16, 466-475.

[180] Hagelberg, S., Hult, K., Fuchs, R., (1989). Toxicokinetics of ochratoxin A in several species and its plasma-binding properties. *Journal of applied toxicology : JAT*. 9, 91-96.

[181] Coronel, M.B., Sanchis, V., Ramos, A.J., Marin, S., (2010). Review. Ochratoxin A: presence in human plasma and intake estimation. *Food science and technology international = Ciencia y tecnologia de los alimentos internacional*. 16, 5-18.

[182] Krogh, P., Axelsen, N.H., Elling, F., Gyrd-Hansen, N., Hald, B., Hyldgaard-Jensen, J., Larsen, A.E., Madsen, A., Mortensen, H.P., Moller, T., Petersen, O.K., Ravnskov, U., Rostgaard, M., Aalund, O., (1974). Experimental porcine nephropathy. Changes of renal function and structure induced by ochratoxin A- contaminated feed. *Acta pathologica et microbiologica Scandinavica. Supplement*. 0, 1-21.

[183] Kuiper-Goodman, T., (1996). Risk assessment of ochratoxin A: an update. *Food additives and contaminants*. 13 Suppl, 53-57.

[184] Organization, W.H., (1994). International programme on chemical safety. Assessing human health risks of chemicals: derivation of guidance values for health-based exposure limits. *Environmental health criteria*. 170, 54.

[185] Walker, R. (2002), *Mycotoxins and Food Safety*. Springer, pp. 249-255.

[186] (1997). Worldwide regulations for mycotoxins 1995. A compendium. *FAO food and nutrition paper*. 64, 1-43.

[187] Bui-Klimke, T.R., Wu, F., (2014). Ochratoxin A and human health risk: A review of the evidence Critical Reviews in Food Science and Nutrition. *Critical reviews in food science and nutrition*.

[188] Kuiper-Goodman, T., Hilts, C., Billiard, S., Kiparissis, Y., Richard, I., Hayward, S., (2010). Health risk assessment of ochratoxin A for all age-sex strata in a market economy. *Food Additives and Contaminants*. 27, 212-240.

[189] Meeting, J.F.W.E.C.o.F.A., Organization, W.H. (2007) *Evaluation of certain food additives and contaminants: sixty-eighth report of the Joint FAO/WHO Expert Committee on Food Additives*. World Health Organization.

[190] Olsen, M., Thorup, I., Knudsen, I., Larsen, J., Hald, B., Olsen, J. (1991), *Nordiske seminar-og Arbejds-rapporter*. Nordic Council of Ministers Copenhagen, Vol. 545.

[191] Haighton, L.A., Lynch, B.S., Magnuson, B.A., Nestmann, E.R., (2012). A reassessment of risk associated with dietary intake of ochratoxin A based on a lifetime exposure model. *Critical reviews in toxicology*. 42, 147-168.

[192] Cole, R.J., Cox, R.H. (1981) *Handbook of toxic fungal metabolites*. Academic Press.

[193] Van der Merwe, K., Steyn, P., Fourie, L., (1965). 1304. Mycotoxins. Part II. The constitution of ochratoxins A, B, and C, metabolites of Aspergillus ochraceus wilh. *Journal of the Chemical Society (Resumed)*, 7083-7088.

[194] Li, S., Marquardt, R., Frohlich, A., Vitti, T., Crow, G., (1997). Pharmacokinetics of ochratoxin A and its metabolites in rats. *Toxicol. Appl. Pharmacol.* 145, 82-90.

[195] Marquardt, R.R., Frohlich, A.A., (1992). A review of recent advances in understanding ochratoxicosis. *Journal of animal science*. 70, 3968-3988.

[196] Faucet-Marquis, V., Pont, F., Stormer, F.C., Rizk, T., Castegnaro, M., Pfohl-Leszkowicz, A., (2006). Evidence of a new dechlorinated ochratoxin A derivative formed in opossum kidney cell cultures after pretreatment by modulators of glutathione pathways: Correlation with DNA-adduct formation. *Mol. Nutr. Food Res.* 50, 530-542.

[197] Pfohl-Leszkowicz, A., Manderville, R.A., (2007). Ochratoxin A: An overview on toxicity and carcinogenicity in animals and humans. *Molecular nutrition & food research*. 51, 61-99.

[198] Manderville, R.A., Pfohl-Leszkowicz, A., (2008). Bioactivation and DNA adduction as a rationale for ochratoxin A carcinogenesis. *World Mycotoxin Journal*. 1, 357-367.

[199] Pohland, A., Schuller, P., Steyn, P., Van Egmond, H., (1982). Physicochemical data for some selected mycotoxins. *Pure and Applied Chemistry*. 54, 2219-2284.

[200] Sibanda, L., De Saeger, S., Van Peteghem, C., (2002). Optimization of solid-phase clean-up prior to liquid chromatographic analysis of ochratoxin A in roasted coffee. *Journal of Chromatography A*. 959, 327-330.

[201] Kuiper-Goodman, T., Scott, P., (1989). Risk assessment of the mycotoxin ochratoxin A. *Biomedical and environmental sciences: BES*. 2, 179-248.

[202] Miller, J.D., (1992). Fungi as contaminants in indoor air. *Atmospheric Environment. Part A. General Topics*. 26, 2163-2172.

[203] Betina, V., (1989). Ochratoxins and related dihydroisocoumarins. *Mycotoxins: Chemical, Biological and Environmental Aspects*. 9, 151-173.

[204] de Jesus, A.E., Steyn, P.S., Vleggaar, R., Wessels, P.L., (1980). Carbon-13 nuclear magnetic resonance assignments and biosynthesis of the mycotoxin ochratoxin A. *J. Chem. Soc., Perkin Trans. 1*, 52-54.

[205] El Khoury, A. (2007). EL Khoury, A. (2007). *Champignons Mycotoxinogènes et Ochratoxine A (OTA) et Aflatoxine B1 (AFB1) dans les Vignobles Libanais: Occurrence et Origine;* Ph.D. Thesis, INP: Toulouse, France, 2007.

[206] Lillehoj, E., Goransson, B., (1980). Occurrence of Ochratoxin‐and Citrinin‐producing fungi on developing danish barley grain. *Acta Pathologica Microbiologica Scandinavica Section B Microbiology*. 88, 133-137.

[207] Abramson, D., (1987). Measurement of ochratoxin A in barley extracts by liquid chromatography-mass spectrometry. *J. Chromatogr. A*. 391, 315-320.

[208] Bauer, J., Gareis, M., (1987). Ochratoxin A in der Nahrungsmittelkette. *Journal of Veterinary Medicine, Series B*. 34, 613-627.

[209] Chu, F., Wilson, B., (1973). Studies on ochratoxins. *CRC Crit. Rev. Toxicol.* 2, 499-524.

[210] Müller, H., (1982). Decontamination of mycotoxins. I. Physical process. *Ubersicht. Tierernähr*. 10, 95-122.

[211] Trivedi, A.B., Doi, E., Kitabatake, N., (1992). Detoxification of ochratoxin A on heating under acidic and alkaline conditions. *Biosci. Biotechnol. Biochem.* 56, 741-745.

[212] Boudra, H., Le Bars, P., Le Bars, J., (1995). Thermostability of Ochratoxin A in wheat under two moisture conditions. *Appl. Environ. Microbiol.* 61, 1156-1158.

[213] Maaroufi, K., Zakhama, A., Baudrimont, I., Achour, A., Abid, S., Ellouz, F., Dhouib, S., Creppy, E., Bacha, H., (1999). Karyomegaly of tubular cells as early stage marker of

the nephrotoxicity induced by ochratoxin A in rats. *Human & experimental toxicology*. 18, 410-415.

[214] Levi, C., Trenk, H., Mohr, H., (1974). Study of the occurrence of ochratoxin A in green coffee beans. *J. Assoc. Off. Anal. Chem*. 57, 866-870.

[215] Urga, K., Gonfa, A., Ochratoxin A and Coffee.

[216] Micco, C., Grossi, M., Miraglia, M., Brera, C., (1989). A study of the contamination by ochratoxin A of green and roasted coffee beans. *Food additives and contaminants*. 6, 333-339.

[217] Tsubouchi, H., Yamamoto, K., Hisada, K., Sakabe, Y., Udagawa, S., (1987). Effect of roasting on ochratoxin A level in green coffee beans inoculated with Aspergillus ochraceus. *Mycopathologia*. 97, 111-115.

[218] Urbano, G., Freitas Leitão, M., Vicentini, M., Taniwaki, M. (2001), *Proc. 19th ASIC Coffee Conf.*

[219] Varga, J., Kocsubé, S., Péteri, Z., Vágvölgyi, C., Tóth, B., (2010). Chemical, physical and biological approaches to prevent ochratoxin induced toxicoses in humans and animals. *Toxins*. 2, 1718-1750.

[220] Deberghes, P., Betbeder, A., Boisard, F., Blanc, R., Delaby, J., Krivobok, S., Steiman, R., Seigle-Murandi, F., Creppy, E., (1995). Detoxification of ochratoxin A, a food contaminant: prevention of growth of Aspergillus ochraceus and its production of ochratoxin A. *Mycotoxin research*. 11, 37-47.

[221] Deberghes, P., Deffieux, G., Gharbi, A., Betbeder, A., Boisard, F., Blanc, R., Delaby, J., Creppy, E., (1993). Detoxification de l'ochratoxine A par des moyens physiques, chimiques et enzymatiques. *Hum. Ochratoxicosis Pathol*. 231, 75-82.

[222] Castegnaro, M., Barek, J., Fremy, J.-M., Lafontaine, M., Miraglia, M., Sansone, E., Telling, G. (1991) *Laboratory decontamination and destruction of carcinogens in laboratory wastes: some mycotoxins*. International Agency for Research on Cancer.

[223] Deberghes, P., Betbeder, A.M., Boisard, F., Blanc, R., Delaby, J.F., Krivobok, S., Steiman, R., Seigle-Murandi, F., Creppy, E., (1995). Detoxification of ochratoxin A, a food contaminant: Prevention of growth of Aspergillus ochraceus and its production of ochratoxin A. *Mycotoxin research*. 11, 37-47.

[224] Peng, C., Ding, Y., An, F., Wang, L., Li, S., Nie, Y., Zhou, L., Li, Y., Wang, C., Li, S., (2015). Degradation of ochratoxin A in aqueous solutions by electron beam irradiation. *Journal of Radioanalytical and Nuclear Chemistry*. 306, 39-46.

[225] Scott, P.M., (1996). Effects of processing and detoxification treatments on ochratoxin A: Introduction. *Food Additives and Contaminants Part a-Chemistry Analysis Control Exposure & Risk Assessment*. 13, 19-21.

[226] Bazin, I., Faucet-Marquis, V., Monje, M.-C., El Khoury, M., Marty, J.-L., Pfohl-Leszkowicz, A., (2013). Impact of pH on the Stability and the Cross-Reactivity of Ochratoxin A and Citrinin. *Toxins*. 5, 2324-2340.

[227] Heilmann, W., Rehfeldt, A., Rotzoll, F., (1999). Behaviour and reduction of ochratoxin A in green coffee beans in response to various processing methods. *European Food Research and Technology*. 209, 297-300.

[228] Bortoli, G., Fabian, M. (2001), *ASIC Congress, Trieste*.

[229] Denvir, A.J., McKenzie, K.S., Rogers, T.D., Miller, D.R., Hitchens, G.D., Andrews, C.C. (2001) US 6171625.

[230] McKenzie, K.S., Sarr, A.B., Mayura, K., Bailey, R.H., Miller, D.R., Rogers, T.D., Norred, W.P., Voss, K.A., Plattner, R.D., Kubena, L.F., Phillips, T.D., (1997). Oxidative degradation and detoxification of mycotoxins using a novel source of ozone. *Food and Chemical Toxicology*. 35, 807-820.

[231] Castegnaro, M., Barek, J., Fremy, J.-M., Lafontaine, M., Miraglia, M., Sansone, E., Telling, G. (1991) *Laboratory decontamination and destruction of carcinogens in laboratory wastes: some mycotoxins.* International Agency for Research on Cancer.

Chapter 2

BIOSYNTHESIS, METABOLISM AND DETOXIFICATION OF OCHRATOXIN A

ABSTRACT

Since the discovery of ochratoxin A (OTA) in 1965, many species of Aspergillus, as well as several species of Penicillium, have been found to produce OTA. The first part of this chapter summarized the species of reported OTA-producing fungi and the appropriate conditions for three representative species that are also major OTA producers, A. ochraceus, A. carbonarius and P. verrucosum. The biosynthetic pathway of OTA and key genes for OTA production are further reviewed. Briefly, in the presence of the polyketide synthase enzyme, malonate and acetate can form a pentaketide, the isocoumarin moiety of OTA, through the polyketide synthesis pathway. Then, a chlorine atom is added by a halogenation reaction. Through the shikimic acid pathway, the phenylalanine group of OTA is synthesized and is likely linked to the chlorinated dihydroisocoumarin at carbon 10 by a non-ribosomal peptide synthase.

After the elucidation of OTA biosynthesis, its biological metabolism and transformation were presented. Kidney, liver, and intestines are the main metabolic organs for OTA, while urine and feces are important excretory pathways. The fate of OTA differs in different species. OTB, 4-OH-OTB, OTα, OTβ, OTHQ, 4(R)- and 4(S)-OH-OTA, 10-OH-OTA, and OP-OTA are the metabolites of OTA in rodents. OTα, the major metabolite of OTA, is regarded as nontoxic. Carboxypeptidases catalyze its formation and degradation. OTB, which can be eliminated more rapidly, can reduce the toxic effects of OTA. OTC, which can be transformed to OTA quickly, has similar toxicity to OTA. Hydroxylation, catalyzed by CYP 450, is an inactivation pathway. In rats, 4(R)-OH-OTA is the main hydroxy metabolite of OTA; however in rabbits, only 10-OH-OTA is found. OP-OTA, because it is more toxic than OTA, can be considered a "toxic product."

Lastly, the reported methods for the detoxification of OTA are summarized, including physical methods such as the OTA-absorbing materials aluminosilicate, bentonite and charcoal, as well as radiation; chemical methods such as hydrogen peroxide, vitamin C, hydrochloric acid, formaldehyde, sodium hydroxide and ammonium hydroxide; and biological methods such as detoxifying bacteria and fungi. These important methods provide a good foundation for OTA control.

Keywords: Ochratoxin A, producing fungi, biosynthesis, biological metabolism, transformation, detoxification

1. Production of Ochratoxin A

1.1. Fungi that Produce Ochratoxin A

Ochratoxin A (OTA) was initially described in 1965. Scott and his colleagues isolated a strain of *Aspergillus ochraceus* from sorghum grain that produced this toxic metabolite [1]. Since then, many more species of *Aspergillus* as well as several species of *Penicillium* were found to produce OTA.

1.1.1. Aspergillus

In 1972, Hesseltine's group investigated 44 strains representative of 9 species of Aspergillus to identify the OTA-producing species. OTA was found in some strains of *A. alliaceus*, *A. sclerotiorum*, *A. sulphureus*, *A. ostianus*, *A. melleus*, and *A. petrakii* [2]. Varga et al. first reported another three species, *A. albertensis*, *A. auricomus*, and *A. wentii*, to be producers of ochratoxin A in 1996 [3].

In 2004, Frisvad et al. reported new ochratoxin A-producing species of Aspergillus: *A. cretensis*, *A. flocculosus*, *A. neobridgeri*, *A. pseudoelegans*, *A. roseoglobulosus*, *A. steynii* and *A. westerdijkiae* [4]. All the species mentioned above belong to the Aspergillus section Circumdati.

Some species of the *Aspergillus* section *Nigri* can also produce ochratoxin A. OTA production by strains of *A. niger* var. *niger* was found by Abarca et al. in 1994 [5]. Horie discovered the production of OTA in *A. carbonarius* in 1995 [6]. Furthermore, Téren and his group tested 150 strains belonging to *Aspergillus* section *Nigri* for ochratoxin A production using three different methods, ELISA, TLC and HPLC, and reported that OTA was detected in the culture filtrates of 5 of the 12 A. *carbonarius* strains in 1996 [7]. Samson and his colleagues in 2004 reported that OTA-producing species of section *Nigri* found in grapes and grape products, such as raisins and wine, included *A. carbonarius* and *A. niger*, with the latter producing lower levels of OTA. Moreover, 4 species discovered from coffee, namely *A. carbonarius, A. niger, A. lacticoffeatus* and *A. sclerotioniger,* all were able to produce OTA [8].

1.1.2. Penicillium

Several species of the genus *Penicillium* were also ochratoxin A producers. As early as 1969, *Penicillium viridicatum*, a new source of ochratoxin A, was introduced by Walbeek et al [9]. Ciegler's group further classified mycotoxin-producing strains of *P. viridicatum* into subgroups based on ability to produce ochratoxin [10]. Moreover, *P. cyclopium* was another OTA-producing species according to Northolt's research [11]. Other species producing OTA were reported to be *P. verrucosum* and *P. nordicum*. Larsen et al. screened 48 *P. verrucosum* isolates and found that 24 isolates mainly originating from plant sources and 20 isolates mainly originating from cheese and meat products had the ability to produce OTA [12]. In addition, isolates from the *P. nordicum* group could produce much larger amounts of OTA compared with that from the *P. verrucosum* group.

In summary, fungi which can produce ochratoxin A are shown in Table 1 (revision based on Abrunhosa et al. [13]).

Table 1. OTA producing fungi

Fungi species			
Aspergillus section Circumdati		*Aspergillus section Nigri*	*Penicillium*
A. ochraceus	*A. wentii*	*A. carbonarius*	*P. viridicatum*
A. sulphureus	*A. cretensis*	*A. niger*	*P. cyclopium*
A. sclerotiorum	*A. flocculosus*	*A. lacticoffeatus*	*P. verrucosum*
A. melleus	*A. neobridgeri*	*A. sclerotioniger*	*P. nordicum*
A. ostianus	*A. pseudoelegans*		
A. petrakii	*A. roseoglobulosus*		
A. albertensis	*A. steynii*		
A. auricomus	*A. westerdijkiae*		

According to Frisvad [4], the following species produce large amounts of ochratoxin A consistently: *A. westerdijkiae, A. cretensis, A. roseoglobulosus, A. pseudoelegans, A. flocculosus, and A. sulphurous. A. ochraceus and A. sclerotiorum* could produce large or small amounts of ochratoxin A but less consistently.

The major producers of OTA in food as well as feed products are considered to be *A. ochraceus, A. carbonarius, A. westerdijkiae, A. alliaceus, A. steynii, P. verrucosum and P. nordicum* [4], which are significantly correlated with the pre-harvest or the post-harvest storage conditions of agricultural crops. *P. nordicum*, a high OTA producer, is isolated primarily from certain fermented meats and cheeses. *A. niger*, a familiar species, is less relevant because most of its isolates are non-ochratoxigenic [14, 15], and the others generally produce small amounts of OTA [16]. However, it can pose some safety concerns in the manufacture of food grade enzymes and organic acids because it is a widely used species in many biotechnological processes. The other OTA producers are rare and therefore do not draw much attention for food safety [13].

1.2. Conditions for Fungi to Produce Ochratoxin A

Environmental conditions such as moisture, temperature, incubation time and substrate type play important roles in the ochratoxin A production by fungi, among which storage temperature and water availability (water activity, a_w) are the most important factors influencing fungal development in stored grain ecosystems [17, 18].

Optimal conditions for fungal growth and ochratoxin A synthesis differ among various species of OTA-producing fungi. We chose representative species of *Aspergillus* section *Circumdati*, *Aspergillus* section *Nigri* and *Penicillium* as examples and summarize the studies on the optimal conditions for production of ochratoxin A.

1.2.1. Conditions for Aspergillus Section Circumdati (A. Ochraceus) to Produce Ochratoxin A

Production of ochratoxin A was originally found in *A. ochraceus*, and many studies investigating its production conditions have been published. In the tropical and subtropical regions, OTA contamination of crops in the fields or in storage is most often caused by *A. ochraceus*, due to the suitable conditions.

In Ramos's study, three *A. ochraceus* isolates were investigated (NRRL 3.113, 3174 and 3.38). The optimal a_w levels for growth on agar media were in the range 0.96–0.98, with a temperature optimum of 30°C for two of the isolates while it was 25–30°C for the other isolate [17]. In addition, maximum growth on barley grains was reached at 30°C, at both a_w levels assayed (0.96 and 0.98). The maximum quantity of OTA was produced at the highest a_w condition (0.98 a_w) and after a three-week incubation time at 25–30°C. Moreover, the tolerated pH values were 3-10.

In Mühlencoert's review [18], *A. ochraceus* grows on several substrates between 8 and 37°C with an optimum for growth at approximately 24°C depending on water activity of the medium. Tolerated a_w values are 0.8-0.99. The temperature and water activity optima for OTA production are different from those optima for growth, but the values reported by several authors differ greatly: OTA production within a range of 12-37°C with an optimum at 31°C [19]. Pardo et al. revealed in 2004 that optimal conditions for the growth and OTA production of *A. ochraceus* were 0.99 a_w and 30°C.

1.2.2. Conditions for Aspergillus Section Nigri (A. Carbonarius) to Produce Ochratoxin A

A. carbonarius was recently discovered to produce ochratoxin A; it mainly infects fruits, especially grapes, thereby largely contributing to the OTA contamination in fresh grapes, raisins, wine and coffee [20].

Two similar studies on the conditions for *A. carbonarius* to grow and produce ochratoxin A were published in 2004. Mitchell et al. used modified synthetic grape juice medium with glycerol or glucose and found that, regardless of a_w modifying solute, the optimal temperature for the growth of most *A. carbonarius* strains was at 30–35°C, while none grew at $<$15°C. The optimum a_w for growth varied from 0.93 to 0.987 depending on the strain. The optimal conditions for OTA production varied with strains. Some strains produced maximum amounts of OTA at 15–20°C and 0.95–0.98 a_w [21]. In Bellí's study [22], two *A. carbonarius* as well as two *A. niger* strains isolated from grapes were studied for *in vitro* OTA production at 25°C on synthetic nutrient medium, over a period of 20 days at different a_w levels. OTA accumulation by these molds favored high levels of a_w. Maximum amounts of OTA were found at the earlier growth states (5 days for *A. carbonarius* and 7–13 days for *A. niger*).

Astoreca clearly analyzed the water activity and temperature conditions for *A. carbonarius* growth and OTA production on irradiated dried grapes in 2010. Over the whole range of 15°C to 30°C, OTA production occurred. The favorable temperature was 25 or 30 °C, depending on the a_w level. Generally, OTA concentration increased at the determined incubation times as the a_w level increased, although there were no statistically significant differences [23].

1.2.3. Conditions for Penicillium (P. Verrucosum) to Produce Ochratoxin A

In areas with temperate climates such as Europe and Canada, *P. verrucosum* is the main source of OTA contamination in grains and their products due to its preference for relatively low temperatures [24].

Lindblad et al. demonstrated that the minimum a_w level for the growth of *P. verrucosum* was approximately 0.81–0.83, and for OTA production, an a_w of 0.83–0.90 was required [25]. Similar results came from Cairns-Fuller's research: three isolates from *P. verrucosum* grew on a milled wheat medium, and the favorable conditions for their growth and OTA production were tested, including the temperature (10–25°C), water activity (0.75–0.995 a_w) and gas composition (air, 25 or 50% CO_2). The optimum growth of *P. verrucosum* was found at 25°C and 0.98 a_w *in vitro* but at 25°C and 0.95 a_w on wheat grain. Significant suppression of growth and OTA production occurred with 50% CO_2 and 0.90–0.995 a_w at 25°C [26].

Arroyo's group examined the effect of water activity (a_w, 0.97, 0.95, 0.93) and pH (4.5, 6.0) on colonization and OTA production by three strains of *P. verrucosum* on bread analogues. At 25°C, optimum growth appeared at 0.97 a_w and pH 6.0. Maximum OTA was produced at approximately 28–36 days for two of the strains with an optimum of 0.93 a_w at pH 6.0 [27].

In summary, for OTA production, *Penicillium* prefer lower temperatures than *Aspergillus* (optimum 20-25°C for *Penicillium* and >30°C for *Aspergillus*), while all OTA-producing fungi need high water activity (normally >0.9).

2. Biosynthesis of Ochratoxin A

Ochratoxin A (OTA) is an important mycotoxin produced as a secondary metabolite by the fungal species of *Aspergillus* and *Penicillium* [1, 28]. OTA is chemically named 7-carboxyl-5-chloro-8-hydroxyl-3,4-dihydro-3-R-methylisocomarin-7-L-βββ-phenylalanine (see Figure 1) and consists of a *para*-chlorophenolic moiety that contains a dihydroiso-coumarin group, which is amide-linked to L-phenylalanine.

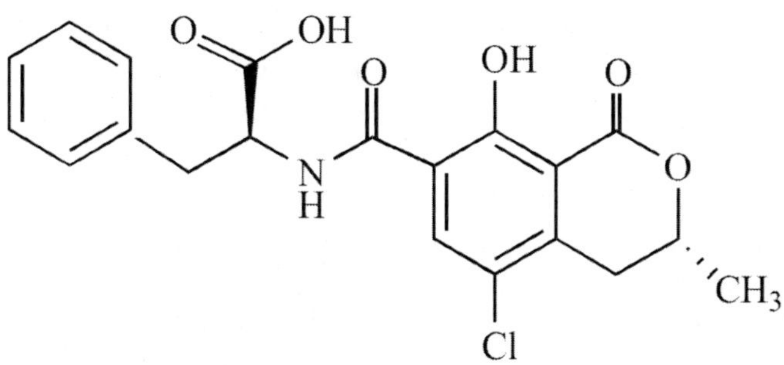

Figure 1. Structure of ochratoxin A.

The empirical formula of ochratoxin A is $C_{20}H_{18}O_6NCl$, and its molecular weight is 403.82 [29]. The IUPAC developed formula for ochratoxin A is L-phenylalanine-N-[(5-chloro-3,4-dihydro-8-hydroxy-3-methyl-1-oxo-1H-2-benzopyran-7-yl)carbonyl]-(*R*)-isocoumarin. The chemical abstract specification (CAS) of ochratoxin A is 303-47-9 [30].

OTA has a weak acidic property. The pK_a values are in the ranges 7.0–7.3 and 4.2–4.4 for the phenolic hydroxyl group of the isocoumarin part and the carboxyl group of the phenylalanine moiety, respectively [31-33]. Under UV light, OTA can exhibit intense green fluorescence in acid conditions and blue fluorescence in alkaline medium, with its crystalline structure varying from colorless to white [33]. OTA is soluble in polar organic solvents (alcohols, chloroform, ketones) at acid and neutral pH but slightly soluble in water and insoluble in saturated hydrocarbons and petroleum ethers. Regarding alkaline conditions, OTA is soluble in general alkaline solutions as well as aqueous sodium bicarbonate solution.

The melting point of OTA when crystallized from benzene as a solvate is approximately 90°C, whereas the melting point of non-solvated crystals from xylene is 169°C. OTA is optically active, and its spectral characteristics are shown in Table 2.

Table 2. Spectral characteristics of ochratoxin A [34]

Spectral	Solvents	Characteristics	References
UV-VIS	ETOH	λmax = 213nm (ε36.800) λmax = 332nm (ε 6.400)	[35]
Fluorescence	ETOH 96%, ETOH /ABS	λmax = 467nm λmax = 428nm	[35]
IR	CHC13	3380; 2988; 1723; 1674; 1528; 1425; 1381; 1304; 1260; 1170; 1140; 1107; 827 cm^{-1}	[36]
NMR ^{1}H250-MHZ	CDC13	δ 12,70; δ 10,80; δ 8,55 (3H); δ 7,23; δ 7,15 (H Aromatic); δ 4,71; δ 5,07 (CH); δ 2,78; δ 3,2 (CH2); δ 1,55 (CH3)	[37-39]
MS	—	m/z 239/241, m/z 255/257, molecular ion m/z 403	[40, 41]

The high stability of OTA determines its particularity. It has been shown that OTA can resist high temperatures and acidity. Therefore, once foodstuffs are contaminated with OTA, it is very difficult to remove this molecule completely. Müller [42] in 1982 showed that OTA was only partially degraded under normal cooking conditions. In addition, OTA showed resistance to three hours of high-pressure steam sterilization at 121°C [43], and its destruction was not complete even at 250°C [44]. Gamma irradiation (up to 7.5 Mrad) does not cause any degradation of OTA in ethanol. However, the degradation of OTA can be achieved at low moisture levels when pre-treated with an excess of sodium hypochlorite (NaClO). Moreover, exposure to fluorescent light is another factor in its degradation. The high stability of OTA may be one reason it is the most toxic member in the group [45].

The biosynthetic pathway of OTA has not yet been completely established. However, the origins of the OTA precursors have been reported. Labeling experiments using both ^{14}C and ^{13}C showed that the phenylalanine moiety originates from the shikimate pathway, while the dihydroisocoumarin part comes from the pentaketide pathway (Figure 2) [46]. Figure 2 shows the widely accepted biosynthetic pathway of OTA.

Briefly, one acetate unit, namely acetyl-CoA, condenses to four malonate units in the catalysis of a polyketide synthase, which is the first step of the synthesis of isocoumarin [48]. Despite the key role of polyketide synthase in this process, the genes encoding polyketide synthase proved to be different between OTA-producing *Aspergillus* and *Penicillium* species [48, 49]. The polyketide synthase-encoding gene in *A. ochraceus* is expressed only when permissive conditions for OTA production are provided. More interestingly, this gene is only expressed during the early stages of OTA synthesis. However, this phenomenon is not found in *Penicillium,* according to the available information. In this kind of fungi, Geisen et al. [46] found two different kinds of polyketide synthase for OTA production in *P. nordicum* and *P. verrucosum*. Apart from OTA, another polyketide-based mycotoxin CIT can also be produced by *P. verrucosum* but not by *P. nordicum*, which might account for the mentioned difference between the two kinds of *Penicillium*.

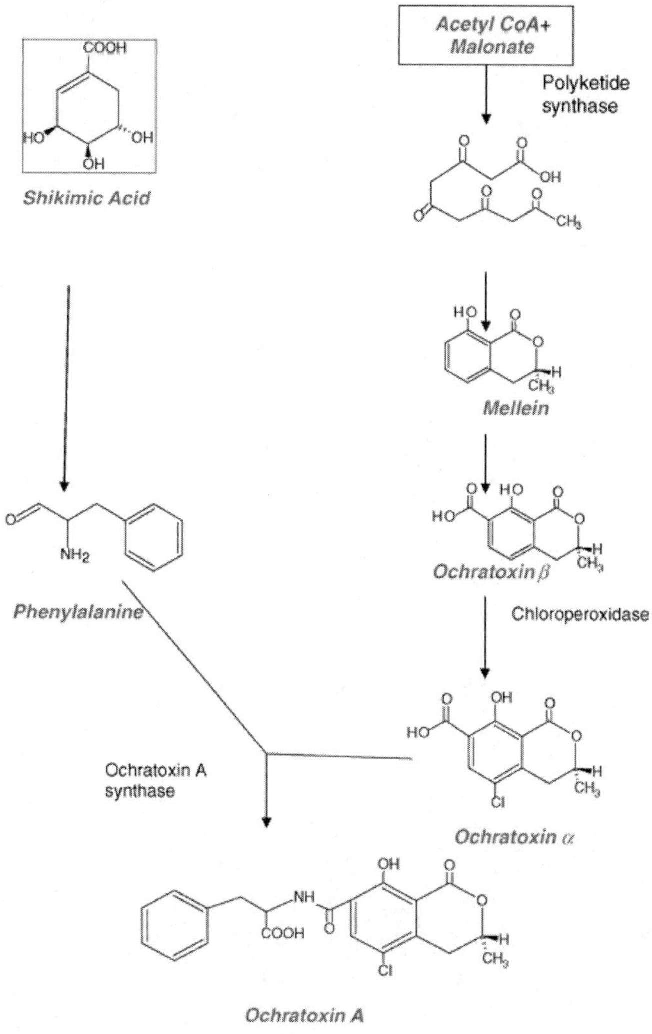

Figure 2. Biosynthesis of ochratoxin A [47].

After the synthesis of isocoumarin, a lactone ring is formed, modifying the polyketide chain into mullein. Subsequently, the structure of ochratoxin β is synthesized by the addition of a carboxyl group derived from the C1 pool, such as sodium formate and S-methylmethionine [50], followed by the formation of ochratoxin α. Chloroperoxidase incorporates a chlorine atom to form ochratoxin β. Finally, ochratoxin A is biosynthesized by the linkage of OTα and phenylalanine [51, 52].

More specifically, the biosynthesis of ochratoxin A in *A. ochraceus*, *A. carbonarius* and *P. verrucosum*, the three species emphatically described in 1.2, will be described.

2.1. Biosynthesis of Ochratoxin A in *A. Ochraceus*

According to a study by Harris, J.P. et al. published in 2001, an isolate from *A. ochraceus* could produce ochratoxin A, ochratoxin B and diaporthin, a similar polyketide, in liquid medium and presented growth-associated production [52]. However, ochratoxin A and ochratoxin B were predominantly produced by the same isolate in a solid substrate, shredded wheat, over 2 weeks' cultivation, in large amounts up to 10 mg/g. They then investigated the possible biosynthetic pathways of ochratoxins utilizing ^{14}C-labelled precursors. As an important intermediate, ochratoxin β could be biotransformed into both ochratoxin A and B efficiently. However, the other essential intermediate, ochratoxin α, could only be biotransformed into ochratoxin A rather than ochratoxin B, suggesting that chlorination is probably the penultimate step in the biosynthesis of ochratoxin A. This point was also supported by the observation of the poor conversion of ^{14}C-labelled ochratoxin B to ochratoxin A, whereas ochratoxin B is the dechlorinated form of ochratoxin A.

Several studies have revealed essential genes for OTA biosynthesis in *A. ochraceus*. In 2003, a portion of polyketide synthase gene (pks) that is indispensable for OTA biosynthesis in *A. ochraceus* was cloned for the first time by O'Callaghan, J and colleagues [53]. Its expression product shared approximately 30% identity with the acyl transferase regions from fungal polyketide synthases. The pks gene was expressed only under OTA-permissive conditions, and a mutant in which the pks gene had been interrupted could not synthesize OTA. They further studied the effect of pH and nutritional stimuli on the expression of OTA biosynthetic genes [54]. A nearly 40-fold increase in OTA occurred when supplemented by potato dextrose broth with yeast extract, with a concomitant increase of pks mRNA transcription in this fungus. The culture pH is another key regulator for OTA production, and *A. ochraceus* seemed to prefer lower pH. Large amounts of OTA were detected at $pH<7.0$; however, the amounts decreased at higher pH values. Excitingly, an OTA biosynthetic gene cluster was proposed because of the discovery of two putative p450 type monooxygenase genes, p450-H11 and p450-B03, and these two genes were detected under all growth conditions. In addition, the expression profile of the p450-B03 gene in particular was very similar to the expression profile of the pks gene.

Several other OTA biosynthesis-related genes have been reported in recent years. Aoyap1 can regulate OTA synthesis by controlling the cell redox balance in *A. ochraceus* [55]. Wang et al. identified two new OTA-related polyketide synthase (PKS) (AoOTApks) genes in *A. ochraceus* strain fc-1 [56]. AoOTApks-1 was considered to be directly involved in the production of OTA; however, AoOTApks-2 was considered an indirect factor in OTA biosynthesis, based on the observation that AoOTApks-1 mutants did not produce OTA, while the AoOTApks-2 mutants produced less OTA than the wild type.

2.2. Biosynthesis of Ochratoxin A in *A. Carbonarius*

A. carbonarius is the main producer of OTA contamination in grapes and wine. Apart from the common pathway of OTA biosynthesis in *Aspergillus* species, which was introduced earlier, Gallo and colleagues provided new insights into the OTA biosynthetic pathway by inactivating a nonribosomal peptide synthetase gene in *A. carbonarius* [57], which revealed for the first time the participation of a NRPS gene product in OTA biosynthesis among

Aspergillus species. In their study, NRPS, which encodes a nonribosomal peptide synthetase, was deleted, eliminating the capacity for OTA production in *A. carbonarius* ITEM 5010. By cultivation of the transformed strain in liquid medium, the absence of OTA as well as the accompanying accumulation of OTα demonstrated that OTα is a product of OTA hydrolysis and that the enzymatic step adding phenylalanine to polyketide dihydroisocoumarin precedes the chlorination step to form OTA in *A. carbonarius*.

A large number of PKS-encoding genes are found in the genome sequence of *A. carbonarius*. Meanwhile, AcOTApks genes also play an essential role in the OTA biosynthetic pathway in *A. carbonarius* [58].

Crespo-Sernpere's group used a suppression subtractive hybridization approach to identify genes whose expression is linked with the ability to produce OTA in *A. carbonarius* [59]. A total of 109 differentially expressed sequences putatively involved in the production of OTA were identified, of which 26% corresponded to regulation processes, 15% to hypothetical proteins, 12% to stress response as well as detoxification, 9% to transport and secretion processes, 7% to amino acid metabolism, another 7% to hydrolysis of energy reserves and 5% to secondary metabolisms.

Moreover, the same group further investigated proteins that might be involved in OTA biosynthesis through proteome analysis of *A. carbonarius* under OTA-producing conditions [60]. Twenty-one differential protein spots showed a two-fold or greater change in expression between highly (W04-40) and weakly (W04-46) OTA-producing *A. carbonarius* strains, among which nine protein spots were identified, including CipC, $NADH^+$ ubiquinone reductase, acetylglutamate, and others. The identified proteins shared isogeny with ones corresponding to amino acid metabolism, regulation, and oxidative stress.

2.3. Biosynthesis of Ochratoxin A in *P. Verrucosum*

Wawrzyniak and Waskiewicz examined the influence of various culture mediums, including maize, rice, wheat, and barley, as well as temperature on the growth and production of OTA by an isolated strain of *P. verrucosum* at a settled humidity [61]. The *P. verrucosum* spores were inoculated into the above cereal substrates, which were previously autoclaved and afterward stored at 10°C, 20°C or 30°C for 40 days. In addition to the detection of OTA accumulation, the biomarker ergosterol content was also tested to evaluate the fungal growth. Among the tested cereals, rice exhibited the highest *P. verrucosum* growth as well as OTA production at all tested temperatures. OTA was produced and accumulated at the temperatures of 20°C and 30°C, with 20°C as the optimal condition. No capacity for OTA biosynthesis was found in *P. verrucosum*.

The brief process of OTA formation in *P. verrucosum* was mentioned above, in which polyketide synthase otapks genes also play a crucial role [62]. A mutant of *P. verrucosum* in which the otapksPV gene was silenced lost the ability to biosynthesize OTA. The protein encoded by the cloned gene is the most similar to the citrinin polyketide synthase CtnpksMa from Monascus anka. Among various nutritional conditions that affect OTA biosynthesis in *P. verrucosum*, glycerol and galactose showed positive regulation: their addition to the culture medium resulted in approximately 19 and 32-fold increases, respectively, in OTA production. Interestingly, the redundant OTA production was related to the upregulated transcription of the otapksPV gene. However, a nearly 10-fold repression of OTA production occurred upon the addition of glucose, with this suppression again mirrored in reduced levels of otapksPV gene transcription [62].

The conclusion regarding the impact of environmental factors on OTA production as well as otapksPV gene expression in *P. verrucosum* coincides with an earlier study. Schmidt-Heydt et al. further found that two circumstances could activate the production of OTA in *P. verrucosum,* either optimal growth conditions or weak stress conditions [63].

To sum up, although it has not yet been fully understood in any fungal species, the acknowledged OTA biosynthetic pathway is elucidated to occur in the presence of polyketide synthase enzyme, malonate and acetate, which can form a pentaketide, the isocoumarin moiety of OTA, through a polyketide synthesis pathway. Then, the chlorine atom is added by a halogenation reaction which is likely to be performed by a chloroperoxidase or a halogenase enzyme, most probably the latter. The phenylalanine part of OTA is synthesized through the shikimic acid pathway and is probably linked to the chlorinated dihydroisocoumarin at carbon 10 by a non-ribosomal peptide synthase [52, 62].

3. Biological Metabolism and Transformation of Ochratoxin A

OTA is rapidly absorbed from both the stomach and the small intestine in rats and mice. The stomach and the small intestine are the major sites of OTA absorption. The duodenum is the third site of absorption, where OTA can be transferred against a concentration gradient [64]. During absorption, organic anion transporters play an important role in the transport of OTA [65-67]. After absorption, many factors (both internal and external), including dose, route and duration of administration as well as species-specific factors such as half-life will determine the concentration of OTA and its metabolites and the affinity of serum binding [45, 68]. The reviews by Ringot et al. [47] and O'Brien et al. [69] could provide more details. Furthermore, another factor in the kinetics of OTA is enterohepatic circulation. The reabsorption of OTA in the kidney is responsible for the residual persistence of the toxin and ultimately results in renal toxicity in rodents [70, 71]; for a review, see Gekle et al. [72].

As mentioned in section 2.1, OTA is a weak binary acid. It can be the monoanion (OTA^-), the dianion (OTA^{2-}) or the fully protonated toxin (OTA^0) at different pH values. Similarly, OTA exists in different form when distributed in different organs: for example, both the monoanion (OTA^-) and the dianion (OTA^{2-}) exist in the duodenal chyme, and the fully protonated toxin (OTA^0) is mainly present in the upper parts of the gastrointestinal tract. For most animal species, the stomach, particularly the proximal jejunum, is the major location of passive OTA absorption in the non-ionized (OTA^0) and monoanion (OTA^-) forms [64, 73, 74]. Due to the high binding affinity of OTA, plasma proteins are mainly responsible for the passive absorption. To illuminate the mechanism of OTA transport across the human intestinal mucosa, Berger et al. [75] cultured CaCo-2 cells in a bicameral system and then developed an *in vitro* model of the human intestinal epithelium. Using simulated *in vivo* duodenum conditions (apical pH 6, basolateral pH 7.4), these experiments showed that the direction of OTA transport is mainly apical to basolateral. Interestingly, carrier systems such as the organic anion polypeptide carrier (Oatp), the organic anion carrier (OAT), and the H^+-dipeptide cotransporter are not involved in OTA transport across CaCo-2 cells, demonstrating that the absorption of OTA is a process of diffusion without carriers. However, the intestinal absorption did not behave in the same way. Both Kumagai and Aibara [64] and Berger et al. [75] concluded that the efflux of the toxin at the apical (AP) pole of the CaCo-2 cells could limit intestinal absorption, which suggested that the ATP-Binding Cassette C2 (ABCC2) participates in the transport of OTA.

These transporters block absorption and indirectly promote the elimination process instead, which contributes to protective tissue barriers. Schrickx et al. found that BCRP is widely expressed in various organs, including the intestines, liver, blood vessels, lactating mammary glands and kidneys [76, 77]. As an ATP-dependent membrane transporter, BCRP also participates in OTA efflux [78].

The percentage of OTA absorption shows species-specific variation: 66% in pigs, 56% in rats and rabbits and 40% in chicken [79]. Subsequently, OTA is transported and distributed to different tissues and organs though the portal system.

3.1. Distribution

The data on equilibrium dialysis, solubility and spectrophotometric analysis show that, once it reaches the blood stream, almost all OTA is bound to serum albumin due to both ionic and hydrophobic forces. The form of OTA in the blood stream could allow its passive absorption in the non-ionized form, which partly explains its long half-life in the body [80, 81]. Meanwhile, only traces of OTA could be found in erythrocytes [73].

To illustrate the role of albumin in OTA distribution, kinetic experiments on OTA were designed by Kumagai [82] using an albumin-deficient rat model. In this study, the OTA concentrations and the rate of OTA excretion in bile and urine in albumin-deficient rats were far higher than in normal rats, which proved that the binding of OTA to albumin in a stable form restricted the distribution of OTA into the hepatic and renal cells, which could be the primary blockage of its elimination.

The interactions between OTA and albumin were explored through both *in vitro* and computational studies [31, 81] An early study reported that bovine serum albumin (BSA) could form stable bonds with several ochratoxin metabolites (OTA, OTB, OTC and OTα). BSA binds to OTA, OTB and OTC at a molar ratio of 2:1; however, it binds to OTα at a ratio of 1:1. Above all, these studies demonstrated that OTA binds to BSA depending on both ionic and hydrophobic forces, based on the isocoumarin part of the toxin and a BSA cationic center for the phenolate ion of the toxin, respectively. Furthermore, McMasters and Vedani [83] more recently explored the interaction sites and form of OTA with human serum albumin (HSA). Based on molecular modeling data, it was concluded that the carboxylic group and the protonated phenol group are involved in the binding to OTA, and the binding sites are lysine (Lys199) and arginine (Arg228), respectively. Recently, Il'ichev et al. [84] reported that optical spectroscopy provided evidence of at least two unique binding sites for OTA, each able to accommodate one dianion.

The binding of OTA to HSA is affected by pH-induced structural changes. No binding is observed between OTA and the denatured protein. Finally, bound OTA strongly quenches the fluorescence of Trp214. These data suggest that the highest-affinity binding site is in subdomain IIA of the protein.

Schwerdt et al. [85, 86] developed an HRP-based system to detect cellular proteins binding OTA. In addition to serum albumin, several other proteins were proved to be capable of binding to OTA, but the specific binding found by this method differs dramatically. Thus, the highly specific binding of OTA might make certain proteins a preferential target of OTA toxicity. Furthermore, binding contributes to the intracellular accumulation of OTA, leading to a prolonged half-life in the mammalian body and emphasizes the toxic potential of its metabolite. However, none of these uncharacterized blood proteins were proven to be related to the transport of OTA.

Using several different species of animal models, Hagelberg et al. [68] studied the toxicokinetics of OTA and the affinity between the plasma and OTA. OTA was administered through the oral or intravenous route at a dose of 50 ng/g body weight. The fraction of OTA shows a species-specific difference, and the plasma clearance was fastest in quail and fish, at 72 and 58 ml/kg.h, respectively, and only 0.17 ml/kg.h in monkey. The bioavailability was as low as 1.6% in fish but as high as 97% in mouse. Meanwhile, the free fraction of toxin in the plasma was found to be less than 0.2% in all species investigated (including man) except fish. Unfortunately, the underlying mechanisms of these species differences in the serum half-life were not completely elucidated (Table 3) [87]. In addition, Stander et al. [88] found that no appreciable conversion of OTA occurred in monkeys after intravenous administration, which could explain the longer half-life time compared to rodents.

Table 3. Elimination half-life of ochratoxin A [87]

Elimination half-life t1/2 (h)	Carp	Quail	Mouse	Pig	Rat	Monkey Macaca	Human
i.v.	8.3	12	48	150	170	840	≈1400
p.o.	0.68	6.7	39	72	150	510	840

They found that the half-life of OTA after oral ingestion (p.o.) was much shorter than after intravenous injection (i.v.), and supposed the main reason to be that before entering the systemic blood circulation, OTA might pass through the liver and be subsequently eliminated through the bile. Several studies on many models and animals suggest that the serum half-life in humans may be the longest.

In animals, the species, the dosage administered, the form of OTA administered, the diet composition and the health status of the animal may influence the concentration of the toxin and its later transformations in tissues and organs. When trying to assess the occurrence of toxin residues in animal tissues, one should focus on these listed factors. The elimination half-life of OTA is much longer in the blood than in tissues, and there is a hypothesis that OTA may bind more easily to blood proteins than tissues ones, supported by several research teams [68, 89, 90]. Some other papers reported that the tissue distribution follows the order kidney > liver > muscle > fat [91] or kidney > muscle > liver > fat [92-94] in pig, rat, chicken and goat. However, another perspective was suggested by Kane et al. [95], who reported a somewhat different tissue distribution after the ingestion of low doses of radioactive OTA, decreasing in the following order: lung, liver, kidney, heart, fat, intestine, testes, muscle, spleen and brain. In 1999, Schwerdt et al. [85] suggested the potential of OTA to bind in a highly specific way to organellar proteins and different cytosolic proteins, through an *in vitro* study using cell cultures of different renal lines and proteins from rat organs. Based on their results, a 62 kDa organellar protein originating from the proximal tubules was considered to be the most prominent. They proposed the hypothesis that this protein contributed substantially to the OTA accumulation in kidney cells, enhancing its nephrotoxic effect despite its low abundance.

The liver and kidney are not the only tissues in which we observed OTA, as be supported by Belmadani, A. et al. and co-workers. They administered a dose of 289 μg OTA/kg/day to male rats for 8 days by gastric intubation and then observed a percentage of 0.022–0.028% in the striatum, cerebellum, brain, ventral mesencephalon and hippocampus [96].

Studies have also been performed on mammalian species such as rat, mice and swine showing the procedure of OTA in utero transfer [97]. In 1983, Appelgren, L.-E conducted whole-body autoradiography studies in mice using high doses of i.v. ^{14}C-labelled OTA. Their results showed that OTA could cross the placenta more readily at days 8 and 9 than at day 10 of gestation, with radioactive labeling appearing within 20 min in the uterine wall and in placental and fetal tissues [98]. From the differences they observed in OTA fetal uptake after different durations of gestation, they concluded that the developmental stage of the placenta, which is considered to be complete after day 9 of gestation, can influence the transfer.

In humans, a study conducted by Zimmerli, B. showed that the OTA concentration in fetal serum was found to be twice the maternal concentration, suggesting the possibility of active placental transfer [99]. Miraglia et al. [100] reported that the level of OTA in the corresponding maternal serum is much lower than that in the placenta and funiculum. Genetic, environmental and pathological factors were suggested to contribute to the placental OTA transport. However, there is no available data on the mechanism of OTA transfer across the placenta. Polarized cells, which possess carrier-mediated transport pathways, like the intestinal epithelial cells and the renal proximal tubules, have the function of separating the fetal blood and the maternal blood circulation. The carrier proteins promote the supply of essential compounds for the growth and development of the fetus. In 2000, Cha et al. [101] isolated hOAT4 from the human placenta, and they found that hOAT4 was probably localized in the trophoblastic membrane, which can transport various organic anions, including OTA.

OTA, known as an accumulative toxic compound, is rapidly absorbed and slowly eliminated. Both fecal and urinary excretions play an important role in the plasma clearance of this toxin in many species. Scientists have found that the route of administration, the dose, the degree of binding with plasma proteins and the enterohepatic circulation of OTA can influence the relative contribution of each excretory route.

3.2. Renal Excretion

As mentioned above, OTA is reported to have a high plasma protein binding potential. Therefore, only a limited amount of OTA could be filtered in the glomeruli. Beyond that amount, OTA instead undergoes tubular elimination in the urine. Moreover, OTA is reabsorbed at all nephron segments. These phenomena are mainly attributed to the transporter proteins, which are responsible for carrier-mediated transport across the tubular membranes.

Organic anion transporters (OATs) from different species have been discovered that mediate the transport of OTA across the basolateral membrane of the proximal tubule [65,102]. Similarly, Tsuda et al. [66] found that OTA is transported by OAT1 and that the accumulation of OTA via OAT1 in mouse proximal tubular cells is the primary event in the development of OTA nephrotoxicity. This evidence demonstrated that piroxicam (PIR), like *para*-aminohippurate (PAH), could inhibit the uptake of OTA by acting as a typical inhibitor of OAT1. Therefore, it is hypothesized that this mechanism might explain the prevention of OTA nephrotoxicity by these compounds. Moreover, human hOAT1 (also present in the liver and brain) and hOAT3 (also present in the liver and brain) have a high affinity with OTA, which may be essentially related to the renal transport of OTA across the basolateral membrane in the human kidney. [67]. In addition, Kusushura et al. [103] isolated rOat3 from the rat brain and demonstrated it to be a similar transporter that mediates the transport of OTA.

As shown by Bahnemann et al. [104], transepithelial secretion is an effective mechanism to accumulate OTA in the tubular lumen and thus effective for its urinary excretion. OTA leaves the cells across the luminal (apical) membrane by facilitated diffusion. These authors suggested that there is a transport system that is sensitive to probenecid (a specific inhibitor of the ABCC superfamily) mediating the facilitated diffusion of OTA at the luminal membrane. To test this hypothesis, Leier et al. [105], using recombinant human MRP2(ABCC2), proved that this unidirectional transport protein contributes to the secretion of OTA into the lumen of kidney proximal tubules. Furthermore, Babu et al. [106] demonstrated that human hOAT4 (localized in the luminal membrane of the proximal tubule) mediates not only the reabsorption but also the efflux of OTA in the apical side of the proximal tubule.

OTA could be reabsorbed in all nephron segments after its excretion. Dahlman et al. [107] discovered that different rat nephron segments show a pH-dependent difference in OTA reabsorption based on *in vivo* microinfusion/microperfusion experiments. However, under physiological conditions, the predominant sites of reabsorption are the proximal straight tubule (PST), the ascending limb of Henle's loop (ALH) and the terminal collecting duct (TCD). Interestingly, reabsorption in the PST and ALH is not pH-dependent but thought to be carrier-mediated. The pH-independent reabsorption in the PST is mediated by the apical organic anion transporter (OAT-K1). It was proven by *in vitro* studies with different cell lines and/or *in vivo* microperfusion experiments in rats that the OTA absorption in PCT is mediated by different transporters. For example, the H^+-dipeptide cotransporter could mediate the OTA reabsorption based on *in vivo* microperfusion experiments in rats [107, 108]; furthermore, transporters such as human hOAT4 and rat kidney-specific rOAT-K2 are also involved in OTA reabsorption at this nephron segment based on *in vitro* data. In short, both the efflux of OTA in the tubule lumen and its reabsorption at the apical side of the proximal tubule are mediated by hOAT4. Similarly, a kidney-specific apical organic anion transporter, rOAT-K1, was found in the rat PST [107] and could mediate OTA reabsorption. Unfortunately, the transport mechanisms of organic anions in ALH remain unclear.

Recently, Kobayashi et al. [109] reported the characterization of OAT2 in the mouse kidney, which involved in OTA uptake. In fact, the expression of OTA2 shows a different distribution in different tissues. The kidney and the apical membranes of the tubules and of the ALH are the predominant tissue in which mOat2 is specifically expressed in mice, whereas in rats, the liver is the predominant organ expressing rOat2. The uptake experiments revealed that mOat2 plays a specific role in ALH reabsorption. However, the role of OAT2 in renal OTA transport in different mammalian species remains largely undetermined. Notably, the localization of OAT2 is reported as apical in the rat and mouse kidney but basolateral in the human kidney [110]. Absorption in the DT and CD occurs mostly by passive non-ionic diffusion [107,108]. An *in vitro* study suggested that OTA is transported across the apical membrane of Madine–Darby canine kidney cells (MDCK) both by non-ionic diffusion and by a H^+-dipeptide cotransporter. The reabsorption of OTA from the CD contributes to the observed long half-life of OTA in the mammalian body [111].

In addition, mouse renal mOat5 has been recently identified as an organic anion transporter that also participates in OTA transport *in vitro* [110]. There is very little information available on the membrane localization of mOat5. Nevertheless, immunohistochemical evidence suggests that OAT5 protein was found in the apical membrane of renal proximal tubules, mainly at the cortico-medullary junction [112].

In summary, the reabsorption of filtered and secreted OTA blocks its excretion, which may cause the accumulation of the toxin in the renal tissue and thus contribute to its renal toxicity.

3.3. Fecal Excretion

Early studies showed that fecal excretion is among the main ways to excrete both OTA and its metabolite OTα. Biliary excretion is another efficient way [64, 74, 113]. The molecular weight of OTA (403.8) is the underlying determinant of the route of excretion. In fact, there are variations in different species of the threshold molecular weight for the biliary excretion of organic anions. Generally, both biliary excretion and renal reabsorption can occur for OTA (403.8), whose molecular weight is between 350 and 500. In addition, intestinal secretion may be another route of OTA excretion [64, 75, 114]. The data from perfusion experiments in rats show that approximately 56% of OTA administered was excreted in both the urine and feces as the unchanged toxin and indirectly as OTα, which suggested that the amount of OTA secreted by the entire gastrointestinal tract may be comparable to the amount excreted through the bile[114]. This intestinal secretion and absorption showed asymmetric transfer of OTA, and it seems that the lower segment of the intestine is the predominant site of OTA secretion [64]. According to the data from the model of the human intestinal epithelium *in vitro*, it is concluded that multidrug resistance-associated protein (MRP2) could be implicated in transepithelial transport. Therefore, the absorption of OTA across the intestinal mucosa would be limited thanks to its excretion through MRP2 at the apical pole of the enterocytes [75]. MRP2 is responsible for the biliary excretion of organic anions, including glutathione, glucuronide and sulfate conjugates of xenobiotics, which means it may also be involved in the hepatobiliary elimination of OTA conjugates [115]. The microbial microflora could hydrolyze OTA to OTα both in the intestine and in the large intestine. Furthermore, the OTA conjugates are hydrolyzed by the intestinal microflora, which release OTA to participate in the enterohepatic circulation.

In goats given a single 0.5 mg/kg oral dose of [^{3}H]-labeled OTA, fecal excretion is the chief route of the toxin's elimination [116]. Sreemannarayana et al. [90] reported that the metabolites of OTA(OT-OH, OTα) are cleared at a much faster rate with much shorter half-lives, while Xiao et al. [117] reported that less than 10% of i.v. administered OTA was excreted in the feces of sheep and that all was hydrolyzed to OTα.

4. Biotransformation Pathways

The biotransformation of OTA has not been fully illuminated in detail. At present, several controversies still exist regarding the intermediates involved in OTA metabolism. Several metabolites have been characterized *in vitro* and/or *in vivo*, while other metabolites appear to exhibit contradictory *in vitro* and/or *in vivo* results. The contribution of metabolites to OTA toxicity still lacks key information.

4.1. Hydrolysis

As mentioned before, OTα, a much less toxic compound than OTA, is formed by the cleavage of the peptide bond, which is a key intermediate in the detoxification of OTA. In rodents, this detoxification pathway appears to occur mainly due to the cecum microbial flora [32, 118, 119]. OTA can be hydrolyzed *in vitro* by carboxypeptidase A, α-chymotrypsin and possibly cathepsin C from lysosomes [120, 121]. Based on experiments involving the *in vitro* incubation of OTA with digesta from rats, Madhyasta et al. [122] discovered that OTA hydrolysis mostly occurs in the presence of digesta from the large intestine or cecum, while little hydrolysis occurs in the stomach contents and digesta from the small intestine even though the pancreatic enzymes, including carboxypeptidase A, are released into the small intestine and have not been reported to be secreted into the large intestine or caecum. Moreover, digesta from rats treated with the antibiotic neomycin sulfate was much less effective. Thus, Madhyasta et al. [122] concluded that the intestinal microbial population not only is able to hydrolyze OTA *in vitro* but also can affect the hydrolysis of orally ingested OTA, reducing its concentration and increasing the concentration of OTα. The hydrolysis of OTA is primarily attributed to the anaerobic microorganisms, particularly bacteria, in the large intestine and caecum. However, it remains unclear which microbial enzymes are responsible for OTA hydrolysis [123]. In a study using homogenate of rat tissues, Suzuki et al. [114] observed that OTα, as a major metabolite of hydrolyzed OTA, occurred at a higher level in the homogenates of the duodenum, ileum and pancreas but a lower level in the liver and kidney. According to other studies, liver has a low capacity for OTA hydrolysis in the rabbit and rat [124-126]. Moreover, Hansen et al. [126] failed to detect any hydrolysis products of OTA in rat hepatocytes. In studies with ^{14}C-labelled OTA in rats, ^{14}C-labelled OTα was the only metabolite detected in the cecum and large intestine, which indicated a lack of efficient OTA hydrolysis in most tissues other than the intestine [127].

Ruminants have a greater resistance to OTA, which has been attributed to the capacity of the rumen microflora to degrade the toxin. In an early study, Hult et al. [128] showed the cleavage of OTA into OTα in all cow ruminal fluids except the abomasum. Two separate studies suggested that rumen protozoa are responsible for this hydrolysis [117, 129]. Using OTA incubated with buffered rumen fluid collected from fistulated cows, Muller et al. [130,131] suggested that the concentration of starch in the diet and the number of protozoa in the rumen are the foremost determining factors of the capacity of the rumen to hydrolyze OTA.

4.2. Phase I Reactions

4.2.1. Detoxification Reactions

Once OTA is absorbed by living creatures, a metabolic pathway is activated to transform this toxic compound. In 1992, Marquardt, R. and his research group reported that through phase I detoxification reactions, a small part of absorbed OTA could be metabolized into hydroxy-ochratoxin A by hydroxylation in the liver. [132]. Recently, Ringot, D. et al completed the description of this pathway, as we discussed before. OTA can be transformed into many other metabolites, as shown in Figure 3 [47].

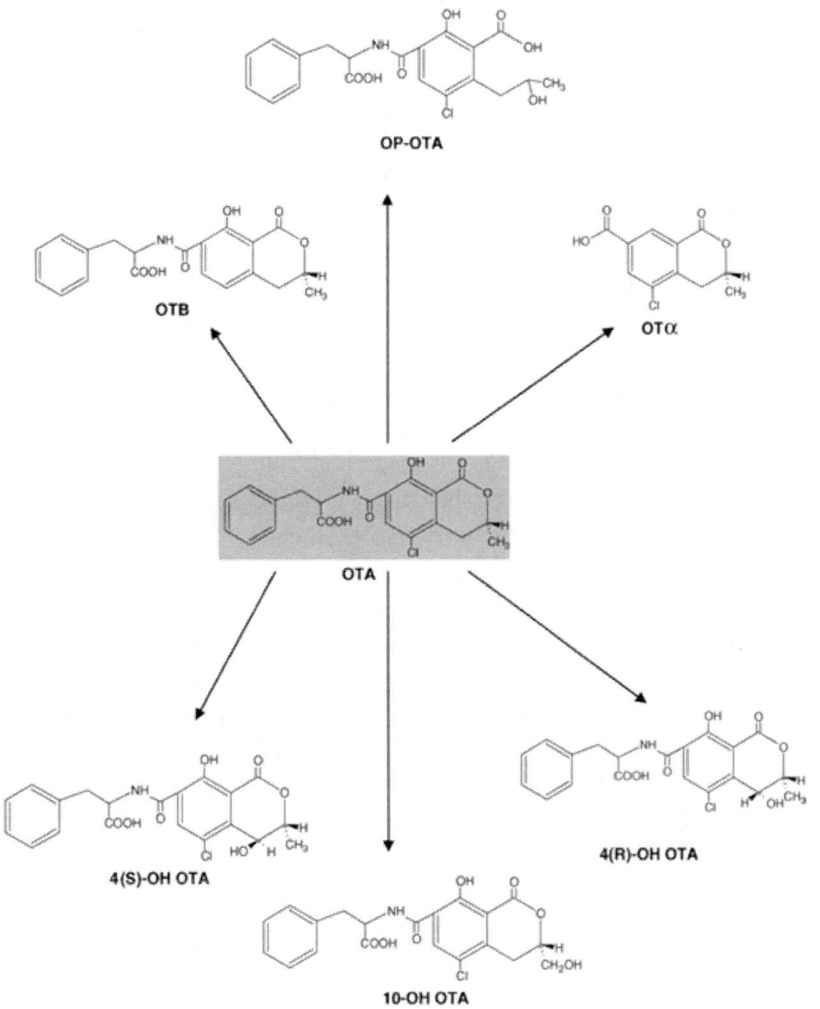

Figure 3. Phase I reactions of ochratoxin A for which the products have been detected *in vivo* [47].

It is well known that the liver is an important detoxifying organ for many kinds of toxicants, and OTA, which can reach the liver via the portal vein [64], is no exception. However, we know little about how OTA is taken up by the hepatocytes or secreted. Two papers together show a mechanism for this procedure in the rat. Oatp1, located in the basolateral cell membrane of hepatocytes, is related to the mechanism of OTA uptake [133]. A later study shows that this protein belongs to the oatp family, which participates in the transportation of multispecific organic anion [134]. In addition, a carrier-mediated mechanism may function in the secretion of OTA via an export system [135, 136].

Some studies have proven that OTA can be metabolized into 4(*R*)-and 4(*S*)-OH OTA, the two epimers of 4-hydroxyochratoxin A (4-OH OTA), in the liver microsomes of human, pig and rat [137-139]. A further study shows that in rat and human liver microsomal systems, the 4(*R*)-OH OTA epimer is mostly formed, whereas in pig liver microsomes, the 4(*S*)-OH OTA epimer is mainly formed In 1996, the optimal pH values for these two isomers in rat liver microsomes were studied by Omar and his research group, showing that 6.0 and 7.0 were the optimal pH values for the ratio between 4(*R*)-OH OTA and 4(*S*)-OH OTA production [140]. It is important to know the toxicity of these two compounds. An early study by Creppy et al. [141,142] showed that 4(*R*)-OH OTA seemed to be as toxic as OTA in yeast, while 4-OH OTA was reported to be non-toxic in rats in an earlier study[143]. The reason for the low toxicity of 4-OH OTA might be its rapid elimination in mammals. No toxicity study on 4(S)-OH OTA has yet been reported. Some studies have been conducted on another metabolite, 10-hydroxy ochratoxin A (10-OH OTA). This compound was first studied in rabbit liver microsomes but not yet *in vivo* [125, 144, 145]. Later *in vitro* studies in cultured human bronchial epithelial cells [146] and in rabbit kidney microsomes [147], together with the earlier studies, concluded that this compound is not genotoxic [146, 147].

4.2.2. Bioactivation Reactions

Many studies have been developed regarding the toxic mechanism of OTA, aiming to unveil the specific mechanism. The DNA damage and lipid peroxidation mediated by OTA are relatively well known. Here we describe the main hypotheses that are currently accepted. Many research teams have studied the interactions between macromolecules and the formation of ochratoxin-derived free radicals. To examine the free radical generation in different cell systems, Hoehler et al. [148, 149] used electronic paramagnetic resonance (EPR) spectroscopy to study this issue, then demonstrated the generation of reactive oxygen species (ROS) by OTA and its analogues. Based on their results, it was concluded that the oxidative damage induced by OTA is indirect rather than direct. Finally, the groups also showed that in *B. brevis* and rat liver microsomes, a strong EPR signal was observed. Considering that OM-OTA, one kind of *O*-methylated derivative, also generated a strong EPR signal, they concluded that the phenolate C8 group is not involved in the ROS generation of OTA. Later studies also proved their findings that when incubating liver mitochondria and hepatocytes with OTA or OM-OTA, no EPR signal could be observed. Further research may be conducted on this issue, as there may be a failure to take up the toxin and/or to a failure to induce free radical production, which would account for the failure to produce EPR signal in hepatocytes and liver mitochondria. A study on the toxicity of OM-OTA by Xiao et al. [144] demonstrated that this compound is toxic in mice; they also hypothesized that its lactone carbonyl group contributed substantially to its toxicity.

Two groups reported other possible reactive metabolites from OTA, known as OTA-derived quinones (OTAQ/OTAHQ) [150, 151]. The reason may be that the chlorine atom is very important for the genotoxicity of OTA, which is consistent with the fact that some chlorinated compounds are known to undergo bioactivation to benzoquinones before inducing oxidative stress and DNA damage [152]. It has also been hypothesized that the genotoxic effect of OTA may be related to its bioactivation from OTA to OTAQ derivatives, which should be further studied in future work. Several studies have demonstrated that the couple (OTAQ/OTAHQ) can be formed *in vitro* through OTA oxidation by photochemical [151, 153] and electrochemical [154] processes; later studies supplied this pathway. The couple can also be generated in a chemical system mimicking cytochromes P450 [150] and in enzymatic systems such as horseradish peroxidase in the microsomes of rat liver [151]. However, recent studies by Zepnick et al. [145] and Gautier et al. [155] using microsomal fractions from the liver and kidney of mice and rats reported that they could not observe the formation of OTAQ/OTAHQ derivatives.

4.3. Phase II Reactions

4.3.1. Glucurono- and Sulfate-Conjugation

OTA is a kind of xenobiotic that can be metabolized into other metabolites in phase II reactions, which are combined with many metabolic steps. First, OTA can be conjugated to the endogenous glucuronic acid with its phenolic group or carboxylic group due to the xenobiotic activity of OTA. In the cell, glucuronosyltransferase (UGT), a cellular membrane-bound microsomal enzyme, is connected to cytochrome P450 (CYP). Both sulfotransferase (ST) and uridine-diphosphate glucuronosyltransferase (UGT), play important roles in the phase II reactions of xenobiotics. The first enzyme catalysis is glucuronjugation, while the second mediates the sulfate conjugation. Among all tissues and organs, the organ most strongly related to xenobiotic glucuronidation is the liver. The intestinal microflora, which plays an important role in human health [156], is also involved in this phase, mainly producing beta-glucuronidase, and this enzyme-hydrolyzed glucuronide conjugate is excreted in the bile as aglycone. Then, the xenobiotic released from this reaction can be reabsorbed by the hepatocytes, and the cycle can repeat.

In the 1980s, Roth et al. [74] conducted study on the distribution profiles of the toxin in mice. They prepared [^{3}H]-labeled OTA and then imported the OTA into the body of mice by the intubation of a single dose of 288 µg/kg. In the bile, they found 28–68% of the conjugated derivatives (glucuronides and sulfate conjugates) metabolized from the labeled OTA, and in the liver and intestinal tissue, the results were 8–17% and 6%, respectively. These findings were supported by a 1995study in which Kühn et al. [157] found that glucuronide conjugates appeared in the bile after feeding pigs with OTA-contaminated fodder. Two years later, Li et al. [90] reported the identification of 18 chromatographic peaks in the samples of bile and urine from the rats raised with some OTA. These peaks could represent OTA conjugates that they explained then, but have remained unknown metabolites. However, in a recent study in 2002, conducted by Gross-Steinmeyer et al. [158], *in vitro* experiments with rat and human hepatocytes showed that no other OTA glucuronide and sulfate conjugates were observed. Based on all the studies focused on the OTA conjugates, it seems that this question needs more research in the future.

4.3.2. Other O-Conjugates

In the early 21st century, Gross-Steinmeyer et al. first found two conjugates with the structures of acyl glycosides, hexose and pentose in their study of the culture of human hepatocytes. These conjugates are now known as *O*-labile ester conjugates of OTA. This result is consistent with the study by [158] Zepnik et al. two years later [159].

Based on this study, little is known on the toxicity of these two conjugate compounds. In some cases, conjugation results in reactions by living beings, despite the classical view that it is a detoxification step. In a few rare cases, other sugars than glucuronic acid can be used by UGT and can then be formed to glycoside conjugates. Acyl glucuronides of xenobiotics are known to be less stable than ether glucuronides and are susceptible to hydrolysis, intramolecular rearrangements, ring opening and covalent binding with target molecules (proteins, nucleic acids). Some hypothesize that these OTA conjugates, as mentioned before, might be related to the OTA toxicity if there is any study proof that they are at least as toxic as acyl glucuronide.

4.3.3. Glutathione Conjugation

There is another important conjugation called glutathione (GSH) conjugation in xenobiotics biotransformation, mainly catalyzed by the glutathione transferase (GST) superfamily, which is rich in the cytoplasm of the liver. In 1994, Malaveille, C. et.al found that a cysteinyl–OTA conjugate is related to the cytotoxicity of OTA [160], and this study provided evidence for the view that GSH conjugation is not always a detoxification step but can instead be a bioactivation step in some cases. It is worth mentioning that this conjugation promotes the formation of peroxyl radicals by forming a thiyl radical. The liver is not observed to produce a high level of the conjugates, mainly because stable glutathione-conjugates, such as quinone-derived glutathione-conjugates, are exported from the liver to other body fluids or other tissues via the MRP2 pump after being produced in the hepatocytes [161]. The latter excretory tissue can receive a relatively high level of glutathione-conjugates because the procedure of the conjugates contained the first excretion via the bile, reabsorption in the gut then and finally transportation to the kidney. In the epithelial cells of the kidney tubules, to cleave the glutathione-conjugates moiety, there are two abundant peptidases: the γ-glutamyl transpeptidase (γ-GT) and the dipeptidase (DPP), which catalyzes the reaction of cleaving off the glutamate residue and removing glycine. Thus, in the kidney, the cysteinyl conjugates are generated last, and then the conjugates are *N*-acetylated and excreted from the kidney. In some other cases, such as quinones, a cytosolic, microsomal and mitochondrial enzyme called β–lyase, which is abundant in the cells of the proximal tubule, can catalyze the bioactivation of cysteinyl conjugates. In 2002, Dai et al. [151] conducted a study on the biological metabolic procedure of the reactivity of OTA toward GSH, showing that OTA undergoes a biotransformation to a reactive intermediate reacted with GSH. In their study, horseradish peroxidase (HRP)/H_2O_2, rat liver microsomes /NADPH or free Fe(II) was incubated with 100 μM OTA (1) in the presence of 5 mM GSH, and afterward they found conjugates demonstrated by the authentic sample of Michael addition conjugate resulting from attachment of GSH to the unsubstituted position of OTQ. The result showed that an OTAHQ-GS conjugate was identified by its groups. In the same year, a OTAHQ-cysteinyl conjugate can be produced by the same group [162], and thus they hypothesized that the OTAHQ-GS conjugate was produced by an oxidative process, perhaps by a OTAHQ-cysteinyl conjugate in kidney, which can then be related to the ROS reaction or participate in covalent binding with the proteins in tissue (Figure 4) However, the glutathione-conjugate produced in this reaction is not more than 1%, making it far more difficult to detect this metabolite.

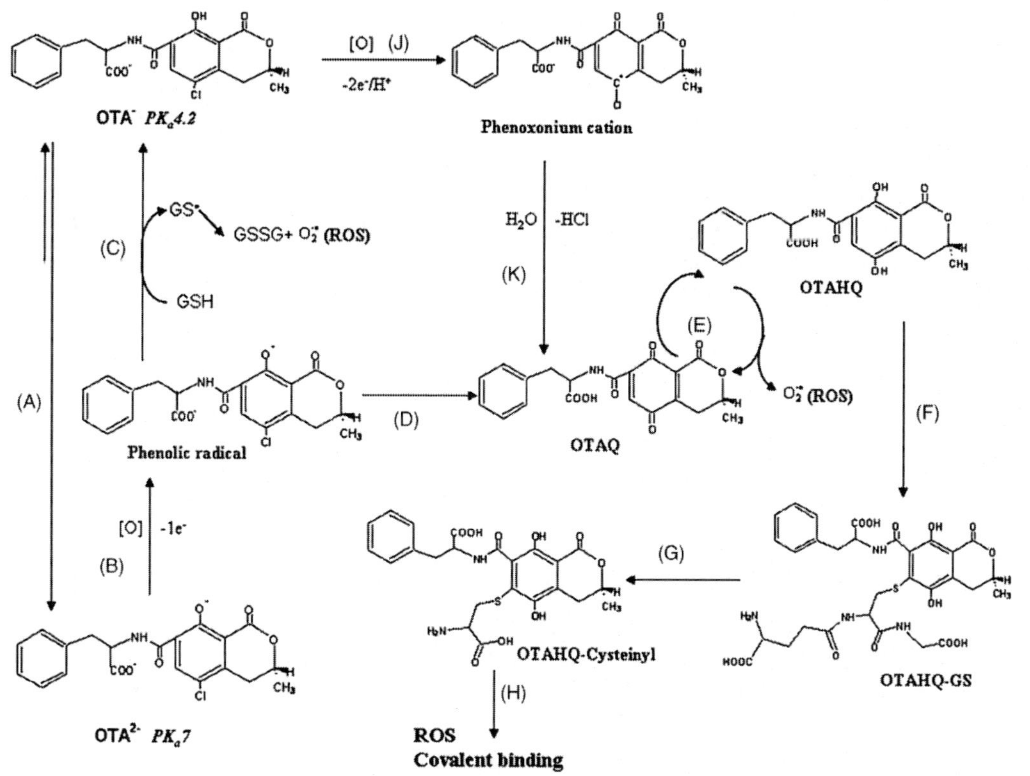

Figure 4. Hypothetic bioactivation pathways of OTA (A) *in vivo* equilibrium between OTA monoanion (OTA^{-}) and dianion (OTA^{2-}) forms; (B-H) proposed ways for OTA bioactivation [47].

5. Biotransformation in Plants

So far as we know, OTA can be metabolized not only in animals but also in plants. Plants can protect themselves from certain chemically reactive compounds, which may be harmful to them, by specializing their organic tissues and cells through physical methods and by producing derived forms of compounds, such as derivatives or conjugates. [163].

In 1994, Ruhland, M. et al. conducted a study on the metabolism of OTA in plants. They incubated wheat and maize cell suspension cultures in Murashige-Skoog medium in the presence of a [U- 14C]-labeled OTA under soft shaking at 28°C within 13 days [164]. They identified several metabolites, including OTα, OH-OTA (I), OH-OTA (II), OH-OTA methyl ester (I), OH-OTA methyl ester (II), OH-OTA--glucoside (I), OH-OTA--glucoside (II), OTA methyl ester, and an oxygenated OTα. These isomeric hydroxylated metabolites should be 4-(S)- and 4-(R)-OTA, mainly because the two compounds' behavior during TLC test with different solvents is comparable to the behavior described for 4-(S)- and 4-(R)-OTA in the study by Stormer et al. [165]. A decline in the concentration of ochratoxin A during the storage of cereals was observed [166], which supports the description above and suggests the transformation of ochratoxin A, possibly by plant enzymes.

In their subsequent study, the metabolism of OTA in plant cells, including vegetables such as carrot, tomato, soybean, and potato and grains such as maize, wheat, and barley, was finally investigated. In their study, these plant-cells were isolated and suspended, then incubated with 2.5 μg/ml of OTA at 28°C for 13 days. Their identifications from these plant cell suspension cultures are consistent with the compounds they identified in previous studies in wheat and maize cells. They also found some polar metabolites, while cannot be identified. In the group of maize, potato, carrot, and tomato, quite many OTα were formed as an intermediate, more polar compound appeared while OT disappeared during the incubation period In contrast, in the other group, including soybean, wheat and barley, only a low amount of OTα was formed, and the main metabolites found were glucoside conjugates and hydroxylation products. Therefore, they concluded that some plant cells do have the ability to catalyze OTA into other compounds.

In 1997, a study was conducted to determine whether intact plants can metabolize OTA, Ruhland, M. et al. used HPLC methods to study this question [167]. In their study, the authors injected OTA into intact paprika, tomatoes, potatoes, and carrots as well as maize, soybean and wheat kernels. The results showed that the OTα was the alone metabolite in tomato and carrots. In maize and wheat, a smaller amount of 4-(S)-OH-OTA was formed, apart from OTα. They did not find the OTαin potato and soybean, but 4-(R)-OH-OTA and 4-(S)-OH-OTA were identified as metabolites. Limited by the knowledge of the OTA residue and its metabolites, they further said that other metabolites or conjugated and non-extractable residues might also exist in these plants. This study demonstrates that intact vegetables and cereals can also exert enzymatic activity to biometabolize OTA into other compounds. The occurrence of OTα and 4-(R)-OH-OTA and 4-(S)-OH-OTA, demonstrated by the research, indicates that the same metabolic pathway as in cell suspension cultures operates in the transformation of OTA in intact plants.

Here is a summary of the metabolism in living creatures. In rodents, there is a variegated metabolism of OTA, especially in rats [168]. In the body, OTA can be metabolized in many organs and tissues such as the kidney, liver, and intestines. Two excretory pathways, urine and feces, are very important in this procedure. In rodents, we have plenty of evidence that the metabolites of OTA can be metabolized into many other compounds, such as OTα, 4(R)- and 4(S)-OH-OTA, 10-OH-OTA, OTB, OTβ, 4-OH-OTB, OTHQ, OP-OTA.. Among all the compounds, OTα, an important major metabolite, is considered nontoxic according to many studies. The nontoxic compounds can be formatted or degraded by carboxypeptidases. In rats, 4(R)-OH-OTA is the main hydroxy metabolite, which is catalyzed mainly by CYP450, while 10-OH-OTA is found to be produced only in rabbits. OTB can be much more rapidly eliminated, and thus the toxic effects of OTA were reduced more or less thereby. OTC can be transformed rapidly into OTA, which is as toxic as OTA. OP-OTA is more toxic than OTA, which is why it is considered as the "toxic product." Regarding the activity of OTA, we think that the lactone carbonyl group of the isocoumaric moiety contributes greatly to its biological activity. So far as we know, some of the metabolites were detected, but we cannot clarify them easily. The formation of OTA-DNA adducts remains a subject of debate. One view of this argument is that DNA adducts can lead to genotoxicity, while another view is that the genotoxicity is caused by OTA or its reactive oxidation species, not by the reaction of the DNA adduction.

6. DETOXIFICATION OF OCHRATOXIN A

Ochratoxin A (OTA), known as a secondary metabolite produced by fungi such as *Aspergillus* spp. and *Penicillium* spp., is a kind of mycotoxin known for its famous carcinogenic and nephrotoxic toxicity. This toxin is very common in variety of foods and feeds, and its presents a serious threat to the health of both animals and human beings, especially their kidney and liver [169]. Its presence has been reported in different foods, such as grain and forage, as well as in grapes and beverages such as grape juices and wines [41]. To date, adequate and sufficient evidence in animals and humans has proven that ochratoxin A might be carcinogenic to animals or humans. Ochratoxin A has been found in many plants, such as wheat, corn, oat and licorice root, threatening o the health of both animals and humans. Detoxifying these products is therefore of prime importance in protecting consumer health [170]. It is crucial to conduct related studies on the control of ochratoxin A. Here, we collected and described detoxification methods for Ochratoxin A, hoping to lead readers to obtain sufficient information on this kind of toxin.

6.1. Physical Methods

6.1.1. Absorption

In traditional industrial areas, aluminosilicate, bentonite and charcoal are often used to absorb some compounds, including ochratoxin A, because the physical structure is porous and their cost is low but produces satisfactory results [171-173]. This physical method of using absorption for the detoxification of OTA is now widely used in agriculture, food procedure and grain storage. Its advantages of low cost and easy operation are clear. *In vitro*, these adsorbents showed good results on the detoxification of OTA, but for *in vivo* tests, only charcoal achieved the desired effect. Moreover, these compounds produced from minerals or industrial activity may cause the retention of essential nutrients and toxin poisoning. Furthermore, this kind of detoxification operates by adsorption only and does not transform toxins to nontoxic compounds, and therefore there is also a problem of secondary pollution.

6.1.2. Radiation

New methods are being applied in the detoxification of OTA. In recent years, gamma-rays have been reported to destroy ochratoxin A and many other aflatoxins in food crops and feed [174]. However, a study conducted in 2014 showed that even at 15 kGy, the gamma-rays were not very effective in the complete destruction of ochratoxin A and aflatoxins, leaving at least 10% ochratoxin A remaining in the tested feed [175]. Moreover, radiation methods are also restricted by huge cost and complex operations.

Physical methods for the detoxification of ochratoxin A are numerous, however, when a more thorough detoxification is required, physical methods may be not effective enough. Combination with other methods is a good solution to increase the effectiveness, including following chemical and biological methods.

6.2. Chemical Methods

Because the physical methods are not very effective, researchers have sought more suitable methods for the detoxification of ochratoxin A. Chemical detoxification of the toxin mainly involves modifying the toxin to its non-toxic derivatives [176]. Usually, hydrogen peroxide, vitamin C, hydrochloric acid, formaldehyde, sodium hydroxide and ammonium hydroxide can effectively reduce the level of ochratoxin A. As early as 1981, scientists had proven that the addition of ammonia to a final concentration 2% inactivates ochratoxin A. The detoxification of contaminated cereal grain like wheat, corn or barley can be performed by ammonization [177]. Electrochemical ozone-generating technology can also be used to reduce ochratoxin A [178], but these methods, containing the compounds listed, will also leave some residual toxins on the food and may destroy other nutritional value and palatability to some degree [179]. Thus, relevant legislation is needed to standardize the use of these compounds.

6.3. Biological Methods

Physical and chemical methods are traditional and widely applied in the detoxification of ochratoxin A. However, there is also a method called microbial detoxification, which means the use of microbes, including bacteria, fungi and yeast, to accomplish detoxification. This procedure is much more complicated than the physical and chemical methods, and scientists have not yet reached consensus on it. There is, however, an agreement that two pathways might be involved in the biodegradation of ochratoxin A. First, as previously described, ochratoxin A can be biodegraded to ochratoxin α and L-β-phenylalanine, which are less toxic than ochratoxin A [180]. Second, ochratoxin A can be degraded into lactone-opened OTA (OP-OTA) [181], which is also considered less toxic than ochratoxin A. However, it is accepted that the most important component in the microbial detoxification of ochratoxin A is the enzyme, meaning a chemical method, that can transform ochratoxin A into other less toxic compounds [182], and physical methods such as adsorption by the cytoderm are also more or less involved. During the microbial degradation of ochratoxin A, acetylation, glucosylation, ring cleavage, hydrolysis, deamination, and decarboxylation are involved [182]. This finding coincides with a study by Stander et al. reporting that some hydrolases were screened for degrading OTA to nontoxic compounds such as OTalpha and phenylalanine, as confirmed by HPLC with fluorescence detection. They also purified the enzyme with the ability to degrade OTA by anion exchange chromatography to homogeneity. [174]. It was reported that some strains of bacteria, fungi and yeasts can biodegrade OTA. Here, we describe several microbial detoxifications based on the microbe species.

6.3.1. Bacterial

(1) *Lactobacillus* spp.

In 2008, Fuchs et al. reported that *Lactobacillus acidophilus* VM 20 decreased ochratoxin A by P95% among thirty different LAB strains, analyzed in a screening trial by HPLC coupled with UV or fluorescence detection, and further study showed that this depended on different parameters, such as the density of the cell, the pH, the concentration of toxins added to the culture fluid, and the variability and species of the bacteria [170]. Vinceneo et al. also found that when fifteen strains belonging to five relevant oenological wine lactic acid bacteria (LAB) species were grown in liquid synthetic culture medium containing ochratoxin A, the portion of OTA removed during the bacterial growth was 8 to 28%. His results showed that the OTA removed from the supernatants was partially recovered (31 to 57%) from the bacterial pellet [183]. The mechanism of OTA degradation by certain *Lactobacillus* spp. strains remains disputed, with the debate between absorption and decomposition. Due to restrictions imposed by technology and methods of detection, the actual mechanism must be further studied and expounded.

(2) *Bacillus* spp.

Recently, Ptechkongkaew et al. identified and found one Bacillus strain with the ability both to restrain the growth of *Aspergillus* strains and to biodegrade 92.5% of OTA. A later study showed that the OTA decrease was due to the degradation of OTalpha by this strain [184]. He proposed that the actual mechanism of OTA degradation by this *Bacillus* strain might be attributed to one efficient enzyme in the bacteria, named CPA. However, this proposal regarding the mechanism is not well accepted and needs more proof and further study.

(3) *Brevibacterium* spp.

Héctor et al. found that certain strains belonging to the *Rhodococcus*, *Pseudomonas*, and *Brevibacterium* genera had the ability to grow in liquid synthetic culture medium containing ochratoxin A. In his study, *Brevibacterium* spp. strains showed 100% degradation of ochratoxin A, and he detected ochratoxin α as a degradation product in the cell-free supernatants, using high-performance liquid chromatography-mass spectrometry (HPLC-MS) [185]. This mechanism also calls for further study.

(4) *Rhizobium*

Ochratoxin A could also be successfully biodegraded by certain strains such as *Rhizopus stolonifer, R. homothallicus, R. microspores* and others. In the results of this study, the kinetics of ochratoxin A biodegradation of the selected strain of *Rhizopus* isolates was then tested and examined. In addition, this strain had the ability to biodegrade nearly all the ochratoxin A, up to 95%, during incubation at 25°C within 16 days [186]. The researchers' opinion on the mechanism of the degradation by these strains also indicates one enzyme, and further study should be conducted.

There are also other bacterial species with the ability to degrade OTA, but the specific molecular mechanism also remains unclear. A reliable explanation of this mechanism may be the absorption of the cell wall and the degradation of the enzyme contained or secreted by the actual bacteria. However, the mechanism remains a significant challenge to determine, and further and deeper study is needed.

6.3.2. Fungi

(1) Mold

In 2006, Bejaoui et al. isolated forty isolated strains from the black *Aspergillus* species *Aspergillus carbonarius, A. niger aggregate* and *A. japonicus*, isolated on French grapes, with the ability to degrade OTA to OTα (ochratoxin alpha). Then, 9 days later, they observed that OTα had decreased and an unknown compound appeared in the culture. [187]. Varga and his co-authors also found that *A. fumigatus* and black *Aspergillus* strains could detoxify ochratoxin A in culture medium, using the detection methods of thin layer chromatography, HPLC and an immunochemical technique. Finally, they identified *A. niger* CBS 120.49, which could effectively transform ochratoxin A into other degradation products, such as ochratoxin α. Their results showed that this strain had the ability to biodegrade 2 mg OTA/L after 10 days of incubation at 30°C [188].

In 2002, Engelhardt reported that one white rot fungus *Pleurotus ostreatus* could biodegrade both OTA (77%) and OTB (97%) when incubated on contaminated barley by solid state fermentation for four weeks, compared to the control state, in which only approximately 40% of OTA was removed [189].

(2) Yeast

Peteri et al. found that *P. rhodozyma,* a strain of yeast, was able to convert OTA to ochratoxin alpha, possibly mediated by an enzyme related to carboxypeptidases, similar to carboxypeptidase A. A later study showed that the enzyme responsible for ochratoxin degradation was cell-bound. Interestingly, the findings also showed that both viable and heat-treated (dead) *P. rhodozyma* cells were still able to adsorb significant OTA, indicating that heat treatment increased the ability of this strain [190]. Schatzmayr et al. selected a strain from the genus Trichosporon, which could biodegrade ochratoxin A into phenylalanine and the non-toxic ochratoxin α (OTα). Further study showed that this strain had the ability to cleave 100% of 0.2 mg OTA/L after five hours of incubation at 35°C and could be fermented and stabilized. [191].

In 2007, it was reported that P. *rhodozyma* CBS 5905 could degrade more than 90% of ochratoxin A in 15 days at 20 degrees, mainly biodegrading OTA to ochratoxin alpha in a manner related to carboxypeptidases. Further study on the enzyme showed that the carboxypeptidase was a metalloprotease, and 30 degrees was the optimum temperature of this enzyme. The enzyme was found to be cell-bound, as heat-treated (dead) *P. rhodozyma* cells were found to be able to adsorb significant amounts of OTA, and it was concluded that heat treatment could enhance the Ochratoxin A detoxification ability of this strain [192]. In the same year, Gil-Serna et al. performed a study on the mechanism of the yeast's ability to degrade OTA. In this study, they found that the genes p450-B03 (cytochrome p450 monooxygenase) and pks (polyketide synthase) in *A. Westerdijkiae* were related to OTA biosynthesis using newly developed and specific real-time RT-PCR protocols. They also observed a high level of extracellular OTA adsorbed to yeast cell walls, particularly at low pH [193].

In summary, the mechanism of OTA detoxification by yeast remains unclear. Different types and strains appear to have different mechanisms to degrade and remove OTA.

6.3.3. Intestinal Microflora

The intestinal microflora are distinct from normal bacteria, as they are combined with the mammalian body, and there is an interaction between the intestinal microflora and human health. Recent studies have shown that intestinal microflora can be related to inflammatory bowel syndrome [194], diabetes [195], cancer [196] and more. There is also the consideration that if we can find a species of intestinal microflora with the ability to degrade OTA, it might lack the potential harmful effects of other bacteria or fungi.

Recently, Upadhaya et al. collected intestinal microflora from swine using aseptic technology. These microflora were then spiked with OTA and incubated at 39 degrees for 6 and 12 hours. It was found that 20% of the OTA toxin was detoxified. Then, they isolated twenty-eight bacterial species aerobically using nutrient broth. The results showed that one anaerobic bacterial isolate, named MM11, detoxified more than 75% of OTA. The bacterium was identified by 16S rDNA sequencing and showed nearly 97% sequence similarity with *Eubacterium biforme*. [197].

A new technology was applied to the study of finding some species degrading OTA in the human intestinal microbiota: semi-automated solid-phase extraction (SPE). Camel et al. used C18-silica as the sorbent and water as the simplest matrix in this method. They found a significant decrease in the level of ochratoxin A because of degradation, with efficiencies ranging from 14% to 47% after 72 h incubation. Furthermore, they also identified three phase I metabolites, ochratoxin α, open ochratoxin A and ochratoxin B, using high resolution mass spectrometry [198].

6.3.4. Others

One kind of protozoa, known as rumen protozoa, was reported to be able to biodegrade OTA [199], which might be why ruminants are resistant to OTA toxicity [200].

In 1996, scientists found by HPLC that some plant cell cultures such as maize, carrot, tomato, potato, soybean, wheat and barley can transform OTA into other compounds.

6.4. Specific Measures

6.4.1. Approach to Handling OTA in the Lab

Because OTA is a nephrotoxic and carcinogenic mycotoxin, its handling requires care and attention. A high-efficiency particulate arrestor (HEPA) or unique filters should be prepared in the laboratory, farm, workshop, or any other location that could be polluted by the OTA. The filters are designed to be transferred into plastic bags without contaminating the maintenance staff. Gloves and masks are needed to protect humans when handling or processing OTA. Additionally, the filters used to contain the poison should be placed in plastic bags immediately after removal of the OTA, and the plastic bag should then be sealed immediately. The sealed bag should be labeled properly to prevent accidental pollution of other people. Waste liquids containing any OTA should be carefully placed and collected in special and suitable containers. The lid should be carefully secured, and a special label should be pasted onto the bottles or the containers in a clearly visible position.. Once the lid is filled with toxin, bottles should be carefully placed in a plastic bag so that the outer surface of the containers cannot directly contaminate other animals or human beings, causing a secondary poisoning. The subsequent steps refer to the section describing methods of detoxification [201].

6.4.2. Prevention of the Mildew from OTA

Different species of grain have different sensitivities to different molds. As OTA is a secondary metabolite produced by *Aspergillus* spp. and *Penicillium* spp., breeding and selecting new crops with mildew resistance can greatly reduce the chance of mycotoxin-contaminated feed and grain, thus achieve the goal of to controlling mycotoxin contamination at the source end, which could be the fundamental way to prevent mildew of the feed and grain [202]. Moreover, genetic modification technology provides a method of introducing exogenous DNA into microorganisms, plants or animals, which endows the target organism with a prospective characteristic, and this technology may be applied to the protection of feed and grain from mildew.

Some strategies can be employed to reduce OTA levels in the storage of feed and grain. We can use better agricultural practices, fungus-resistant crop varieties, the correct application of fungicides and the proper storage of commodities to achieve the control of OTA, which could be a critical control point in minimizing OTA pollution in the food supply chain [180]. It is known that certain important factor are related: [203] cleanliness of harvesters, the weather before and during the harvest of the grain, temperature at harvest, the time before the drying of the grain, the efficiency of the drying machinery, the physical state of the grains,, fungal competition, and transport of the grain.

To create a suitable storage environment, it is important to protect the feed and grain from mildew. It has been reported that temperatures between 20°C and 30°C and relative humidity between 80% and 90% are the best environmental conditions for fungal reproduction. Moreover, the water content in the feed or grain, ranging from 17% to 18%, is also suitable for fungal growth. Therefore, the environmental conditions of the storage must be strict, controlling the raw materials and the raw material moisture content (not higher than 14%); it is also important and sufficient to keep the air in the warehouse dry and clean [204]. If possible, always check and turn the stored feed or grain over to identify mildew. Feed or grain contaminated with mildew should be immediately removed and processed. Related precautions are as described in the previous section.

Fungicides can disrupt the cell walls of many fungi, inhibiting mold cell growth, when storage conditions are too poor or in the rainy season, to prevent the mycotoxin contamination of feed ingredients. Fungicides can also be added as storage agents [181].

If mold has occurred, consider the actual extent of mildew. Seriously moldy feed or grain, in principle, must be abandoned in case of poisoning the animal or human eating it. If the case is less serious, some detoxification methods, as described before, can be performed to remove the OTA.

REFERENCES

[1] Van der Merwe, K., Steyn, P., Fourie, L., Scott, D.B., Theron, J., (1965). Ochratoxin A, a toxic metabolite produced by Aspergillus ochraceus Wilh.

[2] Hesseltine, C., Vandegraft, E.E., Fennell, D.I., Smith, M.L., Shotwell, O.L., (1972). Aspergilli as ochratoxin producers. *Mycologia*, 539-550.

[3] Varga, J., Kevei, E., Rinyu, E., Téren, J., Kozakiewicz, Z., (1996). Ochratoxin production by Aspergillus species. *Applied and Environmental Microbiology*. 62, 4461-4464.

[4] Frisvad, J.C., Frank, J.M., Houbraken, J., Kuijpers, A.F., Samson, R.A., (2004). New ochratoxin A producing species of Aspergillus section Circumdati. *Studies in Mycology*. 50, 23-43.

[5] Abarca, M., Bragulat, M., Castella, G., Cabanes, F., (1994). Ochratoxin A production by strains of Aspergillus niger var. niger. *Applied and Environmental Microbiology*. 60, 2650-2652.

[6] Horie, Y., (1995). Productivity of ochratoxin A of Aspergillus carbonarius in Aspergillus section Nigri. *Nippon Kingakukai Kaiho*. 36, 73-76.

[7] Téren, J., Varga, J., Hamari, Z., Rinyu, E., Kevei, F., (1996). Immunochemical detection of ochratoxin A in black Aspergillus strains. *Mycopathologia*. 134, 171-176.

[8] Samson, R.A., Houbraken, J., Kuijpers, A.F., Frank, J.M., Frisvad, J.C., (2004). New ochratoxin A or sclerotium producing species in Aspergillus section Nigri. *Stud Mycol*. 50, 45-61.

[9] Walbeek, W.v., Scott, P., Harwig, J., Lawrence, J., (1969). Penicillium viridicatum Westling: a new source of ochratoxin A. *Canadian journal of microbiology*. 15, 1281-1285.

[10] Ciegler, A., Fennell, D., Sansing, G., Detroy, R., Bennett, G., (1973). Mycotoxin-producing strains of Penicillium viridicatum: classification into subgroups. *Applied microbiology*. 26, 271-278.

[11] Northolt, M., Van Egmond, H., Paulsch, W., (1979). Ochratoxin A production by some fungal species in relation to water activity and temperature. *Journal of Food Protection®*. 42, 485-490.

[12] Larsen, T.O., Svendsen, A., Smedsgaard, J., (2001). Biochemical characterization of ochratoxin A-producing strains of the genus Penicillium. *Applied and Environmental Microbiology*. 67, 3630-3635.

[13] Abrunhosa, L., Paterson, R.R., Venâncio, A., (2010). Biodegradation of ochratoxin A for food and feed decontamination. *Toxins*. 2, 1078-1099.

[14] Pitt, J., Taniwaki, M., Teixeira, A., Iamanaka, B. (2001), *19ème Colloque Scientifique International sur le Café, Trieste, Italy, 14-18 mai 2001*. Association Scientifique Internationale du Café (ASIC), pp. 1-5.

[15] Serra, R., Mendonça, C., Venâncio, A., (2006). Fungi and ochratoxin A detected in healthy grapes for wine production. *Letters in applied microbiology*. 42, 42-47.

[16] Esteban, A., Abarca, M.L., Bragulat, M.R., Cabañes, F.J., (2004). Effects of temperature and incubation time on production of ochratoxin A by black aspergilli. *Research in Microbiology*. 155, 861-866.

[17] Ramos, A., Labernia, N., Marın, S., Sanchis, V., Magan, N., (1998). Effect of water activity and temperature on growth and ochratoxin production by three strains of Aspergillus ochraceus on a barley extract medium and on barley grains. *International Journal of Food Microbiology*. 44, 133-140.

[18] Mühlencoert, E., (2004). Ochratoxin A production by Aspergillus ochraceus. *Technische Universitat Munchen, Munic, Germanythesis*, 124.

[19] Sweeney, M.J., Dobson, A.D., (1999). Molecular biology of mycotoxin biosynthesis. *FEMS Microbiology Letters*. 175, 149-163.
[20] Abarca, M., Accensi, F., Bragulat, M., Castella, G., Cabanes, F., (2003). Aspergillus carbonarius as the main source of ochratoxin A contamination in dried vine fruits from the Spanish market. *Journal of Food Protection*. 66, 504-506.
[21] Mitchell, D., Parra, R., Aldred, D., Magan, N., (2004). Water and temperature relations of growth and ochratoxin A production by Aspergillus carbonarius strains from grapes in Europe and Israel. *Journal of Applied Microbiology*. 97, 439-445.
[22] Bellí, N., Ramos, A., Sanchis, V., Marín, S., (2004). Incubation time and water activity effects on ochratoxin A production by Aspergillus section Nigri strains isolated from grapes. *Letters in Applied Microbiology*. 38, 72-77.
[23] Astoreca, A., Barberis, C., Magnoli, C., Dalcero, A., (2010). Aspergillus carbonarius growth and ochratoxin A production on irradiated dried grapes under different water activity and temperature conditions. *World Mycotoxin Journal*. 3, 175-182.
[24] Magan, N., Aldred, D., (2005). Conditions of formation of ochratoxin A in drying, transport and in different commodities. *Food additives and contaminants*. 22, 10-16.
[25] Lindblad, M., Johnsson, P., Jonsson, N., Lindqvist, R., Olsen, M., (2004). Predicting noncompliant levels of ochratoxin A in cereal grain from Penicillium verrucosum counts. *Journal of applied microbiology*. 97, 609-616.
[26] Cairns - Fuller, V., Aldred, D., Magan, N., (2005). Water, temperature and gas composition interactions affect growth and ochratoxin A production by isolates of Penicillium verrucosum on wheat grain. *Journal of Applied Microbiology*. 99, 1215-1221.
[27] Arroyo, M., Aldred, D., Magan, N., (2005). Environmental factors and weak organic acid interactions have differential effects on control of growth and ochratoxin A production by Penicillium verrucosum isolates in bread. *International journal of food microbiology*. 98, 223-231.
[28] Frisvad, J., Samson, R., Chełkowski, J., (1991). Mycotoxins produced by species of Penicillium and Aspergillus occurring in cereals. *Cereal grain: mycotoxins, fungi and quality in drying and storage.*, 441-476.
[29] Betina, V. (1989) *Mycotoxins. Chemical, biological and environmental aspects*. Elsevier.
[30] Boorman, G., (1989). NTP Technical report on the toxicology and carcinogenesis studies of ochratoxin A (CAS No. 303-47-9) in F344/N Rats (Gavage Studies)(NIH Publication No. 89-2813). *National Toxicology Program, US Department of Health and Human Services: Research Triangle Park, NC*.
[31] Chu, F., Wilson, B., (1973). Studies on ochratoxins. *CRC critical reviews in toxicology*. 2, 499-524.
[32] Galtier, P., (1990). Pharmacokinetics of ochratoxin A in animals. *IARC scientific publications*, 187-200.
[33] Valenta, H., (1998). Chromatographic methods for the determination of ochratoxin A in animal and human tissues and fluids. *Journal of Chromatography A*. 815, 75-92.
[34] El Khoury, A., Atoui, A., (2010). Ochratoxin A: general overview and actual molecular status. *Toxins*. 2, 461-493.

[35] Miller, J.D., (1992). Fungi as contaminants in indoor air. *Atmospheric Environment. Part A. General Topics*. 26, 2163-2172.

[36] Betina, V., (1984). Mycotoxins. Production, isolation, separation and purification. *Mycotoxins. Production, isolation, separation and purification.*

[37] de Jesus, A.E., Steyn, P.S., Vleggaar, R., Wessels, P.L., (1980). Carbon-13 nuclear magnetic resonance assignments and biosynthesis of the mycotoxin ochratoxin A. *Journal of the Chemical Society, Perkin Transactions 1*, 52-54.

[38] Lillehoj, E., Goransson, B., (1980). Occurrence of Ochratoxin - And Citrinin - Producing Fungi on Developing Danish Barley Grain. *Acta Pathologica Microbiologica Scandinavica Section B Microbiology*. 88, 133-137.

[39] El Khoury, A. (2007). EL Khoury, A. Champignons Mycotoxinogènes et Ochratoxine A (OTA) et Aflatoxine B1 (AFB1) dans les Vignobles Libanais: Occurrence et Origine. Ph.D. Thesis, INP, Toulouse, France, 2007.

[40] Abramson, D., (1987). Measurement of ochratoxin A in barley extracts by liquid chromatography-mass spectrometry. *Journal of Chromatography A*. 391, 315-320.

[41] Bauer, J., Gareis, M., (1987). Ochratoxin A in the food chain. *Zentralblatt für Veterinärmedizin. Reihe B. Journal of veterinary medicine. Series B*. 34, 613.

[42] Müller, H., (1982). Decontamination of mycotoxins. I. Physical process. *Ubersicht. Tierernähr*. 10, 95-122.

[43] Trivedi, A.B., Doi, E., Kitabatake, N., (1992). Detoxification of ochratoxin A on heating under acidic and alkaline conditions. *Bioscience, biotechnology, and biochemistry*. 56, 741-745.

[44] Boudra, H., Le Bars, P., Le Bars, J., (1995). Thermostability of Ochratoxin A in wheat under two moisture conditions. *Applied and Environmental Microbiology*. 61, 1156-1158.

[45] Kuiper-Goodman, T., Scott, P., (1989). Risk assessment of the mycotoxin ochratoxin A. *Biomedical and environmental sciences: BES*. 2, 179-248.

[46] Geisen, R., Mayer, Z., Karolewiez, A., Färber, P., (2004). Development of a real time PCR system for detection of Penicillium nordicum and for monitoring ochratoxin A production in foods by targeting the ochratoxin polyketide synthase gene. *Systematic and applied microbiology*. 27, 501-507.

[47] Ringot, D., Chango, A., Schneider, Y.-J., Larondelle, Y., (2006). Toxicokinetics and toxicodynamics of ochratoxin A, an update. *Chemico-biological interactions*. 159, 18-46.

[48] O'Callaghan, J., Caddick, M., Dobson, A., (2003). A polyketide synthase gene required for ochratoxin A biosynthesis in Aspergillus ochraceus. *Microbiology*. 149, 3485-3491.

[49] Petrak, J., Myslivcova, D., Man, P., Cmejla, R., Cmejlova, J., Vyoral, D., (2006). Proteomic analysis of iron overload in human hepatoma cells. *American Journal of Physiology-Gastrointestinal and Liver Physiology*. 290, G1059-G1066.

[50] Yamazaki, M., Maebayashi, Y., Miyaki, K., (1971). Biosynthesis of ochratoxin A. *Tetrahedron Letters*. 12, 2301-2303.

[51] Steyn, P., (1998). The biosynthesis of mycotoxins. *Revue de médecine vétérinaire*. 149, 469-478.

[52] Harris, J.P., Mantle, P.G., (2001). Biosynthesis of ochratoxins by Aspergillus ochraceus. *Phytochemistry*. 58, 709-716.

[53] O'Callaghan, J., Caddick, M.X., Dobson, A.D.W., (2003). A polyketide synthase gene required for ochratoxin A biosynthesis in Aspergillus ochraceus. *Microbiol-Sgm*. 149, 3485-3491.

[54] O'Callaghan, J., Stapleton, P.C., Dobson, A.D.W., (2006). Ochratoxin A biosynthetic genes in Aspergillus ochraceus are differentially regulated by pH and nutritional stimuli. *Fungal Genet Biol*. 43, 213-221.

[55] Reverberi, M., Gazzetti, K., Punelli, F., Scarpari, M., Zjalic, S., Ricelli, A., Fabbri, A.A., Fanelli, C., (2012). Aoyap1 regulates OTA synthesis by controlling cell redox balance in Aspergillus ochraceus. *Appl Microbiol Biot*. 95, 1293-1304.

[56] Wang, L.Q., Wang, Y., Wang, Q., Liu, F., Selvaraj, J.N., Liu, L.N., Xing, F.G., Zhao, Y.J., Zhou, L., Liu, Y., (2015). Functional Characterization of New Polyketide Synthase Genes Involved in Ochratoxin A Biosynthesis in Aspergillus Ochraceus fc-1. *Toxins*. 7, 2723-2738.

[57] Gallo, A., Bruno, K.S., Solfrizzo, M., Perrone, G., Mule, G., Visconti, A., Baker, S.E., (2012). New Insight into the Ochratoxin A Biosynthetic Pathway through Deletion of a Nonribosomal Peptide Synthetase Gene in Aspergillus carbonarius. *Appl Environ Microb*. 78, 8208-8218.

[58] Gallo, A., Knox, B.P., Bruno, K.S., Solfrizzo, M., Baker, S.E., Perrone, G., (2014). Identification and characterization of the polyketide synthase involved in ochratoxin A biosynthesis in Aspergillus carbonarius. *Int J Food Microbiol*. 179, 10-17.

[59] Crespo-Sernpere, A., Gonzalez-Candelas, L., Martinez-Culebras, P.V., (2010). Genes differentially expressed by Aspergillus carbonarius strains under ochratoxin A producing conditions. *Int J Food Microbiol*. 142, 170-179.

[60] Crespo-Sempere, A., Gil, J.V., Martinez-Culebras, P.V., (2011). Proteome analysis of the fungus Aspergillus carbonarius under ochratoxin A producing conditions. *Int J Food Microbiol*. 147, 162-169.

[61] Wawrzyniak, J., Waskiewicz, A., (2014). Ochratoxin A and citrinin production by Penicillium verrucosum on cereal solid substrates. *Food additives & contaminants. Part A, Chemistry, analysis, control, exposure & risk assessment*. 31, 139-148.

[62] O'Callaghan, J., Coghlan, A., Abbas, A., Garcia-Estrada, C., Martin, J.F., Dobson, A.D.W., (2013). Functional characterization of the polyketide synthase gene required for ochratoxin A biosynthesis in Penicillium verrucosum. *Int J Food Microbiol*. 161, 172-181.

[63] Schmidt-Heydt, M., Baxter, E., Geisen, R., Magan, N., (2007). Physiological relationship between food preservatives, environmental factors, ochratoxin and otapksPV gene expression by Penicillium verrucosum. *Int J Food Microbiol*. 119, 277-283.

[64] Kumagai, S., Aibara, K., (1982). Intestinal absorption and secretion of ochratoxin A in the rat. *Toxicology and applied pharmacology*. 64, 94-102.

[65] Gekle, M., Silbernagl, S., (1994). The role of the proximal tubule in ochratoxin A nephrotoxicity *in vivo*: toxodynamic and toxokinetic aspects. *Kidney and Blood Pressure Research*. 17, 40-49.

[66] Tsuda, M., Sekine, T., Takeda, M., Cha, S.H., Kanai, Y., Kimura, M., Endou, H., (1999). Transport of ochratoxin A by renal multispecific organic anion transporter 1. *Journal of Pharmacology and Experimental Therapeutics*. 289, 1301-1305.

[67] Jung, K.Y., Takeda, M., Kim, D.K., Tojo, A., Narikawa, S., Yoo, B.S., Hosoyamada, M., Cha, S.H., Sekine, T., Endou, H., (2001). Characterization of ochratoxin A transport by human organic anion transporters. *Life sciences*. 69, 2123-2135.

[68] Hagelberg, S., Hult, K., Fuchs, R., (1989). Toxicokinetics of ochratoxin A in several species and its plasma - binding properties. *Journal of Applied Toxicology*. 9, 91-96.

[69] O'Brien, E., Dietrich, D.R., (2005). Ochratoxin A: the continuing enigma. *Critical reviews in toxicology*. 35, 33-60.

[70] Stein, A., Phillips, T., Kubena, L., Harvey, R., (1985). Renal tubular secretion and reabsorption as factors in ochratoxicosis: Effects of probenecid on nephrotoxicity. *Journal of Toxicology and Environmental Health, Part A Current Issues*. 16, 593-605.

[71] Albassam, M., Yong, S., Bhatnagar, R., Sharma, A., Prior, M., (1987). Histopathologic and electron microscopic studies on the acute toxicity of ochratoxin A in rats. *Veterinary Pathology Online*. 24, 427-435.

[72] Gekle, M., Sauvant, C., Schwerdt, G., (2005). Ochratoxin A at nanomolar concentrations: a signal modulator in renal cells. *Molecular nutrition & food research*. 49, 118-130.

[73] Galtier, P., (1977). Contribution of pharmacokinetic studies to mycotoxicology—Ochratoxin A. *Veterinary Science Communications*. 1, 349-358.

[74] Roth, A., Chakor, K., EkuéCreepy, E., Kane, A., Roschenthaler, R., Dirheimer, G., (1988). Evidence for an enterohepatic circulation of ochratoxin A in mice. *Toxicology*. 48, 293-308.

[75] Berger, V., Gabriel, A.-F., Sergent, T., Trouet, A., Larondelle, Y., Schneider, Y.-J., (2003). Interaction of ochratoxin A with human intestinal Caco-2 cells: possible implication of a multidrug resistance-associated protein (MRP2). *Toxicology letters*. 140, 465-476.

[76] Jonker, J.W., Merino, G., Musters, S., van Herwaarden, A.E., Bolscher, E., Wagenaar, E., Mesman, E., Dale, T.C., Schinkel, A.H., (2005). The breast cancer resistance protein BCRP (ABCG2) concentrates drugs and carcinogenic xenotoxins into milk. *Nature medicine*. 11, 127-129.

[77] Maliepaard, M., Scheffer, G.L., Faneyte, I.F., van Gastelen, M.A., Pijnenborg, A.C., Schinkel, A.H., van de Vijver, M.J., Scheper, R.J., Schellens, J.H., (2001). Subcellular localization and distribution of the breast cancer resistance protein transporter in normal human tissues. *Cancer research*. 61, 3458-3464.

[78] Schrickx, J., Lektarau, Y., Fink-Gremmels, J., (2006). Ochratoxin A secretion by ATP-dependent membrane transporters in Caco-2 cells. *Archives of toxicology*. 80, 243-249.

[79] Galtier, P., Alvinerie, M., Charpenteau, J., (1981). The pharmacokinetic profiles of ochratoxin A in pigs, rabbits and chickens. *Food and cosmetics toxicology*. 19, 735-738.

[80] Chu, F.S., (1971). Interaction of ochratoxin A with bovine serum albumin. *Archives of biochemistry and biophysics*. 147, 359-366.

[81] Chu, F.S., (1974). A comparative study of the interaction of ochratoxins with bovine serum albumin. *Biochemical pharmacology*. 23, 1105-1113.

[82] Kumagai, S., (1985). Ochratoxin A: plasma concentration and excretion into bile and urine in albumin-deficient rats. *Food and chemical toxicology*. 23, 941-943.

[83] McMasters, D.R., Vedani, A., (1999). Ochratoxin binding to phenylalanyl-tRNA synthetase: computational approach to the mechanism of ochratoxicosis and its antagonism. *Journal of medicinal chemistry*. 42, 3075-3086.

[84] Il'ichev, Y.V., Perry, J.L., Simon, J.D., (2002). Interaction of ochratoxin A with human serum albumin. Preferential binding of the dianion and pH effects. *The Journal of Physical Chemistry B*. 106, 452-459.

[85] Schwerdt, G., Freudinger, R., Silbernagl, S., Gekle, M., (1999). Ochratoxin A-binding proteins in rat organs and plasma and in different cell lines of the kidney. *Toxicology*. 135, 1-10.

[86] Stojković, R., Hult, K., Gamulin, S., PleStina, R., (1984). High affinity binding of ochratoxin A to plasma constituents. *Biochemistry international*. 9, 33-38.

[87] Petzinger, E., Ziegler, K., (2000). Ochratoxin A from a toxicological perspective. *Journal of Veterinary Pharmacology and Therapeutics*. 23, 91-98.

[88] Stander, M.A., Nieuwoudt, T.W., Steyn, P.S., Shephard, G.S., Creppy, E.E., Sewram, V., (2001). Toxicokinetics of ochratoxin A in vervet monkeys (Cercopithecus aethiops). *Archives of toxicology*. 75, 262-269.

[89] Studer-Rohr, I. (1995), Swiss Federal Institute of Technology Zürich.

[90] Li, S., Marquardt, R., Frohlich, A., Vitti, T., Crow, G., (1997). Pharmacokinetics of ochratoxin A and its metabolites in rats. *Toxicology and applied pharmacology*. 145, 82-90.

[91] Ferrufino-Guardia, E., Tangni, E., Larondelle, Y., Ponchaut, S., (2000). Transfer of ochratoxin A during lactation: exposure of suckling via the milk of rabbit does fed a naturally-contaminated feed. *Food Additives & Contaminants*. 17, 167-175.

[92] Madsen, A., Hald, B., LillehØ j, E., Mortensen, H.P., (1982). Feeding Experiments with Ochratoxin a Contaminated Barley for Bacon Pigs: 2. Naturally Contaminated Barley given for 6 weeks from 20 kg compared with Normal Barley Supplemented with Crystalline Ochratoxin A and/or Citrinin. *Acta Agriculturae Scandinavica*. 32, 369-372.

[93] Mortensen, H.P., Hald, B., Larsen, A.E., Madsen, A., (1983). Ochratoxin A contaminated barley for sows and piglets: pig performance and residues in milk and pigs. *Acta Agriculturae Scandinavica*. 33, 349-352.

[94] Hald, B., (1990). Porcine nephropathy in Europe. *IARC scientific publications*, 49-56.

[95] Kane, A., Creppy, E.E., Roth, A., Röschenthaler, R., Dirheimer, G., (1986). Distribution of the [3H]-label from low doses of radioactive ochratoxin A ingested by rats, and evidence for DNA single-strand breaks caused in liver and kidneys. *Archives of toxicology*. 58, 219-224.

[96] Belmadani, A., Tramu, G., Betbeder, A., Steyn, P., Creppy, E., (1998). Regional selectivity to ochratoxin A, distribution and cytotoxicity in rat brain. *Archives of toxicology*. 72, 656-662.

[97] Hallén, I.P., Breitholtz - Emanuelsson, A., Hult, K., Olsen, M., Oskarsson, A., (1998). Placental and lactational transfer of ochratoxin A in rats. *Natural toxins*. 6, 43-49.

[98] Appelgren, L.-E., Arora, R., (1983). Distribution of 14 C-labelled ochratoxin A in pregnant mice. *Food and Chemical Toxicology*. 21, 563-568.

[99] Zimmerli, B., Dick, R., (1995). Determination of ochratoxin A at the ppt level in human blood, serum, milk and some foodstuffs by high-performance liquid chromatography with enhanced fluorescence detection and immunoaffinity column cleanup:

methodology and Swiss data. *Journal of Chromatography B: Biomedical Sciences and Applications*. 666, 85-99.

[100] Miraglia, M., Brera, C., Corneli, S., Cava, E., Montanino, G., Miraglia, E. (1998), *Proceedings of IX International IUPAC Symposium on Mycotoxins and Phycotoxins–Developments in Chemistry, Toxicology and Food Safety, Alaken Inc., USA*, pp. 165-179.

[101] Cha, S.H., Sekine, T., Kusuhara, H., Yu, E., Kim, J.Y., Kim, D.K., Sugiyama, Y., Kanai, Y., Endou, H., (2000). Molecular cloning and characterization of multispecific organic anion transporter 4 expressed in the placenta. *Journal of Biological Chemistry*. 275, 4507-4512.

[102] Groves, C.E., Morales, M., Wright, S.H., (1998). Peritubular transport of ochratoxin A in rabbit renal proximal tubules. *Journal of Pharmacology and Experimental Therapeutics*. 284, 943-948.

[103] Kusuhara, H., Sekine, T., Utsunomiya-Tate, N., Tsuda, M., Kojima, R., Cha, S.H., Sugiyama, Y., Kanai, Y., Endou, H., (1999). Molecular cloning and characterization of a new multispecific organic anion transporter from rat brain. *Journal of Biological Chemistry*. 274, 13675-13680.

[104] Bahnemann, E., Kerling, H.-P., Ensminger, S., Schwerdt, G., Silbernagl, S., Gekle, M., (1997). Renal transepithelial secretion of ochratoxin A in the non-filtering toad kidney. *Toxicology*. 120, 11-17.

[105] Leier, I., Hummel-Eisenbeiss, J., Cui, Y., Keppler, D., (2000). ATP-dependent para-aminohippurate transport by apical multidrug resistance protein MRP2. *Kidney international*. 57, 1636-1642.

[106] Babu, E., Takeda, M., Narikawa, S., Kobayashi, Y., Enomoto, A., Tojo, A., Cha, S.H., Sekine, T., Sakthisekaran, D., Endou, H., (2002). Role of human organic anion transporter 4 in the transport of ochratoxin A. *Biochimica et Biophysica Acta (BBA)-Molecular Cell Research*. 1590, 64-75.

[107] Dahlmann, A., Dantzler, W.H., Silbernagl, S., Gekle, M., (1998). Detailed mapping of ochratoxin A reabsorption along the rat nephron *in vivo*: the nephrotoxin can be reabsorbed in all nephron segments by different mechanisms. *Journal of pharmacology and experimental therapeutics*. 286, 157-162.

[108] Zingerle, M., Silbernagl, S., Gekle, M., (1997). Reabsorption of the nephrotoxin ochratoxin A along the rat nephron *in vivo*. *Journal of Pharmacology and Experimental Therapeutics*. 280, 220-224.

[109] Kobayashi, Y., Ohshiro, N., Shibusawa, A., Sasaki, T., Tokuyama, S., Sekine, T., Endou, H., Yamamoto, T., (2002). Isolation, characterization and differential gene expression of multispecific organic anion transporter 2 in mice. *Molecular pharmacology*. 62, 7-14.

[110] Youngblood, G.L., Sweet, D.H., (2004). Identification and functional assessment of the novel murine organic anion transporter Oat5 (Slc22a19) expressed in kidney. *American Journal of Physiology-Renal Physiology*. 287, F236-F244.

[111] Schwerdt, G., Gekle, M., Freudinger, R., Mildenberger, S., Silbernagl, S., (1997). Apical-to-basolateral transepithelial transport of ochratoxin A by two subtypes of Madin-Darby canine kidney cells. *Biochimica et Biophysica Acta (BBA)-Biomembranes*. 1324, 191-199.

[112] Koepsell, H., Endou, H., (2004). The SLC22 drug transporter family. *Pflügers Archiv*. 447, 666-676.
[113] Fuchs, R., Radic, B., Peraica, M., Hult, K., Plestina, R., (1988). Enterohepatic circulation of ochratoxin A in rats. *Period. Biol*. 90, 39.
[114] Suzuki, S., Satoh, T., Yamazaki, M., (1977). The pharmacokinetics of ochratoxin a in rats. *The Japanese Journal of Pharmacology*. 27, 735-744.
[115] Chandra, P., Brouwer, K.L., (2004). The complexities of hepatic drug transport: current knowledge and emerging concepts. *Pharmaceutical research*. 21, 719-735.
[116] Nip, W.K., Chu, F.S., (1979). Fate of ochratoxin A in goats. *Journal of Environmental Science & Health Part B*. 14, 319-333.
[117] Xiao, H., Marquardt, R., Frohlich, A., Phillips, G., Vitti, T., (1991). Effect of a hay and a grain diet on the rate of hydrolysis of ochratoxin A in the rumen of sheep. *Journal of animal science*. 69, 3706-3714.
[118] Galtier, P., Alvinerie, M. (1976), *Annales de Recherches Vétérinaires*, Vol. 7, pp. 91-98.
[119] Galtier, P., (1999). Biotransformation and fate of mycotoxins. *Toxin Reviews*. 18, 295-312.
[120] Pitout, M., (1969). The hydrolysis of ochratoxin A by some proteolytic enzymes. *Biochemical Pharmacology*. 18, 485-491.
[121] Doster, R., Sinnhuber, R., (1972). Comparative rates of hydrolysis of ochratoxins A and B *in vitro*. *Food and cosmetics toxicology*. 10, 389-394.
[122] Madhyastha, M., Marquardt, R., Frohlich, A., (1992). Hydrolysis of ochratoxin A by the microbial activity of digesta in the gastrointestinal tract of rats. *Archives of environmental contamination and toxicology*. 23, 468-472.
[123] Höhler, D., Südekum, K., Wolffram, S., Frohlich, A., Marquardt, R., (1999). Metabolism and excretion of ochratoxin A fed to sheep. *Journal of animal science*. 77, 1217-1223.
[124] Kanisawa, M., Suzuki, S., (1978). Induction of renal and hepatic tumors in mice by ochratoxin A, a mycotoxin. *Gan*. 69, 599.
[125] Størmer, F., Støren, O., Hansen, C.E., Pedersen, J., Aasen, A., (1983). Formation of (4R)-and (4S)-4-hydroxyochratoxin A and 10-hydroxyochratoxin A from Ochratoxin A by rabbit liver microsomes. *Applied and environmental microbiology*. 45, 1183-1187.
[126] Hansen, C.E., Dueland, S., Drevon, C.A., Størmer, F., (1982). Metabolism of ochratoxin A by primary cultures of rat hepatocytes. *Applied and environmental microbiology*. 43, 1267-1271.
[127] Galtier, P., Charpenteau, J.-L., Alvinerie, M., Labouche, C., (1979). The pharmacokinetic profile of ochratoxin A in the rat after oral and intravenous administration. *Drug Metabolism and Disposition*. 7, 429-434.
[128] Hult, K., Teiling, A., Gatenbeck, S., (1976). Degradation of ochratoxin A by a ruminant. *Applied and Environmental Microbiology*. 32, 443-444.
[129] Kiessling, K.-H., Pettersson, H., Sandholm, K., Olsen, M., (1984). Metabolism of aflatoxin, ochratoxin, zearalenone, and three trichothecenes by intact rumen fluid, rumen protozoa, and rumen bacteria. *Applied and environmental microbiology*. 47, 1070-1073.

[130] Müller, H.M., Lerch, C., Müller, K., Eggert, W., (1998). Kinetic profiles of ochratoxin A and ochratoxin α during *in vitro* incubation in buffered forestomach and abomasal contents from cows. *Natural toxins*. 6, 251-258.

[131] Müller, H.M., Müller, K., Steingass, H., (2001). Effect of feeding regime on the metabolism of ochratoxin A during the *in vitro* incubation in buffered rumen fluid from cows. *Archives of Animal Nutrition*. 54, 265-279.

[132] Marquardt, R.R., Frohlich, A.A., (1992). A review of recent advances in understanding ochratoxicosis. *Journal of animal science*. 70, 3968-3988.

[133] Kontaxi, M., Echkardt, U., Hagenbuch, B., Stieger, B., Meier, P.J., Petzinger, E., (1996). Uptake of the mycotoxin ochratoxin A in liver cells occurs via the cloned organic anion transporting polypeptide. *The Journal of pharmacology and experimental therapeutics*. 279, 1507-1513.

[134] Eckhardt, U., Schroeder, A., Stieger, B., Hochli, M., Landmann, L., Tynes, R., Meier, P.J., Hagenbuch, B., (1999). Polyspecific substrate uptake by the hepatic organic anion transporter Oatp1 in stably transfected CHO cells. *The American journal of physiology*. 276, G1037-1042.

[135] Kontaxi, M., Echkardt, U., Hagenbuch, B., Stieger, B., Meier, P.J., Petzinger, E., (1996). Uptake of the mycotoxin ochratoxin A in liver cells occurs via the cloned organic anion transporting polypeptide. *Journal of Pharmacology and Experimental Therapeutics*. 279, 1507-1513.

[136] Eckhardt, U., Schroeder, A., Stieger, B., Höchli, M., Landmann, L., Tynes, R., Meier, P.J., Hagenbuch, B., (1999). Polyspecific substrate uptake by the hepatic organic anion transporter Oatp1 in stably transfected CHO cells. *American Journal of Physiology-Gastrointestinal and Liver Physiology*. 276, G1037-G1042.

[137] Størmer, F., Pedersen, J., (1980). Formation of 4-hydroxyochratoxin A from ochratoxin A by rat liver microsomes. *Applied and environmental microbiology*. 39, 971-975.

[138] Størmer, F., Hansen, C.E., Pedersen, J., Hvistendahl, G., Aasen, A.J., (1981). Formation of (4R)-and (4S)-4-hydroxyochratoxin A from ochratoxin A by liver microsomes from various species. *Applied and environmental microbiology*. 42, 1051-1056.

[139] Oster, T., Jayyosi, Z., Creppy, E.E., El Amri, H.S., Batt, A.-M., (1991). Characterization of pig liver purified cytochrome P-450 isoenzymes for Ochratoxin A metabolism studies. *Toxicology letters*. 57, 203-214.

[140] Omar, R.F., Gelboin, H.V., Rahimtula, A.D., (1996). Effect of cytochrome P450 induction on the metabolism and toxicity of ochratoxin A. *Biochemical pharmacology*. 51, 207-216.

[141] Creppy, E., Kern, D., Steyn, P., Vleggaar, R., Röschenthaler, R., Dirheimer, G., (1983). Comparative study of the effect of ochratoxin A analogues on yeast aminoacyl-tRNA synthetases and on the growth and protein synthesis of hepatoma cells. *Toxicology letters*. 19, 217-224.

[142] Creppy, E., Kane, A., Giessen-Crouse, E., Roth, A., Röschenthaler, R., Dirheimer, G. (1986), *Toxic Interfaces of Neurones, Smoke and Genes*. Springer, pp. 310-314.

[143] Hutchison, R., Steyn, P., Thompson, D., (1971). The isolation and structure of 4-hydroxyochratoxin A and 7-carboxy-3, 4-dihydro-8-hydroxy-3-methylisocoumarin from penicilliumviridicatum. *Tetrahedron letters*. 12, 4033-4036.

[144] Xiao, H., Marquardt, R.R., Abramson, D., Frohlich, A.A., (1996). Metabolites of ochratoxins in rat urine and in a culture of Aspergillus ochraceus. *Applied and environmental microbiology*. 62, 648-655.

[145] Zepnik, H., Pähler, A., Schauer, U., Dekant, W., (2001). Ochratoxin A-induced tumor formation: is there a role of reactive ochratoxin A metabolites? *Toxicological Sciences*. 59, 59-67.

[146] Pinelli, E., El Adlouni, C., Pipy, B., Quartulli, F., Pfohl-Leszkowicz, A., (1999). Roles of cyclooxygenase and lipoxygenases in ochratoxin A genotoxicity in human epithelial lung cells. *Environmental toxicology and pharmacology*. 7, 95-107.

[147] El Adlouni, C., Pinelli, E., Azemar, B., Zaoui, D., Beaune, P., Pfohl - Leszkowicz, A., (2000). Phenobarbital increases DNA adduct and metabolites formed by ochratoxin A: Role of CYP 2C9 and microsomal glutathione - S - transferase. *Environmental and molecular mutagenesis*. 35, 123-131.

[148] Hoehler, D., Marquardt, R.R., McIntosh, A.R., Xiao, H., (1996). Free radical generation as induced by ochratoxin A and its analogs in bacteria (Bacillus brevis). *Journal of Biological Chemistry*. 271, 27388-27394.

[149] Hoehler, D., Marquardt, R.R., McIntosh, A.R., Hatch, G.M., (1997). Induction of free radicals in hepatocytes, mitochondria and microsomes of rats by ochratoxin A and its analogs. *Biochimica et Biophysica Acta (BBA)-Molecular Cell Research*. 1357, 225-233.

[150] Gillman, I.G., Clark, T.N., Manderville, R.A., (1999). Oxidation of ochratoxin A by an Fe-porphyrin system: model for enzymatic activation and DNA cleavage. *Chemical research in toxicology*. 12, 1066-1076.

[151] Dai, J., Park, G., Wright, M.W., Adams, M., Akman, S.A., Manderville, R.A., (2002). Detection and characterization of a glutathione conjugate of ochratoxin A. *Chemical research in toxicology*. 15, 1581-1588.

[152] Wiese, F., Chang, H., Lloyd, R., Freeman, J., Samokyszyn, V., (1998). Peroxidase-catalyzed oxidation of 2, 4, 6-trichlorophenol. *Archives of environmental contamination and toxicology*. 34, 217-222.

[153] Gillman, I., Yezek, J., Manderville, R., (1998). Ochratoxin A acts as a photoactivatable DNA cleaving agent. *Chemical Communications*, 647-648.

[154] Calcutt, M.W., Gillman, I.G., Noftle, R.E., Manderville, R.A., (2001). Electrochemical oxidation of ochratoxin A: correlation with 4-chlorophenol. *Chemical research in toxicology*. 14, 1266-1272.

[155] Gautier, J.-C., Holzhaeuser, D., Markovic, J., Gremaud, E., Schilter, B.t., Turesky, R.J., (2001). Oxidative damage and stress response from ochratoxin A exposure in rats. *Free Radical Biology and Medicine*. 30, 1089-1098.

[156] Human Microbiome Project, C., (2012). Structure, function and diversity of the healthy human microbiome. *Nature*. 486, 207-214.

[157] Kühn, I., Valenta, H., Rohr, K., (1995). Determination of ochratoxin A in bile of swine by high-performance liquid chromatography. *Journal of Chromatography B: Biomedical Sciences and Applications*. 668, 333-337.

[158] Gross-Steinmeyer, K., Weymann, J., Hege, H.-G., Metzler, M., (2002). Metabolism and lack of DNA reactivity of the mycotoxin ochratoxin A in cultured rat and human primary hepatocytes. *Journal of agricultural and food chemistry*. 50, 938-945.

[159] Zepnik, H., Völkel, W., Dekant, W., (2003). Toxicokinetics of the mycotoxin ochratoxin A in F 344 rats after oral administration. *Toxicology and applied pharmacology*. 192, 36-44.

[160] Malaveille, C., Brun, G., Bartsch, H., (1994). Structure-activity studies in E. coli strains on ochratoxin A (OTA) and its analogues implicate a genotoxic free radical and a cytotoxic thiol derivative as reactive metabolites. *Mutation Research/Fundamental and Molecular Mechanisms of Mutagenesis*. 307, 141-147.

[161] Monks, T.J., Lau, S.S., (1997). Biological reactivity of polyphenolic-glutathione conjugates. *Chemical research in toxicology*. 10, 1296-1313.

[162] Brow, M.E., Dai, J., Park, G., Wright, M.W., Gillman, I.G., Manderville, R.A., (2002). Photochemically Catalyzed Reaction of Ochratoxin A with d - and l - cysteine. *Photochemistry and photobiology*. 76, 649-656.

[163] Conn, E.E. (1985), *UCLA symposia on molecular and cellular biology (USA).*

[164] Ruhland, M., Engelhardt, G., Wallnofer, P.R., Schafer, W., (1994). Transformation of the Mycotoxin Ochratoxin-a in Wheat and Maize Cell-Suspension Cultures. *Naturwissenschaften*. 81, 453-454.

[165] Stormer, F.C., Storen, O., Hansen, C.E., Pedersen, J.I., Aasen, A.J., (1983). Formation of (4r)-4-Hydroxyochratoxin-a and (4s)-4-Hydroxyochratoxin-a and 10-Hydroxyochratoxin-a from Ochratoxin-a by Rabbit Liver-Microsomes. *Appl Environ Microb*. 45, 1183-1187.

[166] Rotter, R.G., Marquardt, R.R., Frohlich, A.A., Abramson, D., (1990). Ensiling as a Means of Reducing Ochratoxin-a Concentrations in Contaminated Barley. *J Sci Food Agr*. 50, 155-166.

[167] Ruhland, M., Engelhardt, G., Wallnofer, P., (1997). Transformation of the mycotoxin ochratoxin A in artificially contaminated vegetables and cereals. *Mycotoxin research*. 13, 54-60.

[168] Wu, Q., Dohnal, V., Huang, L., Kuca, K., Wang, X., Chen, G., Yuan, Z., (2011). Metabolic pathways of ochratoxin A. *Current drug metabolism*. 12, 1-10.

[169] Stoev, S.D., (2010). Studies on carcinogenic and toxic effects of ochratoxin A in chicks. *Toxins*. 2, 649-664.

[170] Fuchs, S., Sontag, G., Stidl, R., Ehrlich, V., Kundi, M., Knasmüller, S., (2008). Detoxification of patulin and ochratoxin A, two abundant mycotoxins, by lactic acid bacteria. *Food and Chemical Toxicology*. 46, 1398-1407.

[171] Chan, B.S., Sellors, K., Chiew, A.L., Buckley, N.A., (2015). Use of multi-dose activated charcoal in phenytoin toxicity secondary to genetic polymorphism. *Clinical Toxicology*. 53, 131-133.

[172] Sturino, J.M., Pokusaeva, K., Carpenter, R., (2015). Effective sequestration of Clostridium difficile protein toxins by calcium aluminosilicate. *Antimicrobial agents and chemotherapy*, AAC. 05050-05014.

[173] Anjos, F.D., Ledoux, D., Rottinghaus, G., Chimonyo, M., (2015). Efficacy of adsorbents (bentonite and diatomaceous earth) and turmeric (Curcuma longa) in alleviating the toxic effects of aflatoxin in chicks. *British poultry science*.

[174] Kumar, S., Kunwar, A., Gautam, S., Sharma, A., (2012). Inactivation of A. ochraceus spores and detoxification of ochratoxin A in coffee beans by gamma irradiation. *Journal of food science*. 77, T44-T51.

[175] Di Stefano, V., Pitonzo, R., Cicero, N., D'Oca, M.C., (2014). Mycotoxin contamination of animal feedingstuff: detoxification by gamma-irradiation and reduction of aflatoxins and ochratoxin A concentrations. *Food Additives & Contaminants: Part A*. 31, 2034-2039.

[176] Raushel, F.M., (2011). Chemical biology: Catalytic detoxification. *Nature*. 469, 310-311.

[177] Chelkowski, J., Goliński, P., Godlewska, B., Radomyska, W., Szebiotko, K., Wiewiorowska, M., (1981). Mycotoxins in cereal grain. Part IV. Inactivation of ochratoxin A and other mycotoxins during ammoniation. *Food/Nahrung*. 25, 631-637.

[178] McKenzie, K., Sarr, A., Mayura, K., Bailey, R., Miller, D., Rogers, T., Norred, W., Voss, K., Plattner, R., Kubena, L., (1997). Oxidative degradation and detoxification of mycotoxins using a novel source of ozone. *Food and Chemical Toxicology*. 35, 807-820.

[179] Lanier, C., Heutte, N., Richard, E., Bouchart, V., Lebailly, P., Garon, D., (2009). Mycoflora and mycotoxin production in oilseed cakes during farm storage. *Journal of agricultural and food chemistry*. 57, 1640-1645.

[180] Abrunhosa, L., Paterson, R.R., Venancio, A., (2010). Biodegradation of ochratoxin a for food and feed decontamination. *Toxins*. 2, 1078-1099.

[181] Karlovsky, P., (1999). Biological detoxification of fungal toxins and its use in plant breeding, feed and food production. *Natural toxins*. 7, 1-23.

[182] McCormick, S.P., (2013). Microbial detoxification of mycotoxins. *Journal of chemical ecology*. 39, 907-918.

[183] Del Prete, V., Rodriguez, H., Carrascosa, A.V., de las Rivas, B., Garcia-Moruno, E., Munoz, R., (2007). *In vitro* removal of ochratoxin A by wine lactic acid bacteria. *Journal of Food Protection®*. 70, 2155-2160.

[184] Petchkongkaew, A., Taillandier, P., Gasaluck, P., Lebrihi, A., (2008). Isolation of Bacillus spp. from Thai fermented soybean (Thua - nao): screening for aflatoxin B1 and ochratoxin A detoxification. *Journal of applied microbiology*. 104, 1495-1502.

[185] Rodriguez, H., Reveron, I., Doria, F., Costantini, A., De Las Rivas, B., Munoz, R., Garcia-Moruno, E., (2011). Degradation of ochratoxin a by Brevibacterium species. *Journal of agricultural and food chemistry*. 59, 10755-10760.

[186] Varga, J., Péteri, Z., Tábori, K., Téren, J., Vágvölgyi, C., (2005). Degradation of ochratoxin A and other mycotoxins by Rhizopus isolates. *International journal of food microbiology*. 99, 321-328.

[187] Bejaoui, H., Mathieu, F., Taillandier, P., Lebrihi, A., (2006). Biodegradation of ochratoxin A by Aspergillus section Nigri species isolated from French grapes: a potential means of ochratoxin A decontamination in grape juices and musts. *FEMS microbiology letters*. 255, 203-208.

[188] Varga, J., Rigo, K., Teren, J., (2000). Degradation of ochratoxin A by Aspergillus species. *International journal of food microbiology*. 59, 1-7.

[189] Engelhardt, G., (2002). Degradation of ochratoxin a and b by the white rot funguspleurotus ostreatus. *Mycotoxin research*. 18, 37-43.

[190] Péteri, Z., Téren, J., Vágvölgyi, C., Varga, J., (2007). Ochratoxin degradation and adsorption caused by astaxanthin-producing yeasts. *Food microbiology*. 24, 205-210.

[191] Schatzmayr, G., Heidler, D., Fuchs, E., Nitsch, S., Mohnl, M., Taubel, M., Loibner, A.P., Braun, R., Binder, E.M., (2003). Investigation of different yeast strains for the detoxification of ochratoxin A. *Mycotoxin research*. 19, 124-128.

[192] Peteri, Z., Teren, J., Vagvolgyi, C., Varga, J., (2007). Ochratoxin degradation and adsorption caused by astaxanthin-producing yeasts. *Food microbiology*. 24, 205-210.

[193] Gil-Serna, J., Patiño, B., Cortés, L., González-Jaén, M.T., Vázquez, C., (2011). Mechanisms involved in reduction of ochratoxin A produced by Aspergillus westerdijkiae using Debaryomyces hansenii CYC 1244. *International journal of food microbiology*. 151, 113-118.

[194] Guslandi, M., (2006). Targeting intestinal microflora in inflammatory bowel disease. *World journal of gastroenterology: WJG*. 12, 7225.

[195] Bondarenko, V.M., Maleev, V.V., Likhoded, V.G., (2014). [Intestinal microflora, obesity and type 2 diabetes]. *Zhurnal mikrobiologii, epidemiologii, i immunobiologii*, 42-49.

[196] Wasilewska, E., Złotkowska, D., Pijagin, M.E., (2012). [The role of intestinal microflora and probiotic bacteria in prophylactic and development of colorectal cancer]. *Postepy higieny i medycyny doswiadczalnej (Online)*. 67, 837-847.

[197] Upadhaya, S.D., Song, J.Y., Park, M.A., Seo, J.K., Yang, L., Lee, C.H., Cho, K.J., Ha, J.K., (2012). Isolation, screening and identification of swine gut microbiota with ochratoxin A biodegradation ability. *Asian-Australasian journal of animal sciences*. 25, 114.

[198] Camel, V., Ouethrani, M., Coudray, C., Philippe, C., Rabot, S., (2012). Semi-automated solid-phase extraction method for studying the biodegradation of ochratoxin A by human intestinal microbiota. *Journal of Chromatography B*. 893, 63-68.

[199] Kiessling, K.H., Pettersson, H., Sandholm, K., Olsen, M., (1984). Metabolism of aflatoxin, ochratoxin, zearalenone, and three trichothecenes by intact rumen fluid, rumen protozoa, and rumen bacteria. *Applied and environmental microbiology*. 47, 1070-1073.

[200] Hult, K., Teiling, A., Gatenbeck, S., (1976). Degradation of ochratoxin A by a ruminant. *Applied and environmental microbiology*. 32, 443-444.

[201] Montesano, R. (1979) *Handling chemical carcinogens in the laboratory, problems of safety*. International Agency for Research on Cancer ;obtainable from Who Publications Center, Lyon Albany, N. Y.

[202] Roland, A., Bros, P., Bouisseau, A., Cavelier, F., Schneider, R., (2014). Analysis of ochratoxin A in grapes, musts and wines by LC-MS/MS: first comparison of stable isotope dilution assay and diastereomeric dilution assay methods. *Analytica chimica acta*. 818, 39-45.

[203] Scudamore, K.A., (2005). Prevention of ochratoxin A in commodities and likely effects of processing fractionation and animal feeds. *Food additives and contaminants*. 22 Suppl 1, 17-25.

[204] Varga, J., Kocsube, S., Peteri, Z., Vagvolgyi, C., Toth, B., (2010). Chemical, physical and biological approaches to prevent ochratoxin induced toxicoses in humans and animals. *Toxins*. 2, 1718-1750.

Chapter 3

DETECTING METHODS OF OCHRATOXIN A

ABSTRACT

Ochratoxin A(OTA) is a significance mycotoxin, and contaminated materials are monitored to exclude Ochratoxin A from animal feed and food to avoid therisk of OTA consumption by animals and humans. Many analytical methods have been employed to detect OTA. The detection of trace mycotoxins in complex matrices, such as food, often requires sample preparation and extraction before analysis. Optimal sample preparation conditions can reduce errors and analysis time as well as enhance analysis sensitivity and enable definitive identification, confirmation and quantification.

This section discusses three aspects of OTA detection, including sample preparation, separation techniques and analytical techniques. The discussion also introduces techniques for the extraction and detection of other mycotoxins. Finally, future development trends of mycotoxin analysis technology are discussed.

Keywords: Ochratoxin A, sample preparation, immunoassay, aptamer, biosensor

1. INTRODUCTION

Many structurally different mycotoxins have been discovered to date. Most of them are both thermally and chemically stable, making it difficult for them to be removed during food processing procedures or other suitable detoxification procedures [1]. To date, aflatoxins (AFTs), zearalenone (ZON), ochratoxin A (OTA), trichothecenes and fumonisins have received the greatest attention because of their frequent occurrence and their severe effects on human health [1-3].

Surveillance surveys have demonstrated that mycotoxin is a world-wide contamination problem [4, 5]. It is estimated that 20% of crop production and 25% of the world's crop production in the European Union may be contaminated with mycotoxins [3]. Therefore, mycotoxin contamination results in tremendous economic losses, including a reduction of livestock and agricultural production, and increased veterinary, health care and regulatory costs. The situation is less severe in Europe and Northern America due to technical, climatic and educational reasons [1]. Many countries have established limits to control and monitor mycotoxin levels, especially for agricultural import products. Maximum tolerable levels and guideline levels have been established for OTA, AFTs, ZON and deoxynivalenol (DON) at the ppb to ppt level in different feed and food products [1, 6-9]. The maximum tolerable levels for other mycotoxins are under debate.

The Food and Drug Administration (FDA) and European Commission have promoted the development of extensive analytical efforts to enable reliable and fast analysis of large quantities of contaminated samples to monitor ppb and ppt levels in complex food environments. In particular, fast and highly sensitive (ppt level) screening assays are necessary for OTA, as it is toxic and carcinogenic.

Analysis of trace mycotoxins in complex food matrices often requires extensive sample extraction and preparation prior to detection analysis, except with ELISA-based assays and other new emerging biosensors, which may not require cleanup. Sample preparation steps are time-consuming (it is often the main time factor in any analysis method and takes approximately two-thirds of total analysis time) and affects the final choice of detection procedure. Thus, sample preparation optimization is an important parameter in any mycotoxin analysis system. The following section considers some aspects of sample preparation, covering general extraction techniques and pressurized liquid extraction, microextraction techniques and more selective techniques, such as solid phase extraction (SPE). The applicability of these techniques in food analysis is also discussed.

2. The Principle and Method of Sample Pre-Treatment

2.1. The Principle of Sampling and Sample Preparation

Sample size, sample extraction and clean-up determine the length of the analytical procedure, accuracy, recovery, and achievable detection limits. The determination of trace contaminants in complex matrices such as food often requires sample extraction and preparation strategies prior to instrumental analysis. The necessary sample size depends on the matrix and the properties and levels of analyte. The typical steps of sample preparation include sampling/homogenization, extraction, clean-up and concentration, followed by final analysis. To detect trace organics, the final analysis method utilizes a powerful separation technique, typically chromatographic, combined with an appropriate detector.

The final analysis is largely determined by the sample preparation strategy in consideration of the instrumentation to be used, the degree of accuracy required and quantitative or qualitative requirements. Notably, sample preparation requires a small fraction of samples that are representative of those being analyzed. The portion or aliquot taken for analysis must represent the original bulk sample. For food sample analysis, smaller sample sizes are easily prepared, and relatively larger sample sizes are easy to obtain for the low detection limits during trace level analysis. However, larger sample size increases the need for more efficient clean-up techniques and possibly higher costs [10].

Sample preparation optimization can reduce the time required for analysis and potential sources of error and is critical for low concentrations analytes in ultra-trace level analyses [10]. According to a statistical survey, the level of uncertainty in a given analysis method is related to the number of procedures performed in the analysis. Therefore, uncertainty could be reduced by the decreasing the number of steps, such as the usage of automated techniques.

2.2. Principles of Extraction and Separation

The selective extraction of analyte targets utilizes the differences in their physical and chemical properties. These properties typically include charge, molecular weight, polarity, solubility (hydrophobicity) and differences in volatility. The extraction of polar mycotoxins, such as fumonisins (FUMs), often requires the presence of water and is aided by organic solvents [11]. However, hydrophobic mycotoxin extraction, such as that for AFTs, largely depend on organic solvents [11, 12]. Matrix effect is another vital factors to determine the choice of extraction solvent due to the differing chemical mixtures can affect it [13]. Different sample extraction techniques are suited to different analytes and matrix types. Some extraction methods, such as immunoaffinity and imprinted polymers, are based on the selectivity for specific structural groupings or mimic a biological selectivity. This section gives a brief review of the extraction techniques available for trace organic analysis with examples taken for their application to solid and liquid foodstuffs [14].

Ochratoxin A (OTA) is a colorless crystalline compound with weak acidic character and blue fluorescence under UV light. The pKA values of OTA is ca 7.5 (phenolic group) and 4.4 (carboxyl group of phenylalanine). The detailed physical and chemical characters of ochratoxin A were discussed in chapter 1. A good extraction and separation method is full consideration of the physical characters of ochratoxins.

2.3. Sample Preparation Techniques

Particle size is a key parameter for reproducible results in solid samples, such as meat, cereals, and fats. The matrix crush extent can influence extraction rates. The analyte target is first separated from the matrix and dissolved into a solvent or fluid. Extraction rates are closely related to solvent penetration of the sample (mass transfer), solubility and matrix effects [15]. Solid samples can generally be prepared by directly grinding or grinding after drying, then extracting the liquid or solvent. After extraction, the analytes are obtained in an organic or aqueous solution, which then requires additional clean-up or concentration. These extract solutions can be treated as liquid samples and handled directly by sorption methods or solvent–solvent extraction methods.

2.3.1. Matrix Solid Phase Dispersion (MSPD)

Matrix solid phase dispersion (MSPD) is an effective technique that was first reported in 1989 [16]. This method provides a porous structure that allows the solvent to penetrate the matrix and extract the analytes. The sample is usually mixed with a matrix, such as C18 bonded silica, followed by washing and elution with a small volume of eluting solvent.

Josep et al. [17] designed a MSPD-based method for examining five mycotoxins (aflatoxins B and G and ochratoxin A) in different cereals. This paper optimized dispersants, eluents and ratios for obtaining the best analysis results. The extraction rates were tested by liquid chromatography–triple quadrupole-tandem mass spectrometry (LC–QqQ-MS/MS). They found that the recovery of the extraction treatment ranged from 64% to 91%, with a relative standard deviation (RSD) lower than 19%.

This technique eliminated the need for multiple extractions, as the sample was mixed with the sorbent material and packed into a mini-column for analyte elution. A layer of selective sorbent can be packed into the bottom of the column as a further 'clean-up' procedure to allow for a more selective extraction.

Although MSPD is useful for trace contaminant analysis in food, it is not easily automated and could be time-consuming for a large number of samples. Some MSPD extracts also require a further clean-up step, particularly with fatty matrices.

2.3.2. Solid Sample Extraction Techniques

Solid samples often require pre-treatment, such as sieving, grinding and drying. Dispersion is used to avoid sample particle aggregation and to ensure good solvent penetration. Drying is particularly important when using non-polar solvents, as any moisture can reduce extraction efficiency. Desiccants such as sodium sulfate, cellulose or diatomaceous earth can help facilitate drying. Several methods have been established to extract samples that can be analyzed or examined as powders or after absorption onto a solid porous matrix.

(1) Super Critical Fluid Extraction (SFE)

The supercritical fluid (SF), such as CO2, can use as an extraction solvent to selective extract the required compound from the matrix. This works well due to the high solvating power, and density of the solvating liquid. It showed faster reaction kinetics than most liquids. Supercritical fluid chromatography on fused silica capillary columns has been applied previously for separating toxins [51].

This method reduces the usage of organic solvent and the carbondioxide is non-flammable, inexpensive and environmentally friendly. The CO_2 is easily removed from the sample matrix after extraction by reducing the pressure.

One of the main disadvantage with SFE is the less robustness of the method compared to other techniques and conditions must be consistent for reproducible extractions. The presence of fat and water in food samples can require extensive sample preparation, and more on-line clean-up procedures in SFE could allow for its further application in food analysis. Furthermore, the SFE technique is not suitable for routine analysis due to high costs and the need for specialized equipment.

(2) Microwave-Assisted Extraction (MAE)

MAE stirs and heats the sample during extraction. This strategy is particularly sutiable for achieving efficient extraction from solid samples. It is often applied to thermally stable compounds to increase the temperature during extraction. As non-polar solvents do not absorb microwave energy, polar solvents such as water must be used. Compared to dialysis, MAE can significantly reduce both the time and amount of solvent used.

Paula et al. [18] designed an analytical method using microwave-assisted extraction (MAE) to assit liquid chromatography with fluorescence detection (LC-FD) detect of ochratoxin A (OTA) in bread samples. With the optimized MAE conditions, the validation of the overall methodology was 0.1–3.00 ng/g. The quantification limit was 0.005 ng/g.

One problem of MAE is that it is not easily automated and it is not suitable for volatile analytes.

2.3.3. Extraction of Liquid Samples

Distillation is a basic form of 'separation' or extraction methods. The separation principle of target components of a mixture is based on boiling points and relative vapour pressures. Therefore, distillation is most popular in high boiling point mixtures or pure solutions. Most extraction methods for liquids are based on partitioning into an immiscible extracting phase.

(1) Liquid–Liquid Extraction (LLE)

Analytes in solution or liquid samples can be extracted by direct subregion using an immiscible solvent. Liquid–liquid extraction is based on the solubility difference of an analyte in two immiscible phases and is governed by the equilibrium distribution/partition coefficient. Analyte extraction is achieved through the differences in solubilizing power (polarity) of the two immiscible liquid phases.

Liquid–liquid extraction (LLE) for toxin separation is based on the different solubilities of the aqueous phase and immiscible organic phase. This technique extracts the compound into one solvent, leaving the rest of the matrix in the other solvent. A separating funnel is typically used, and the two immiscible phases are mixed by shaking followed by standing to separate. In some cases, salt is added to avoid emulsions, and centrifugation can be used if necessary. Either layer can be collected for further treatment. Hexane and cyclohexane are often used to remove non-polar contaminants, e.g., lipids and cholesterol. This method is effective for several mycotoxins and can be used in small-scale preparations [50].

The major disadvantage of bulk liquid-liquid extraction is the need for large volumes of organic solvents. Additionally, this method has limited selectivity. For trace level analysis, it requires future clean-up or analyte enrichment/concentration steps prior to instrumental analysis.

(2) Miniaturised Liquid-Liquid Extraction Techniques

Microextraction techniques can reduce solvent usage and was first reported in 1996. Recently, a novel single-drop microextraction (SDME) technique based on a single droplet of solvent suspended at the tip of a needle and exposed to the sample solution have been reported [19]. To date, most applications have applied to environmental or water samples, and the detection limit is comparable to traditional liquid–liquid extraction.

Although the method has potential for liquid samples, its use in food applications is limited. Recently, a research group reported the use of SDME for the analysis of organophosphorus pesticides in orange juice.

2.3.4. Solid Phase Extraction Techniques (SPE)

Solid phase extraction (SPE) is a liquid-solid partition method using for removing and concentrating trace organic analytes from solutions or liquid samples. The extracting phase of SPE is a solid sorbent. The use of sorbents is different based on the difference in mechanisms for analyte retention/extraction.

A wide range of sorbents have been used, including C18 and C8-bonded phases on silica, Florisil (activated magnesium silicate), polymeric resins (divinyl benzene copolymer or polystyrene), polar sorbents such as alumina, cyano and amino-bonded. Ionic functional groups such as amino groups can also be bonded to polymeric sorbent or silica to form ion-exchange sorbents. Mixed-mode sorbents include different phases, which enable interactions based on H-bonding, adsorption, polar/non-polar interactions, cation/anion exchange or size exclusion.

Jornet et al. [20] established a reversed-phase HPLC method to analyze Ochratoxin A at low concentrations (21 mg L^{-1}) in wines. This method involves solid-phase extraction (SPE) of samples using octadecylsilane cartridges prior to chromatography. The accuracy and precision of this method was very high, and its relative standard deviations were lower than10%. The recovery rate of Ochratoxin A in artificially contaminated wines over 0.1–3.0 mg $L^{-1}$was higher than 80%.

SPE requires uniform packing to avoid poor efficiency. Although reliable pre-packed commercial cartridges are now available, automated systems with reproducibility for some sample types still require optimization. The matrix effect influences sorbent extractability due to the competition for retention. Many traditional sorbents have shown poor selectivity and insufficient retention of very polar compounds. Hydrophilic materials can enhance the extraction rates of the most polar compounds by SPE.

(1) Selective Sorbents in SPE

1) Immunosorbents

A highly selective immunosorbent can be developed by employing the antigen-antibody molecular recognition using antibodies [21]. The principle of immuno-based sample preparation is molecular recognition and the high selectivity and affinity of the antigen–antibody interaction. There are various immuno-based sample preparation techniques, such as immunoaffinity extraction sorbents, ultrafiltration or membrane-based techniques. It is particularly suited to complex biological and environmental samples. Immunosorbents can be designed for a single target based on specific antibody [22] or multiplex target based on antibodies which bind to the analogue to the antigen (termed as crossreactivity). However, this technique is needed the initially production of the antibody, which increases the analyses cost. Matrix effect influences the target-antibody interaction, leading to poor extraction recoveries. A review written by Hennion and Pichon [23], describes antibody-based extraction immunosorbents and the artificial antibodies-based extraction sorbents in details.

2) Molecularly Imprinted Polymers (MIPs)

To avoid depending on antibody production, recent reports have used artificial antibodies to pretreat samples. This technology is referred to as MIP, as these molecules have been synthesized to mimic the specificity of immunological products. Molecularly imprinted polymers (MIPs) with selective binding properties for ochratoxin A (OTA) in Polar/protic media have been developed, and methods for analyzing mycotoxins are now commercially available [24].

Factors that are crucial to the success of retention targets on the sorbent include shape recognition in the imprints or cavities and other physicochemical properties, such as hydrophobic interactions, ion-pairing, steric interactions and hydrogen-bonding [22, 25]. To maximize template-functional monomer, it is important to have rationally designed OTA mimics as templates and a set of novel basic and neutral functional monomers. Selectivity MIPs can often be used for a number of matrices, even though the interactions may be different.

MIP-SPE sorbents can be used in large sample volumes because of their selectivity. They can be heated and kept stable in strong acids, organic solvents and bases. However, MIP needs to be fully custom for each analyte, and stringent cleaning of the sorbent must be performed. It is necessary to remove the target used as a template before trace analysis. To avoid this 'carry-over,' structural analogues of the analyte can be used to produce the sorbent cavity [26, 27]. Compounds labeled with stable isotope have also been used as MIP templates [28].

Selective interactions generally occur in organic solvent, but do not work as well in water or for samples containing a large amount of water. However, this problem can be overcome by initially retaining the sample via non-selective interactions then washing with organic solvent to trap the analyte through selective binding or by transferring the sample and analytes into an organic solvent prior to MIP-SPE [29].

MIPs have been used as selective sorbents for a wide range of targets and matrices [26, 28]. Justus et al. designed a selective molecularly imprinted polymers for solid-phase extraction of OTA [30]. The use of selective sorbents is a vital factor in SPE for the selective and sensitive analysis of trace targets. However, preparing and producing such sorbents is time-consuming for new and emerging targets.

2.3.5. Immuno-Based Sample Preparation

(1) Characteristics of the Immuno-Basedextraction Sorbents

Immunosorbents (IS) are usually prepared by linking antibodies to a solid phase carrier. The solid phase carrier is then packed in an SPE precolumn orcartridge. Use of an immunosorbent in SPE is very similar to using a conventional C_{18} cartridge. However, IS is generally stored in a wet media, such as PBS (phosphate-buffered saline) buffer. Moreover, there are several parameters that have to be carefully controlled and optimized when packing an IS to develop a robust SPE approach. These parameters include the affinity between antibodies (Abs), and a close relationship between extraction recovery and the amount of immobilized antibodies. In addition, sample matrices and additives in the samples can block target–antibody interaction to some extent, resulting in decreased extraction recovery.

(2) Antibodies

Antibodies are vital factors that determine the immunosorbent potential. Both polyclonal (PAbs) and monoclonal (MAbs) antibodies can be used for immobilization. In general, MAb production is more expensive than PAb production, but it guarantees the long-term availability of reproducible antibodies without requiring animals for batch-to-batch production. As a result, the use of MAbs is preferred over PAbs in IS. Both MAbs and PAbs are commercially available.

(3) Effect of the Solid Support and Antibody Immobilization Procedures

The solid support is another critical parameter in IS. Variables include basic properties such as biochemical and chemical inertness, good mechanical stability and uniformity of particle size. Moreover, the solid support used for antibody immobilization should be easily activated to allow antibody adsorption. It should also have a large pore size, as antibodies are large molecules. To avoid any non-specific interactions, the supports should be hydrophilic.

Finally, the bio specific activity of antibodies must be retained during the immobilization procedure.

Covalent binding to the antibodies is achieved by reacting free amino groups of the antibody with reactive aldehyde or epoxide groups on the support surface. Alternatively, carbonyldiimidazole, cyanogen bromide or N-hydrosuccinimide can also be used to activate the groups. Traditional supports, such as agarose, cellulose, silica and synthetic polymers, are commonly used in immunoaffinity chromatography. Silica- and agarose-based supports are commercially available and are the most popular sorbents. Agarose-based ISs are generally not pressure-resistant and cannot be operated at high flow-rates. As a result, there has been an increase in silica-based ISs for on-line direct coupling to chromatographic separation techniques.

Antibody immobilization through oriented bonding deserves attention. Because lysine-amino groups exist throughout antibody molecules, covalent coupling of antibodies to activated silica-based and agarose-based supports through these groups is considered random due to their different orientations. Alternatively, streptavidin and protein G- or A- based sorbents can interact with these proteins through non-covalent binding.

As protein G- and the related protein A-based supports bind antibodies at their Fc sites, the antibody binding sites are oriented away from the surface and towards the solution. These types of IS were successfully used in small compound separation, such as Ochratoxin A. Reducing the pH of the surrounding solution can easily disrupt binding, which is quite strong under physiological conditions [29]. The reusability and durability of IS are important elements for application in an automatic analytical system. Home-made ISs have retained consistent target binding capabilities even after being used with hundreds of samples. They can be kept for at least 1 year by storing them at 4°C in PBS containing 0.02% sodium azide without any special treatment.

(4) Capacity and Bonding Density

Bonding density is an important parameter because it influences the future antigen-binding capacity. It is determined by the number of antibodies linked to the sorbent surface and is usually measured in mg/ml or mg/g. The measurement of the amount of antibody immobilized onto a support is usually performed by detecting the antibody concentration in the binding solution by UV spectrophotometry. The specific surface area of the solid support accessible to antibodies is largely influenced by bonding density. Another important parameter is antibody purity. Use of non-purified antisera for binding reduced the bonding density due to the co-immobilization of other proteins.

(5) Elution Conditions

Effective elution solutions ideally break the analyte–antibody interactions but do not adversely affect the activity of immobilized Abs. The most commonly used methods include chaotropic agents or water-miscible organic solvents, or a combination of both to reduce the pH to 2 or 3. The choice of elution solution first depends on the affinity between the antibodies and targets and second depends on the nature of the target because of the ratio between hydrophobic and electronic interactions involved in the antigen–antibody interactions [31].

2.4. Separation Methods

Chromatographic methods are commonly available for mycotoxin analysis. As discussed in the above section, sample pre-treatment plays a key point in the analysis and this is normally done by LLE or by SPE methods. In this section, we discuss several chromatographic separation methods, including their advantages and challenges in mycotoxin analysis and detection.

2.4.1. Thin Layer Chromatography (TLC)

The most common method traditionally used for mycotoxin analysis is TLC, which can economically screen large numbers of samples. There are several previous reviews about the application of TLC in mycotoxin analysis [32-36]. TLC analysis of mycotoxins is still popular for both semi-quantitative and quantitative purposes. This is because it is high-throughput, has a low detection cost and ease of target compound identification through UV–vis spectral analysis. As with HPLC, several methods have been established to obtain the best results with each separate mycotoxin class, and both one-dimensional and two-dimensional analyses are frequently used [36]. TLC inherently requires sample preparation. The combination of an efficient and robust immuno-affinity column clean-up with the low-cost TLC procedure is a promising analytical approach for detecting mycotoxins in food by improving the sensitivity, separation, resolution and visualization of mycotoxins on TLC plates. Silica gel columns have also been used to purifymycotoxins [35]. Modified or unmodified gel layers have also been applied in TLC. Octadecyl bonded layers seem to be the most widely used [37-41], however, others in vogue are phenyl non-polar bonded [33], silanized [42], impregnated with organic acid, and polyamide based.

TLC has been routinely used as a cost efficient method of mycotoxin analysis with a fair amount of accuracy but poor repeatability. Upon the emergence of HPLC, the use of TLC in mycotoxin analysis has decreased, as HPLC has superseded this technique in accuracy. TLC also requires large quantities of solvent, intensive laboratory procedures and poor automation. However, it is possible to rapidly screen a variety of samples.

While TLC continues to be used, efficient sample preparation could further increase the spectrum of mycotoxins that can be detected using this method. Further development of more sensitive analytical systems and enhanced automation in terms of improving its performance could also make TLC a more popular method in the future.

2.4.2. High Performance Liquid Chromatography (HPLC)

One instrument heavily used in mycotoxin analysis is HPLC. It employs various adsorbents based on the chemical and physical structure of the specific mycotoxin. The most important characteristic of normal and reversed-phase columns for separation and toxin purification is their polarity. Generally, small mini columns have been used for sample pretreatment, and large scale preparative columns have been used to prepare mycotoxin standards [29, 43, 44].

The analysis of different mycotoxins using HPLC is similar. UV or fluorescence detectors that rely on the presence of a chromophore in the target molecules are the most popular detection methods. Several toxins (e.g., OTA, AFTs, CIT) can be directly detected by HPLC-FD due to their natural fluorescence [45]. However, many mycotoxins lack a suitable chromophore, thus their determination requires derivatization. Examples of derivatizing agents are 9-(fluorenylmethyl) chloroformate and o-phthaldialdehyde. Derivatization can be employed either pre- or post-column. HPLC is the industry standard for mycotoxin detection, and numerous protocols exist for OTA analysis [45].

In contaminated dried figs, OTA was extracted by orthophosphoric acid and methanol following clean up by an IAC and then studied using HPLC–FD [46]. The LOD for OTA was 0.12g kg^{-1}.

2.4.3. Gas Chromatography (GC)

GC is generally used to identify and quantify the presence of mycotoxins in food samples. In order to detect the volatile products, usually the GC is linked to flame ionisation detector (FID), Fourier transform infrared spectroscopy (FTIR) or MS detection techniques [47-49]. Most of the mycotoxins have to be derivatised for GC analysis in consider of their poor volatility [50]. Chemical derivatisation of mycotoxins such as polyfluoroacylationor silylation have been estabblised for obtaining a volatile material [50].

Trichothecenes have been examined in chemotaxonomic studies of Stachybotrys, Fusarium and Trichoderma derivatization with pentafluoropropionic anhydride [56]. GC–MS detection allowed simultaneous monitoring of up to four compounds during a 23-min GC run. The volatile fungal metabolites were measuredin grain as fungal contamination indicators [51]. The samples were evaluated for OTA, DON and ergosterol using GC–MS. The results showed that it is possible to measure OTA level below the maximum limit of 5 mg kg^{-1} developed by the Swedish National Food Administration for grain. It has also been reported that gas chromatography–mass spectrometry can simultaneous determination of trichothecene mycotoxins and zearalenone in cereals at ppb levels [52]. George et al. designed a assay of Ochratoxin A in wine and beer by Gas Chromatography Mass Selective Detection, and the total run time is 28 min and the limits of detection (LOD) and quantitation (LOQ) are 0.05 and 0.10 ug/L, respectively.

Several examples of OTA detection utilizing derivatisation have been published [52]. This approach gave a LOD of 0.1gkg^{-1}. However, GC detection is not expected for commercial protocols due to HPLC are cheaper and faster alternatives. However, there are some challenges or using GC in analysis of mycotoxins.

2.4.4. Capillary Electrophoresis (CE)

CE is an excellent tool for the separation of biological molecules, especially macromolecules. There are two significant advantages for its high performance. First, CE exhibits a nearly flat plug flow profile, and the antibodies have small diffusion constants. Both of these characteristics eliminate band broadening. This is because the efficiency of CE separation is inversely proportional to the analyte diffusion constant. Thus, CE is highly efficient in separating biopolymers with small diffusion constants. The specific detection of closely related mycotoxins requires a sophisticated separation technique and high sensitivity. CE has been used in mycotoxin separation in recent years. Effective separation can be based on mass- and charge-dependent migration in an electrical field. Capillary electrophoresis can be applied for fast separation in aqueous buffer solutions, eliminating the use of organic solvents. The combination of CE with sensitive fluorescence-based detection methods have been published for AFT [53], and fuminosin [54]. Other separation modes such as capillary isoelectrofocusing (CLEF) , capillaryisotechophoresis (CITP), micellar electrokinetic capillary chromatography (MECC), capillary gel electrophoresis (CGE) and capillary zone electrophoresis (CZE), have also been developed. However, they are barely used for the separation of mycotoxins.

3. THE DETECTION METHOD FOR OCHRATOXIN A

3.1. Conventional Ochratoxin Analysis

Because of the rigorous regulations, high toxicity and established OTA maximum levels, its sensitive and precise detection is increasingly required down to the sub ppb range in a broad range of food, feed and biological matrices. There have been several recent summaries and evaluations of the relevant analytical techniques and protocols.

OTA has strong native fluorescence activity, thus HPLC-FL is preferred as the routine analysis method for monitoring OTA and their metabolites due to it shows advantage in selectivity and sensitivity compared to other detectors. Generally, positive findings are confirmed by HPLC experiment [55]. In consideration of thin-layer chromatography (TLC) with FL detection is robustness and easy and cost-effective handling; it is still routinely used in countries outside Europe and North America. However, this technique is showed low sensitivity towards OTA (ppb range) and easy disturbed by the sample matrix [56]. Due to GC-based methods commonly suffer from a time-consuming and error prone derivatisation protocol (methylation), it is less attractive for mycotoxins analysis. Nevertheless, GC/MS can be used to confirm ambiguously positive findings [55]. In general, the above mentioned method is needed a careful sample clean up and analyte enrichment by liquid/liquid partitioning or increasingly by SPE with immunosorbents or silica [55,57]. Low recovery rates are also a problem due to OTA binds very strongly to proteins and DNA affecting decreasing the extraction/purification efficiency in urine, plasma, tissue and milk samples [55].

3.1.1. LC/MS for OTA

The early LC/MS for OTA detecting has rarely applying direct liquid injection and thermospray ionization[58] They are major focused on the mass spectrometric properties of OTA, such as ionisation efficiency and in-source fragmentation. The recovery rate of spiked cereal samples demonstrated detection limits of LC/MS are three orders of magnitude less sensitive than HPLC-FL. Despite its low sensitivity and robustness, it was already recognized that LC/MS might be extremely useful for structural elucidation and confirmation of OTA and distinguish OTA from its metabolites in animals and plants [55].

Although a lot of routine analysis and qualitative/quantitative LC/MS/(MS) method development, the analysis of ochratoxins is still exclusively focused on OTA since it exhibits the highest toxicity and most frequent occurrence of all known ochratoxins in biological and food environments. Due to its high toxicity, OTA analysis has to be performed down to the ppt range in agricultural commodities and feed, food and beverages thereof, such as cereals, coffee, wine, beer, milk and cheese to estimate the risk of OTA intake [57]. In addition, plasma and urine samples have been repeatedly analysed to monitor actual OTA contamination, distribution and excretion kinetics in humans and animals [59-64]. In this context, LC/MS is predominantly applied to confirm OTA positive results obtained by ELISA or HPLC-FL [65]. LC/MS is also an excellent tool to elucidate the structure of *in vivo* metabolites and OTA adducts in biological fluids and tissues, leading to a better understanding about the mode of action in OTA-induced mutagenesis and carcinogenesis in animals and humans [66-70].

Other ochratoxins, like OTB and OTC as well as products of the metabolic pathways of OTA, as ochratoxin, 4-hydroxyOTA, 10-hydroxy OTA and glutathione, 3'monophosphate desoxyguanidine and carbohydrate conjugates of OTA have been only a matter of investigation in the course of a couple of metabolism and biological mode of action studies when LC/MS was used for structural confirmation or elucidation.

3.1.2. Typical LC Conditions for LC/MS Analysis

Chromatographic separation prior to MS detection is exclusively achieved on REVERSED-PHASE (RP) stationary phases (RP-18, RP-8) with methanol/water, acetonitrile/water and methanol/acetonitrile/water mixtures as mobile phases, both in the isocratic and gradual mode. RP chromatography of acidic ochratoxins is highly pH dependent and acids, like formic acid, acetic acid and trifluoroacetic acid (TFA) are used as additives to achieve sufficient chromatographic retention. Typically, LC/MS run times are less than10 min on conventional RP columns since the analysis is focused on OTA in most applications [71] To obtain high MS selectivity, chromatographic separation is frequently used in conjunction with a suitable sample preparation procedure. Therefore, it is critical to remove matrix components from the analyte to reinforce MS sensitivity, validity, and repeatability. Optimal LC/MS separation characteristics can also be achieved in metabolism studies using a mixture of OTA and its oligonucleotide conjugates orits glycosylated and hydroxylated metabolites. Representative separation run times rangefrom20 minto35 min.

(1) Mass Spectrometry

In general and independent of interface geometries and ionization polarities, OTA demonstrates abundant fragmentation behavior in APCI interfaces, leading to significantly decreased sensitivity compared to ESI. Therefore, ESI has been almost exclusively used in qualitative and quantitative OTA analysis, and it is more robust for routine analyses than APCI. OTA produces enough adduct ions in the electrospray source when sodium or potassium ions are present in the mobile phase [72, 73], whereas one group detected acetonitrile adducts of OTA in the negative ion mode [74].

(2) Method Validation and Matrix Effects

Quantitative LC/MS analysis provides remarkable sensitivity down to the ppt range in far-ranging biological environments and different foods. Although this is less sensitive than LC-FL by a factor of 5–10, LC/MS can be applied to confirm OTA-positive samples that have been analyzed by ELISA or HPLC-FL [73, 75] or it can be used as an alternative detection technique to LC-FL [76]. Method recovery rates ranging predominantly from 70 to 121% were highly dependent on the sample matrix. Roasted coffee has been shown to offer such a complicated matrix that adequate LC/MS sensitivity and recovery values can only be achieved when the extracts are subjected to SPE or to IAC [77] with a polymeric sorbent material that offers RP along with anion exchange properties. Similarly, an adequate ionization efficiency, and thus, a dramatic decrease insensitivity have been reported for the analysis and detection of OTA in dark beer brands [78].

Representative linear ranges have been obtained with one to three orders of magnitude and with good connection factors, although these values were not always dependent on the matrix. OTB and Zearalenone (ZAN) were applied as internal standards to improve the reliability and accuracy of LC/MS analysis. The two compounds did not resemble OTA by either MS or LC, and one particular study in wine showed that due to its constructional imparity, ZAN could only partially make up for matrix effects, as it was impossible to gain one representative calibration curve for distinct brands of wine. It is interesting that the performance of this internal calibration protocol was strikingly enhanced (dramatic decreasein the dispersion of calibration curves) after IAC was removed from the main part of the sample matrix. In contrast, the use of OTB is not critical in tissue samples and plasma, although OTB elutes earlier than OTA from the LC column.

Several authors have carefully evaluated the validation data and provided evidence that the LC/MS methodology in OTA analysis offers great precision and accuracy. Accuracy was determined by spiking samples with defined amounts of OTA. Elbaek and Sorensen suggested that pH conditions can reduce matrix effects and improve accuracy during SPE clean-up of milk samples [76].

3.2. Immunochemical Methods for Rapid Mycotoxin Detection

As immunoassays have a unique ability to routinely handle a large number of samples and do not require time consuming procedures and sophisticated equipment, many immunoassays for OTA have been developed. Absolute leaders in the second group of analytical methods for mycotoxins are immunochemical methods with high sensitivity and selectivity provided by specific antibodies. Application of immunoassays for individual mycotoxin rapid screening is now a common analytical practice. Main trends for research in this field are sensitivity improvement, matrix effect reduction, simplification and shorter time of analysis. Immunomethods for rapid detection of individual mycotoxins are summarized in recent reviews concerning immunochemical methods [44] and analytical methods in general [79]. Moreover, reviews about single mycotoxins or groups of related toxins, such as ochratoxin A (OTA), have been published [75].

An increasing trend is the development of multi-mycotoxin screening methods in similar with chromatographic methods. In this review, we summarize the methods for rapid determination of mycotoxin and focus on multiple analyte detection by immunochemical techniques compared to existing multi-methods for other groups of low molecular analytes. We describe state-of-the-art approaches and perspectives for multiple mycotoxin detection and present both instrumental and non-instrumental immunochemical methods.

3.2.1. OTA Conjugation with BSA or KLH to Prepare Artificial Antigen

Antigens (any target or foreign entity) used for immunization may be highly purified preparations, molecules containing a mixture, a single molecular species, or an extremely complex "antigen" (e.g., microorganism or cell). Because mycotoxins are low molecular mass compounds, they do not evoke an immune response and have to be modified by binding to a larger carrier molecule before immunization, such as bovine serum albumin (BSA). It is necessary to introduce a functional group into the selected molecule to allow for coupling. When trying to obtain antibodies with the required specificity, design of the hapten is an important variable, especially when antibodies are targeted to recognize an entire class of structurally related compounds [80].

Conjugation of OTA with KLH or BSA has been accomplished through several methods: a carbodiimide method and an active ester method. BSA,OVA,KLH and g-globulin have been used as the carrier protein. One study has shown that the ratio of hapten conjugation to BSA was higher than to OVA using the same synthesis method [81]. The ratio of OTA to KLH incomplete antigen was evaluated as 1250:1 [82], but one study showed that the sensitivity of antibody with g-globulin as the carrier protein was higher than KLH as the carrier protein [83]. However, KLH and BSA have been used as carrier proteins for OTA conjugation for the coating antigen and immunogen, respectively. The use of different conjugate proteins and synthesis methods for the coating antigen (OTA-KLH) and the immunization antigen (OTA-BSA) was beneficial for screening for specific antibodies [84]. This approach could be used to reduce the cross reactivity (CR) of the antibody to conjugate protein or to OTA analogs, which partially came from incomplete by-product antigen synthesis [85], such as the CR of anti-OTA antibody to OTA and BSA.

3.2.2. Antibody Preparation

Non-covalent binding-based biomolecular recognition can form interactions between all species. During the biorecognition process, the complementary shape plays a pivotal role, in addition to ionic bonds, hydrogen bonds, and hydrophobic interactions. Paul Ehrlich invented the term "antibodies," and he was the first to identify the function of a particular class of proteins over a century ago. Knowledge of antibody structure and function is required to understand the fundamental principles of antibody-based detection systems. In this section, we describe antibody structure and types and discuss the antibody preparation process.

In the immune system defense mechanism, antibodies are highly soluble serum glycoproteins. All antibodies are built from the pairing of heavy and light polypeptide chains, and all of these protein molecules use the generic term "immunoglobulin" (Ig). There are three types of antibodies. Monoclonal antibodies (mAbs) from each B-cell clone have exactly the same affinity and specificity for the antigen used for immunization. However, many different B cells respond and secrete antibodies upon immune system challenge with antigen. In a polyclonal response, the B cells all have the potential to recognize the antigen, but in slightly different ways, with different affinities, specificities and cross-reactivities. Polyclonal antibodies are derived from the challenged immune system, and these antibodies have batch-to-batch variation and a tendency to cross react with conformationally similar compounds, limiting their use in defined hapten-specific assays. However, pAbs are easier to produce and often have a high sensitivity for capturing their targets.

In 1975, Milstein and Kohler [86] first described the discovery of hybridoma technology, which revolutionized the production of true antigen-specific antibodies. Primed B cells immunized with the desired antigen were obtained from experimental animals (e.g., mice or rats), and myeloma cells were fused to form hybridoma cells, which produce the desired specific antibodies. Thus, mAbs provide potentially limitless amounts of identical antibody with single-epitope specificity. Although rat- or mouse-derived mAb production is now a routine process, it can also be an expensive and time-consuming process.

Recombinant antibody phage-display has been developed during the past two decades and has changedthe way we generate antibodies for specific analyte detection. Recombinant antibodies can be generated by genetic engineering or by a combinatorial approach from the spleens of immunized animals and by assembling scFv, hybridoma cells or Fab fragments. It is achieved by PCR amplification of the rearranged antibody genes from B-lymphocytes to construct an antibody-gene library. The combinatorial libraries can be generated from non-immunized animals (native libraries) or from the B-cells of immunized animals (immune libraries). Because the libraries contain a combination of random light-chains and heavy-chains, the combination of antibody libraries may contain antibody fragments that are not found in nature, so the libraries can be used to separate antibodies specific for the immune antigen (e.g., highly toxic substances or recognition of its own antigen). The following section will focus on the latest developments in immunochemical detection of mycotoxins, with a particular emphasis on multiple analyte determination. These approaches include the rapid visual test which can carry out insitu (lateral-flow dipstick, circulation and column test), and laboratory high-throughput analysis instruments (microplate enzyme-linked immunosorbent assay (ELISA), several capillary electrophoretic immunoassays, fluorescence polarization immunoassay, and immunosensors). For each type of immunoassay, we discuss their perspective in multiple analyte application and cite examples [87].

3.2.3. Instrumental Methods

Immunochemical methods using relatively simple equipment (microplate reader, luminometer, capillary electrophoresis system, etc.) are often used to acquire quantitative or semi-quantitative results. In contrast to chromatographic methods, immunoassays are usually combined with simple extraction, pre-concentration and clean-up procedures, or often do not require pre-concentration or clean-up. Binding of specific antibodies to a solid support (direct competitive enzyme-linked immunosorbent assay (ELISA)) or coated antigens (indirect competitive ELISA) are the most frequently used approaches for mycotoxin immunoassays. These formats can be used in non-homogenic assays, such as microtiter-plate immunoassays and sensors. Capillary electrophoretic immunoassays and fluorescent polarization immunoassays are examples of homogenic assays.

(1) Microtiter Plate Enzyme-Linked Immunosorbent Assays

The ELISA microplate format is often used for rapid mycotoxin detection.ELISA tests are commercially available for the detection of individual mycotoxins, aflatoxin M1 (AFM1) (aflatoxin B1 (AFB1), deoxynivalenol (DON), ZEA, OTA, T-2 toxin, citrinin) in some foods and for groups of mycotoxins. Therefore, they are useful tools in screening and quantification, and they provide the benefits of sensitivity and speed. Depending on antibody cross-reactivity, standard ELISA procedures can be adapted for individual or groups of related mycotoxins. ELISAs are still being developed for uncommon mycotoxins in special matrices. ELISA assays have been developed for other macrocyclic trichothecenes and satratoxin G, which are associated with indoor air [88]. Compared to other methods, the multiple label approach for simultaneous analyte detection is still not widespread because the optimum conditions for enzymatic reactions are different [89].

There are currently two primary methods: the use of antibodies and related compounds to determine the appropriate antibody cross-reaction, and the use of an enzyme-labeled plate in a separate line of compounds. The use of an antibody with high cross-reactivity against mycotoxins on behalf of all target groups can be used to determine the total concentration of related compounds (total aflatoxin, total fumonisin).The use of a monoclonal antibody specific for detecting acetylated DON was developed in both competitive direct and indirect ELISA formats, but it also demonstrated cross-reactivity and sensitivity for nivalenol (NIV), DON and some DON derivatives [90].We were unable to find examples of the identification of multiple non-cross-reacting mycotoxins by ELISA using the second multi-assay approach.

(2) Immunosensors

Immunosensors are devices rooted in the detection of analyte–antibody interactions. Changes in physicochemical parameters are generated and are electronically sensed when specific biological molecules interact [91,92]. There are two groups of sensors: luminescent/colorimetric/electrochemical sensors and surface plasmon resonance sensors. These sensors transform this interaction to an analytical signal suitable for analyte concentration measurements. Most immunosensors are based on different ELISA formats and use automatic flow devices. The term "biosensor" often refers to the real-time read-out or reversibility. The analytical cycle consists of four stages: competition (notin the non-competitive format), capture, display (enzyme labeling), and sensor regeneration, all of which automatically operate [93]. Other rapid immunoassay sensors have been combined with a simple methanol (or acetonitrile)-water-extraction toxin from food samples.

(3) Colorimetric and Luminescent Sensors

Colorimetric and luminescent sensors can onvertvisible or ultraviolet light transformation into analytical signals. These sensors require different support, including wave guides, microbeads or optical fibers. Color development is measured with a spectrometer by reading the absorbance across the beads at 620 nm, as an enzymatic label alkaline phosphatase was used. Compared to the sensitivity of a microtiter plate ELISA, this method detected AFB1 at a level as low as 0.2 ng ml^{-1} in artificially contaminated food materials.

Fluorescent sensors are more commonly used to detect mycotoxins. They have been implemented in three different formats to detect fungal mycotoxin: a direct competition format, an indirect competitive format, and a non-competitive format. Fluorescent labeling shortened decreased the analysis time by eliminating the need to add the substrate and allow for color development compared to colorimetric sensors.

A competitive direct ELISA-based sensor has been developed to measure AFB1 in corn. Monoclonal antibodies were covalently bound to an etched 800 mm core optical fiber through a heterobifunctional silane. The analysis of the fluorescence intensity of fluoresce in isothiocyanate (FITC)-labeled fumonisin derivatives has been analyzed. The analytical sensor signal had a range of 10~1000 ng mL^{-1} of AFB1, the limit of detection was 10 ng mL^{-1}, and the midpoint of dynamic range was 70 ng mL^{-1} [94]. A completely automated flow-through immunosensor using fluorescence detection based on direct competitive ELISA was developed to detect ZEA. The standard immunosensor solutions exhibited a dynamic range from 0.019 to 0.422 ng mL^{-1}, and the detection limit was 0.007 ml^{-1} ng. After ZEA extraction using accelerated solvent extraction, wheat, corn, and swine feed samples were analyzed by this approach [95].

Using a sensor like horseradish peroxidase (HRP) could simultaneously detect mycotoxin in the form of a multi-component fluorescent label. This approach allows for the determination of the total analyte concentration with one fluorescent label.

Fluorescent biosensors were developed for AFB1 [96] and ZEA [97] using the indirect competitive ELISA principle to evaluate FITC fluorescence intensity. In reference food materials such as peanuts, nut puree, and pistachios, the sensor could detect AFB1 to a level of 4 ng g^{-1}. If separate zones of beads with conjugates for different analytes are used, this method can be applied to multiple mycotoxin assays.

An array biosensor system has been developed for simultaneous, multiplexed detection of AFB1, OTA, and DON, AFB1. A one-step indirect competitive assay in the flow cell was achieved by using NeutrAvidin-coated slides patterned with biotinylated mycotoxin analogues. A biotinylated secondary antibody lane was added to the planar waveguides as a positive control. A mixture of different combinations of mycotoxin standards and Cy5-labeled anti-mycotoxin antibodies were then used to expose the slides. The system could run both sandwich and competitive immunoassay formats using a single waveguide surface. Both small and large molecular weight food contaminants could be simultaneously detected [98]. An analogous array biosensor was developed using a single waveguide to detect multiple toxins (staphylococcal enterotoxin B, botulinum toxoids, trinitrotoluene, ricin, choleratoxin, and AFB1). Various immunoassay formats have been used for the different analytes. For AFB1 detection, biotinylated AFB1 was immobilized on the waveguide surface. The immobilized AFB1 competed with AFB1 in solution to bind to Cy5-labeled fluorescent antibodies. In the presence of specific fumonisins, the sample fluorescence decreased. The sandwich format was used for large analytes [99].

Non-competitive formats utilized the binding of target mycotoxins without competition of a labeled molecule. Native mycotoxin fluorescence is the main limitation. This approach had a lower LOD for AFB1 detection compared to the sensitivity of several fiber optic methods [100]. The use of FITC can increase the sensitivity of the method. The most sensitive format was the native fluorescence-based sensor; as many mycotoxins have their own fluorescence (aflatoxins, ZEA, OTA, citrinin), this approach can be used to simultaneously monitor these mycotoxins. Separate signals can be obtained by using specific antibody locations on the waveguide or spectral selection. However, one disadvantage of this approach is the sensitivity of the signal to the refractive index of the sample solution.

Chemiluminescence can provide higher sensitivity without the need for an excitation source or additional optical requirements, making it a good tool for multi-analyte detection. Using chemiluminescent labels (peroxidase/luminol),a parallel affinity sensor array (PASA) has been developed to detecttriazine herbicides with detection limits at 20 ng^{-1} [101].

(4) Surface Plasmon Resonance (SPR) Sensors

Surface Plasmon resonance (SPR) is an optical phenomenon under conditions of total internal reflection used to measure changes on the surface of thin metal films (usually Au or Ag) [101]. Nearly all present-day SPR sensors for mycotoxins are based on Biacore sensors (Biacore AB, Sweden, Uppsala), unlike fluorescence sensors that require different type of equipment. The immunoassay formats of the SPR sensor are similar to those of indirect competitive ELISA assays, wherein the antigen is attached to the metal surface, which is attached directly to an ELISA or other non-competitive format. Because the majority of mycotoxins have a low molecular mass, the change from the combination to the surface of the mass may be too small, resulting in a significant change in the refractive properties. Therefore, quantitative indirect methods of the mycotoxin with antigen-modified surfaces have primarily been used with the same immunoreagents (polyclonal antibody and AFB1-

BSA conjugate), and the sensitivity of SPR sensors and SPR sensors have been compared for AfB1. In contrast to ELISA, the SPR sensor (12–25000 ng ml^{-1}) (3.0-49 ng/ml) has a more sensitive but narrow linear range of detection [102]. Yu et al. investigated a simple ochratoxin A biosensor that was comprised of a polypyrrole film doped with chloride on a gold surface on a miniaturized surface plasmon resonance sensor. The SPR, thickness of each film and angle were monitored. In 2005, the same authors synthesized a molecularly imprinted polypyrrole film in place of a gold surface on the surface plasmon resonance sensor to detect ochratoxin A. The molecularly imprinted polypyrrole film was electrochemically polymerized on the sensor surface from a solution of ochratoxin A and pyrrole in ethanol/water (9:1 v/v). Film growth was monitored in situ by measuring the increasing SPR angle.

The authors investigated the binding properties of the molecularly imprinted polypyrrole film by loading ochratoxin A standard solutions into the integrated 20-μL flow cell [103]. The study primarily focused on the concerns of sensor development in toxin measurement, and the practicality of these sensors in real analyses is still questionable. In practice, toxins usually coexist in a multiplexing complex matrix, but the detection of multiple analytes is an important prerequisite for small molecular toxins. Ligler et al. developed a portable array biosensor that utilized planar wave guides and total internal reflection fluorescence excitation patterned with antibody capture to monitor a wide variety of analytes. Competitive immunoassays have been successfully developed to detect small molecular toxins in complex matrices such as food.

One of the critical characteristics of SPR sensors is the regeneration process. If the antibody is high affinity, the bound antibody cannot be completely removed from the sensor surface immobilized with hapten-protein conjugates by the regeneration solution. Rather than using antibodies to examine AFB1, the application of single chain antibody fragments allowed for less stringent regeneration conditions and the simplification of the regeneration procedure [104]. A similar method allowed for the detection of AFB1 at a level of 3 ng g^{-1} in grain sample extracts [105].

In contrast with common immunoassay and luminescent sensor methods, SPR does not require special labeling, which allows for non-specific interactions of the adsorbed molecules during sample analysis. This requires additional clean-up steps, making analysis more extensive and complicated. Moreover, SPR test sensitivity is relatively low compared to enzyme or fluorescence immunoassays, as SPR does not involve signal enhancement.

(5) Fluorescence Polarization Immunoassay

Fluorescence polarization immunoassay (FPIA) is a homogeneous method, in contrast to the methods described above. Its function is not attached to the solid surface of the immunoreagents. Analysis of this approach is often associated with fluorescence, the specific antibody-binding site competition in the solution of labeled analyte (tracer) and free analyte. The analytical signal, which corresponds to the rate of its rotation in solution, is the value of fluoresce in fluorescence polarization. Free tracer is a small moving molecule with a high rotational speed. The tracer has a lower rotational speed, and therefore, a higher fluorescence polarization value. Consequently, the amount of free analyte present in the sample is inversely proportional to the polarization value.

The test procedure is relatively simple, as it does not require any washing steps because FPIA does not require the separation of free and antibody-bound analyte. Furthermore, the equilibration time is very short. Polarization measurements may be performed using portable equipment. The incubation time can sometimes affect the assay sensitivity and the kinetics of interactions can be comparatively slow. The midpoint of the competition curves ranged from 0.03 mg mL^{-1} (15 s incubation) to 41 mg mL^{-1} (12-min incubation) via the monoclonal antibody to DON [106]. The kinetics improved with the use of another antibody, which was cloned with innovative DON-fluoresce in tracer, which achieved shorter response times. A technique for the rapid quantitative determination of DON was developed by Lippolis et al. [107] in durum wheat kernels, pasta and semolina using the same antibody.

A background signal has been observed in both naturally contaminated and spiked samples, strictly relying on the testing matrix. Quantitative determination of DON was possibly more accurate than 0.10 mg $g^{-1}$for all agricultural productsafter subtracting the background DON level. Comparative analyses of naturally contaminated samples performed by HPLC-FL and FPIA methods showed a good correlation ($r > 0.995$) [107].

In barley, the dynamic FPIA ranged from 5 to 200 ng mL^{-1} for OTA determination, with a midpoint at 39 ng mL^{-1} and an LOD of 3 ng mL^{-1} [108].

The FPIA readout signal is an estimate, and it resembles a mean characteristic of the solution (namely the fluorescence polarization value). Therefore, it cannot be used for the simultaneous inspection in a single tube for some individual targets. FPIA was only applied to detect a signal, concerning the existence of one or more compounds with no differentiation. In this aspect, FPIA was used in the urine screening of multiple drugs (e.g., cocaine metabolite, several barbiturates and amphetamine) [109]. However, FPIA can potentially and simultaneously detect parallel several analytes for use in a multi-immunoassay with separate tubes.

(6) Capillary Electrophoretic Immunoassay

Capillary electrophoresis (CE) has proven to be a powerful separation technique for macromolecule separation. CE-based immunoassays are called capillary electrophoretic immunoassays (CEIAs). When combined with sensitive detection methods such as laser-induced fluorescence (LIF), a capillary electrophoretic immunoassay (CEIA) offers several advantages over conventional immunoassays [110, 111]. CEIAs are performed without a supporting phase, e.g., insolution, such as FPIA. In addition, a CEIA uses a similar set of reagents as for FPIA, e.g., specific antisera and fluorescently labeled tracers. The assay procedure includes the incubation of the liquid sample with antibody solution and tracer or followed by the application of a tiny aliquot (nL to pL) of this mixture onto the capillary. Separation of the antibody-tracer complex and the unbound fluorescent tracer occurs as a result of electrophoretic forces. LIF detection allows for the acquisition of peak areas proportional to the fluorescein-labeled target concentrations. All tracers are bound to the antibody in the case of absent of free target, as the signal (peak area) of non-bound tracer is minimal, while the signal of the complex is maximal. If target is present in the sample, the bound tracer signal decreases, while the non-bound tracer peak area increases.

One advantage of CEIAs is the integration of the separation step. A CEIA has other advantages over conventional immunoassays, as it requires much less sample and reagents, has a simplified procedure and easy automation, allows for the simultaneous measurement of multiple analytes, permits the direct visualization of immunocomplex formation and dissociation, and can combine CE detection with UV, LIF, and mass spectrometry (MS).

3.2.4. Non-instrumental Methods

The development of rapid methods for mycotoxin analysis is a major trend in the last decade. Moreover, non-instrumental-based rapid screening techniques, which could be used in other areas, are becoming increasingly important. In general, the quantification of toxin concentration is not always required as the first step in an analytical survey, and a simple qualitative test is often sufficient.

The non-instrumental evaluation of results is based on visual estimation. Consequently, different visual labels are used, such as colloidal gold, enzymes for catalytic enzymatic reactions, encapsulating a visible dye, fluorescent labels and liposomes. The basic immunoassay formats used in non-instrumental tests to measure mycotoxins are indirect or direct competitive ELISAs. Non-instrumental tests usually provide qualitative results, as negative or positive, characterizing the absence (or presence) of the target at concentrations higher than the fixed cutoff level. The cut-off level can be determined based on either complete color suppression or a noticeable decrease in color development. Several tests use this type of semiquantitative estimation based on optical intensity. Special control zones have been included in the majority of current tests, to allow for easier data interpretation. All rapid tests with visual results estimation can be combined with a simple sample preparation procedure, such as extraction with methanol or methanol/water (or buffer) mixture, filtering and dilution in buffer.

AOAC International (http://www.aoac.org) and the European Mycotoxin Awareness Network EMAN (http://www.mycotoxins.org) have generated a list of commercially available test kits for mycotoxin determination.

(1) Lateral Flow Tests

Lateral flow tests, also termed immunochromatographic strip (ICS) tests, are visual, rapid and user-friendly test formats. The lateral flow strip usually consists of four parts: sample pad, conjugate pad, porous membrane and absorbent pad. After application onto the sample pad, the liquid components of the assay move along the membrane by capillary flow to the absorbent pad. Lateral flow devices can be constructed in several forms depending on which reagent is labelled (the antibody or the antigen) and the label type.

Colloidal gold is a common label for the ICS test, and the colloidal gold particles range from 20~40-nmin diameter. They provide binding zones for red antibodies. In the lateral flow strip, colloidal gold-labeled specific antibodies are pre-deposited onto the conjugate pad (Figure 1). The test line (or lines in the case of multi-assay) is commonly coated with mycotoxin-carrier protein (such as BSA/OVA) conjugate onto the porous membrane. A control line above the test line is coated with secondary antibody (e.g., anti-mouse/rabbit IgG) to confirm strip validity. In a typical assay procedure, a liquid sample or sample extract solution is first applied to the sample pad. The sample then dissolves the colloidal gold-labeled specific antibodies and carries the mixtures along the porous membrane. The specific antibodies bind to the mycotoxin in the sample such that it cannot bind to the mycotoxin–BSA conjugate on the test line. The entire antibody-mycotoxin complex migrates along the membrane and binds to the secondary antibodies on the control line. Therefore, the presence of analyte above the cutoff value results in a colorless test line and a red control line. However.in the absence of mycotoxin, the colloidal gold-labelled specific antibodies bind to the mycotoxin-carrier protein conjugates on the test line. Therefore, the absence of mycotoxin results in red color on both the test line and control line.

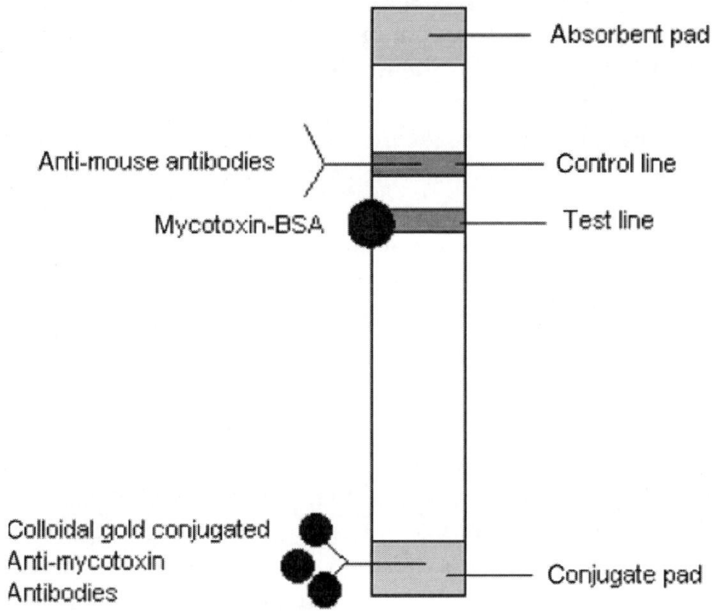

Figure 1. Principle of a lateral flow test.

Lateral flow devices for T-2 toxin, AFB1, DON, OTA and ZEA are commercially available. The usage of a portable hand-held quantification instrument enhances assay sensitivity. An ICS based on monoclonal antibodies has been established for OTA detection with a detection limit of 500 ng mL^{-1} [112]. The same format was used to detectAFB1 in pig feed with a detection limit of 5 ng g^{-1} AFB1 [113].

A colloidal carbon-antibody conjugate was used as a label for sporidesmin A determination. This results in black lines, and the sensitivity was less (25 ng ml^{-1}) than the colloidal gold conjugate (4 $ngml^{-1}$) [114]. Polystyrene microspheres are unsuitable for ICS because they form non-specific attachments to the ICS membrane.

A lateral flow ELISA, which uses an enzymatic marker, showed a ten times higher sensitivity than colloidal gold labels. However, it requires additional washing steps [115].

(2) Dipstick Tests

Dipstick, dot-ELISA or strip assays are based on a membrane with specific antibody spots and possesses several assay steps. According to the direct competitive ELISA test approach, the dipstick with a fixed antibody is immersed in different solutions, including the sample, chromogenic substrate and the analyte–enzyme conjugate.

A two-steps dipstick was established for15-acetylDON (15-acDON). The dipstick with a specific antibody zone was applied to a mixture of 15-acDON-HRP conjugate and sample solution, which was then transferred to the chromogenic substrate solution. A separate dipstick was used as negative control. The LOD in buffer was5 ng mL^{-1}, more than ten times higher than that obtained by microtiter plate ELISA (0.35 ng mL^{-1}). At this toxin level, the dot color reaction significantly decreased compared to the negative control. There was no color response at 20-25 ng mL^{-1} [116]. A similar dipstick was used to assay T-2 toxin in wheat. A visible LOD could be detected at 0.25 ng ml^{-1} in buffer solution under optimal conditions. The color response was completely inhibited at 3 ng mL^{-1} T-2 toxins. It was possible to make a visually clear distinction between wheat extract spiked with12 ng g^{-1} T-2 toxin and the negative control [117]. An anti-HRP antibody can be used as a negative control. A dipstick for AFB1 detection included two lines: a control line with anti-HRP antibodies, and the other contained anti-AFB1 antibodies. The AFB1-HRP bound anti-HRP antibodies in a non-competitive format. After substrate application, a color reaction occurredon two lines [118].

Dipsticks for multi-mycotoxin detection have been frequently compared to other non-instrumental immunoassays. Five specific antibodies (T-2toxin, 3-acDON,AFB1, roridin A and ZEA) were immobilized onto one dipstick membrane. To design the direct competitive dipstick assay, five different toxin-HRP conjugates were added to the mycotoxin solution, which was then incubated with the dipstick for 30 min. For each dipstick, corresponding negative control dipsticks should be prepared. The LOD for spiked wheat samples was 600 ng $g^{-1}$for 3-acDON,30 ng $g^{-1}$for AFB1,100 ng $g^{-1}$for T-2 toxin, 500 ng $g^{-1}$for roridin A, and 60 ng $g^{-1}$for ZEA [119]. The sensitivity of this multi-mycotoxin detection dipstick was less than single target. This may be because only one spot color on each dipstick was considered positive to simplify the results. Furthermore, immunoreagent concentrations were higher than those used in the single analyte test to obtain equal color intensities for all five negative spots.

(3) Flow-Through Tests

Enzyme-linked immunofiltration assay (ELIFA) or flow-through assay (IFA) utilizes a direct competitive ELISA format where specific antibodies are absorbed by a membrane. First. the sample is added to the upper surface of the membrane. Then, it flows through the membrane into an absorbent pad. At the same time, analyte was immobilized to the antibody spot on the membrane. The analyte–enzyme conjugate is added to bind the remaining dissociative antibodies. Finally, the chromogenic substrate is applied to achieve a colored product in the presence of retained enzyme. Colloidal gold can be used as a visual IFA label to avoid the use of a chromogenic substrate [113]. However, contrary to lateral flow tests, colloidal gold is not frequently used as a label for flow-through assays.

Liberty et al. [120] developed a membrane-based flow-through enzyme immunoassay to detect OTA in roasted coffee. This method used methanol/5% aqueous $NaHCO_3$ to extract the sample and as a cleanup solvent to achieve a high partition coefficient for OTA in the mobile phase. Then, developing the solid-phase (aminopropyl) cleanup to chromatographically elute OTA. The use of aminopropyl cleanup dramatically decreases the incidence of false positives for both flow-through enzyme immunoassay and HPLC methods. The flow-through results can be visually evaluated, and the sensitivity was 4 μg kg^{-1} in spiked roasted coffee.

IFA was used to measure OTA in wheat using the above mentioned format. The LOD for OTA was 4 ng g^{-1} [121]. A collaborative study of five laboratories showed that these flow-through kits effectively detected the presence of OTA [119]. However, the IFAs were 50 times less sensitive than ELISA with instrumental determination [122]. To improve the sensitivity of IFA, some modifications have been applied to flow-through tests. First, the absorbent body was not immobilized to the response membrane to avoid limitations related to the volumes of reagents, sample and washing buffer. Second, it was possible to wash with a wash bottle. Third, the target analyte can be pre-concentrated in the antibody spot by replacing the adsorbent body and using several sample portions. Fourth, AFB1 was detected using signal amplification with avidin-HRP conjugate and biotinylated tyramine for detection, which removes the volume restriction [123]. A prototype eight-well immunofiltration test device was developed for OTA detection. Membranes coated with different anti-toxin antibodies were immobilized to the well, and a cellulose layer support was placed under the membrane as a filter pad. The LOD of OTA-spiked wheat samples was 50 ng g^{-1} [124]. However, the sensitivity was lower than single analyte detection.

3.2.5. An Immunoassay for Ochratoxin a Without the Use of Mycotoxin

Standard immunoassays to detect OTA generally require the use of the mycotoxin itself, either conjugated or in free form. However, it may be toxic to both users and manufacturers, hindering its extensive application. To solve such problems, one rational alternative to using mycotoxins as immunochemical reagents is to develop peptide or protein substitutes to mimic mycotoxins. Phage display peptide libraries allow for the selection of new molecules capable of mimicking the functional and structural features of native chemicals or proteins. This technology can be applied to research new reagents, which are promising substitutes for the original targets. To screen peptides able to mimic ochratoxin A (OTA) in anti-OTA monoclonal antibody (McAb) interactions in establishing an OTA immunoassay, an anti-OTA McAb was used as the target for panning-elution selection from an M13 phage-displayed random seven-peptide library. After four rounds of panning, Liu's group identified 11 phage-displayed peptides able to stimulate OTA. An Enzyme-linked Immunosorbent Assay (ELISA) was established for detecting OTA using the phages. In the most sensitive assay, the range of linearity of the inhibition curve was 200–8000 pg/ml, and the corresponding LOD was 150 pg mL^{-1}. The inserted peptide sequences were deduced by DNA sequencing. The shared conservative amino acid residue sequence was IR(V)PMV(L)XX, where X is any amino acid residue, which was verified by two synthesized peptides. The results suggested that those phage peptides are suitable OTA substitutes in developing the immunoassay. The new method is comparable to the conventional indirect competitive ELISA, and it is specifically useful for OTA detection.

3.3. Electrochemical Affinity Biosensors to Detect Mycotoxins

Electrochemical biosensors to inspect mycotoxins are remarkable due to their simplicity, selectivity, sensitivity, low-cost, and in some circumstances, portability, integration and miniaturization in automated applications [125]. Most of them use high affinity interactions between disparate antigens and corresponding antibodies, but the use of new types of specific ligands (e.g., aptamers) is rising. To convert the target-ligand interaction into a readout signal, several electrochemical techniques (potentiometric, conductimetric, amperometric, impedimetric) have been applied, as discussed below [126-128].

3.3.1. Electrochemical Transduction

Amperometry (and voltammetry) is the most suitable electrochemical transducer because of its high sensitivity and its working response over a wide range of target concentrations. Amperometry is widely used and is based on the measurement of currents under variable (voltammetry) potentials or a fixed potential (potentiostatic control). A great amount of research has focused on finding electrode nanostructurations or configurations. Electrodes are generally made of carbon (graphite, glassy-carbon) or inert metals (Pt, Au), although the main disadvantage is the need for regeneration between measurements, so single-use disposable screen-printed electrodes (SPEs), characterized by mass-production and low-cost, are now being widely used to overcome this problem [129].

Electrochemical impedance spectroscopy (EIS) is an informative, sensitive and nondestructive technique that allows for the tracing of reactions at the interface of the sensor and for studying its electrical properties [127]. Label-free EIS aptasensors and immunosensors survey changes in electron-transfer resistance on the electrode surface due to the complexation of the mycotoxins in the range of 1–20 ng mL^{-1}, with an LOD of approximately 0.5 ng mL^{-1}. One of the main merits of EIS biosensors is that there is no conjugating of antibodies or mycotoxins with an enzyme to generate the electrochemical signal [129]. This suggests that with modification to the electrode surface, for example using an Au electrode surface, and the use of 4-carboxyphenyl (4-CP) film, a covalent linkage can form between the 4-carboxyphenyldiazonium salt and an antibody to build a sensitive impedance immunosensor to measure OTA with an LOD of 0.5 ng mL^{-1} [130].

Although they are used less often, partly due to their lower signal/noise ratio, conductimetric and potentiometric immunosensors have also been developed for aflatoxin [131]. Potentiometric sensors measure variations in potential at zero current, while conductimetric sensors measure variations in conductivity between two electrodes. Chemically sensitized field effect transistors (CHEMFET) have been used to detect DON [132] and is a vital subgroup of the potentiometric biosensors.

3.3.2. Recognition Receptors

Both artificial (antibody fragments, engineered antibodies, aptamers and receptors such as molecularly imprinted polymers) and natural (DNA, antibodies) receptors have been used for specific mycotoxin binding [133]. Specific antibodies raised against the specific mycotoxins are most common, but the use of aptamers to recognize ligands is currently rising in the OTA field.

Some authors have employed the response of enzymatic inhibition generated by mycotoxins to develop enzymatic biosensors for identifying targets. However, the main obstacle in biosensor development is the lack of selectivity, as other molecules or toxins from the sample are also likely to restrain the enzymatic reaction [134]. For instance, amperometric biosensors were developed for Alternaria mycotoxins and were quantified using mushroom tyrosinase on a carbon paste electrode [135] for aflatoxin B1(AFB1) restraining acetylcholinesterase (AChE) [136] and for OTA as the substrate of horseradish peroxidase(HRP) immobilized in a polypyrrole matrix [137]. These sensors were not practical assays but rather “proof of concept” due to the lack of selectivity. For example, acetylcholinesterase is irreversibly repressed by carbamate pesticides and organophosphoruswith a high LOD [138].

(1) Antibodies and Antibody Fragments

Specific antibody synthesis is available for a range of analytical immunoassays based on the selective recognition of any epitope belonging to a specific mycotoxin. Mycotoxins are small molecules with no immunogenicity (haptens), which have to be bound to suitable carrier proteins to elicit an adequate immune reaction. This implies that there is some possible affinity to the corresponding protein, e.g., for BSA [139]. Most of the mycotoxin immunosensors are based on competitive assays with typical formats include indirect and direct schemes [140] but not sandwich formats due to the small molecule size.

There are three types of antibodies based on their generation methods: monoclonal (mAb), polyclonal (pAb) and recombinant (rAb). MAbs are generated from positive hybridomas fused by two cell types from immunized mice, spleen cells and murine myeloma cells. This approach is beneficial due to the unlimited antibody production and uniformity.Polyclonal antibodies are inexpensive, simple and easy to purify from the blood of immunized animals.However, cross-reactivity to the protein used for the conjugate (e.g., BSA) is often a problem. It has been reported that recombinant antibodies (rAbs) is not require the use of animals. The functional gene of interest is cloned and transmitted into eukaryotic or prokaryotic organisms from positive hybridomacells or spleen cells with or without immunization. Finally, rAbs can be expressed, screened and collected [141].

Antibodies can be treated with proteolytic enzymes or reducing agents to produce segments that expose the antigen-binding region orepitope (Fab) to increase specificity and affinity. Moreover, it is possible to separate and purify variable immunoglobulin segments using recombinant technologies primarily involved in toxin binding [140]. Fab fragments are an alternative to the whole antibody, and their smaller size contributes to the minimization of non-specific binding through the Fc fragment and decreases the steric hindrance of heterogeneous systems [133]. Peptide library phage-display methods could be used to authenticate molecules with high specificity and affinity for the mycotoxin, and these techniques could be used to generate antibodies toxic antigens that are difficult to produce and harmful to animals with traditional antibody raising techniques.

(2) DNA and Artificial Receptors

The affinity of aflatoxins to both denatured and native DNA (ssDNA and dsDNA) has been confirmed [142]. The hybridization response between an immobilized probe and a target in solution via the guanine oxidation reaction was established with AFB1, which is nonspecifically captured in the double strand, producing a detection range from 10 to 30 mg L^{-1}. Inspection of variation in the time-dependent current from DNA hybridization initiated by aflatoxin M1 has a detection range from 1.9 to 20.9 nM [143]. An impedimetric AFM1 biosensor based on gold nanoparticles and an ssDNA probe was applied to milk samples and generated a linear response over the concentration range of 1–14 ng mL^{-1}.An electrochemical double stranded DNA biosensor used to measure the electrochemical oxidation response of OTA at a glassy-carbon electrode could also be applied to detect the specific interaction between DNA and OTA [144]. However, most hybridization DNA biosensors have low sensitivity and selectivity for their use in toxicological or environmental detection, and very few studies have reported the use of this technology.

Supramolecular organic structures (e.g., chlorophyllin and cyclodextrins) are well known to strongly bind aflatoxins, but with low selectivity [145]. However, binding peptides and molecularly imprinted polymers (MIPs) generated through combinatorial synthesis have shown great potential for their use as mycotoxin affinity ligands. In addition, MIPs are generated by linking functional polymers in the presence of the target. After UV or thermal polymerization, the target (template molecule) is removed, and a cavity complementary to the target forms, where only the target can bind, acting similar to synthetic antibodies [146]. The affinities of MIPs are lower than antibodies, but have the advantage that the assays can be performed in non-aqueous conditions. Nevertheless, despite some analytical fluorescent applications [147]. MIPs have only been applied to selective cleanup extraction of trichothecene ZEA + DON [148] and OTA [149] or in solid phase columns, where the selectivity of mycotoxin in a complex food matrix was the primary question, rather than the role of synthetic receptors in biosensor applications.

3.3.3. Application of Electrochemical Biosensors in OTA Detection

Several electrochemical immunosensors have been reported with dynamic ranges of mycotoxin concentrations that are approximately 1-100 ng $mL^{-1}$and LODs in the range of 0.5–20 ng mL^{-1} [150]. A recently reported electrochemical aptasensor for OTA has better selectivity and sensitivity than conventional immunoassays and has shorter incubation times. Amperometric biosensors are most common, but EIS is becoming more popular.

OTA contamination occurs at low levels (nanograms per gram), so the development of highly selective and sensitive electrochemical biosensors is urgentto satisfy the needs of the detection limits set by regulatory legislation. Electrochemical OTA biosensors are usually applied to roasted coffee, beer, corn, wine [151], wheat and other cereals, but have also been used in more unusual specimens such as apples [152]. One typical problem in wine samples is the presence of potential electroactive interferents such as polyphenolic compounds, which can be overcome by the use of complexing agents (e.g., poly-(vinylpyrrolidone)) [153]. There have been many studies on OTA immunosensors since 2004, in which Palleschi group fixed polyAntibody of OTA onto screen-printed electrodes (SPCEs). Both polyclonal (pAb-OTA) and monoclonal (mAb-OTA) antibodies against OTA have been compared, with mAbs having at least one-order of magnitude lower IC_{50} values in some cases. OTA immunosensor transducers are usually SPEs, metallic electrodes, or glassy-carbon, and immobilization methods include gold colloid layers, diazonium functionalized gold electrodes for enhancing conjugated OTA surface loading, passive adsorption, and coupling to carbodiimide chemistry covalent procedures or a carboxymethylated dextran hydrogel on a gold electrode. Malhotra et al. performedgeneric rabbit antibody (r-IgGs) fixation to various nanomaterials that successfully generated OTA immunosensors (e.g., nanocrystalline TiO_2–chitosan) , including any line coupled to acacia gum [154], fumed silicaNPs (nano-SiO2) with chitosan (CH) films [154], sol–gel derived cerium oxide films [155]. However, the use of generic non-specific r-IgG antibodies to OTA restricts their selectivity. In addition, competitive indirect and direct electrochemical OTA immunosensors on SPE transducers have been developed with improvements due to nanostructuration with the use of magnetic beads (MBs) and AuNPs in an indirect assay format.

The advantages and intrinsic sensitivity of EIS have led to the development of several impedimetric OTA immunosensors with very low LODs. In most circumstances, a simple ferricanide/ferrocyanide charge-transfer mediator is necessary to achieve impedance measurements with no need for any limited electrochemical mediator enzymatic labels. It is important to identify an efficient method for fixationonto the electrode to produce a range of values for initial electrochemical electron-transfer impedance, producing large variations based on OTA binding to ligands (e.g., aptameror antibody). For this purpose, sulfonated-aniline electropolymerization, co-immobilization with BSA onto C-11 SAMs [156] and diazonium organic salt-modified gold electrodes have been reported by Zamfir et al.

Liu's group [157] reported a convenient, highly sensitive and selective electrochemical immunosensor based on an indirect competitive assay format for ochratoxin A (OTA) detection. The sensing substrate was prepared using a gold electrode-modified with a self-assembled monolayer of 1,6-hexanedithiol, which mediated the assembly of a gold colloid layer able to enhance OTA–ovalbumin conjugate surface-loading to improve sensitivity in electrochemical readouts. After competition of the limited anti-OTA mouse monoclonal antibody between OTA analyte and immobilized happen in sample solution, alkaline phosphatase (ALP)-labeled horse anti-mouse immunoglobulin G (IgG) antibody was selectively combined onto the electrode surface, acting as an indicator of OTA concentration in the sample. The electrochemical response originating from the oxidation of 1-naphthyl phosphate enzymatic products was inversely proportional to OTA concentration from 10 pg/ml to 100 ng/ml, with an LOD as low as 8.2 pg/ml. This method showed good recovery and a negligible matrix effect in corn samples.

3.3.4. Recent Advances in Electrochemical Mycotoxin Biosensors and Future Trends

We discuss some recent advances (use of nanostructuration, MBs and aptamers) in this section. Recent scientific progress of conventional biosensors have overcome some of the difficulties in mycotoxin detection and improved their analytical properties. One area that has advanced is the use of nanomaterial composites to modify electrodes [158]. Nanomaterials have many features important for electrochemical biosensors, such as stability, biocompatibility, easy dispersability, high surface area, unique size and shape, easy functionalization and rapid fabrication, allowing for increased current production and enhanced ligand surface concentration [159]. The main challenge of this technique is that they have composition-dependent chemical and physical properties that require careful optimization. Carbon nanotubes (CNTs) and metallic nanoparticles (NPs), especially gold nanoparticles (AuNPs), are the most frequently used materials [159, 160]. For example, coupling of SAMs (self-assembled monolayers from alkane-thiols) with AuNPs allowed for the generation of OTA immunosensors after hapten-immobilization. Nanomaterials also increase the ability to use ligand (antibodies or aptamer) conjugation to nanomaterials [127] and the charge transfer resistance on the electrode in EIS [160]. For example, DNA probes in conjunction with silver NPs or AuNPs have been generated for AFM1 immunosensors [161]. OTA voltammetric immunosensors have been developed through a synergistic effect of OTA gold nanostructuration conjugated to bovine serum albumin protein [139]. Nanoparticles can be used as spacer matrices to extend the use of OTA–BSA by generating binding sites with increased potential to interact with anti-OTA antibodies. Conjugation of the enzyme tracers with extravidin and biotinylation of the antibody eliminates the incubation step to generate amperometric currents with a labeled secondary antibody [139]. NPs can also act as labels for electrochemical detection, but amplifying enzymes are often preferable in this application [159].

3.4. Aptamer

3.4.1. Structural Basis and Characteristics of Aptamers in Target Identification

Aptamer ligands contain 20–90 oligonucleotides and are single-stranded nucleic acids (DNA, RNA). Aptamers can bind to their targets with high affinity and high specificity, ranging from large proteins to small molecules, such as amino acids or drugs. Aptamers are generated by an *in vitro* selection process named SELEX (systematic evolution of the ligand by the exponential enrichment process), which was first reported in 1990 [162]. The basic process is outlined in Figure 2. (1). Construction of random oligonucleotide libraries. The primary binding region of aptamers is part of the fixed sequences, which are located at both ends of the oligonucleotide library. Generally, 20–60 oligonucleotides compose the random sequences, which can theoretically produce 10^{14-15} random sequences, and the random sequences are in the middle of the aptamer sequence. The oligonucleotide 3-D structure diversity is vital for the diversity of random sequence determination. The library can react with almost all types of naturally occurring molecules due to its large oligonucleotide storage. (2) Separation of the target-nucleic acid complex. The sequence library is incubated with a specific target under the appropriate conditions. The target-nucleic acid complexes are then separated and eluted by magnetic bead separation, centrifugation, capillary electrophoresis, filtration or flow cytometry. (3) Amplification. Polymerase chain reaction (PCR) is the primary amplification technique for the eluted nucleic acids. After 6–20 repeated processes of adsorption, separation, elution, and amplification, aptamers can bind to their target with high specificity and high affinity, with dissociation constants in the nmol or pmol range.

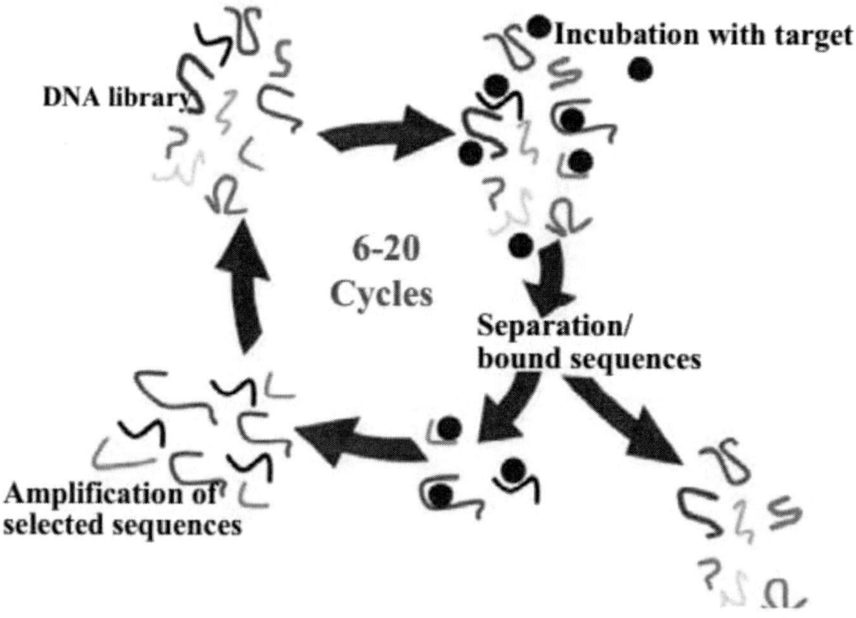

Figure 2. General illustration of the SELEX procedure.

Aptamers can fold into defined three-dimensional conformations under certain physicochemical conditions, allowing it to bind to target molecules through specific interactions with high affinity constants. Aptamer binding to the target occurs through specific interactions, including non-covalent Van der Waals forces, hydrogen bonding, hydrophobic laminate, and static interactions, in addition to other associated factors, including selectivity and directionality. This mutual recognition depends on matching induction aptamer and target spatial structure, which requires a variety of forces, such as hydrogen bonding and electrostatic interactions. To adapt to the targets, aptamers fold into unique pocket-shaped structures. The corresponding target group could react with some aptamer functional groups. Both allow for the aptamer distinguishing molecular structures at the atomic level. Compared to antibodies, which have been the standard for molecular recognition of mycotoxins for several decades, aptamers have clear advantages: they do not require the immunization of animals; they have less variability than pAbs, increased chemical and thermal stability, and inexpensive in-vitro synthesis. Finally, aptamers are not susceptible to denaturation in the presence of solvents, which are commonly used in mycotoxin extraction [163]. The comparison of aptamers to antibodies is presented in Table 1.

Table 1. Comparison between aptamer and antibody

Aptamer	Antibody
Small molecular; Aptamer generation does not depend on *in vivo* conditions allowing the selection of aptamers against toxic or poor immunogenic targets; Manipulation by changing some synthesized conditions *in vitro* including temperature, pH, buffer, etc.	Big molecular; Antibodies are identified through an *in vivo* process, requiring animal immunization, inducing animmune response against a target into a biological system
Storage and transportation at room temperature; Reversible degeneration induced by temperature	Sensitive to temperature and humidity changes; Limited storage time. Irreversible Degeneration induced by temperature.
Easy production; Low-cost; Chemically synthesized without batch variation	Laborious production; High-cost; Significant variations in antibodies from batch to batch.
Low-immunogenic and low-toxic molecules; High affinity and specificity; Broad range of targets (toxins, metal ions, organic dyes, antibiotics, peptides and small organic molecules, etc.)	Strong-immunogenicity; Difficult to elicit toxic and poor-immunogenic targets.
Labeling can not cause loss in affinity	Labeling can cause loss in affinity
Similar structure or cross-reaction substances can be separated	Cross-reaction substances can not be separated
SELEX conditions can be manipulated to select aptamers with the desired properties	During antibody production, binding sites of the target are identified by the animal immune system. The identification is restricted by *in vivo* parameters.

3.4.2. Selection of Aptamers for Mycotoxins

Mycotoxins generate trace exogenous harmful residues, are low molecular weight (< 1 kDa) and have stable structures and unabated toxicity under normal heating conditions. OTA-specific aptamers were defined in 2008, following characterization by equilibrium dialysis [164, 165] and fluorescence polarization. By affinity chromatography coupled to molecular fluorescence detection, OTA-specific aptamers were used to detect OTA in wheat grains. Their binding affinities are in the nanomolar range, as determined by equilibrium dialysis [164], which is comparable to or lower than the binding constants of antibodies to OTA. Consequently, many electrochemical OTA aptamer sensors were developed after 2009, and one was developed in our laboratory using magnetic beads (MBs) as solid support [166]. Aptamer-based electrochemical sensors are very sensitive, displaying detection limits in the range 0.03–0.8 ng mL^{-1}OTA. More work is expected in the near future, due to a new program that will likely connect the aptamer conformational change to the redox mediator. More sensitive (up to 10^{-4}ng mL^{-1}OTA) label-free aptasensors have been described using electrochemical impedance spectroscopy (EIS) [167].

In 2010, six unique aptamer sequences were found to bind AFB1 with high affinity, and one of them (AFB1-39) had a dissociation constant of 0.1 m Mafter 18 rounds of SELEX [168]. This work opens the door to new aptasensors and solid extraction columns, though no studies have been published to date. As an initial study, our laboratory has demonstrated the possibility of using AFB1-39 aptamer in the development of future electrochemical aptasensors. Neoventures Biotechnology Inc. has patented specific aptamers to AB1 and ZEA targets with K_D=10 nM [169]. They have also described commercial aptamer-based devices for the detection of AFB1, OTA and ZEA using lateral-flow technology. Therefore, it is essential to identify new aptamers against mycotoxins. Aptamers against OTA, FB1, AB1 and ZEA have been reported, and a large increase is expected in the near future. The corresponding sequences are listed in Table 2.

Considering the sensitivity of the instrument and complexity of the contaminated matrix, mycotoxins (trace substances) in the matrix must be enriched and purified to be detected. Mycotoxin purification methods primarily utilize liquid-liquid extraction and chromatographic columns. Because mycotoxin concentrations are so low, it is very easy to cause environmental pollution and worsen their recovery. Chromatographic columns require a large amount of organic solvents. Thus, the column has become a widely used technology because of its simple operation and low solvent consumption. It is widely used in the immunoaffinity column (IAC) clean-up technique, which is based on interactions between antibody and antigen. Antibodies can be easily affected by the matrix, solvent, pH and salt concentration. Additionally, IAC is expensive, and cross-reactivity has been reported in the use of antibodies for analyzing mycotoxins and their metabolites [174]. Those deficiencies limit the use of IAC clean-up-based methods in mycotoxin detection. Aptamers can be used repeatedly. To exclude interference of other similar substances, aptamers can be applied in mycotoxin pre-treatment and avoid the use of organic solvents. Compared with IAC, aptamer preparation costs are lower. Aptamers can also be used as alternatives to antibodies, as they have higher affinity, specificity, structural stability, synthesis and repeatability.

Chapuis-Hugon et al. [175] reported that aptamer immobilization via covalent binding (cyanogen bromide-activated sepharose) has higher binding efficiency (29%). Compared with non-covalent binding (streptavidin-activated agarose), aptamers are more conducive to OTA purification from red wine. The effect of aptamer purification by solid phase extraction (SPE) method and IAC are similar. Aptamer columns can be used for different complex matrices and reused after a few minutes (antibody columns generally need 1–2 days before reuse). In the future, aptamers will replace the use of IAC. De Girolamo et al. [176] absorbs aptamers with a coupling gel and used them as a sorbent to prepare SPE columns for OTA removal from wheat. Using high performance liquid chromatography with fluorescence detection (HPLC-FLD), the average recoveries ranged from 74% to 88%, with an LOD of 23 pg g^{-1}. This column could be used five times without any loss in performance. A good correlation was found in comparative HPLC-FLD analyses of 33 wheat samples in which OTA was removed by aptamer-SPE or IAC (r= 0.990). They also detected OTA in 96 wheat samples using aptamer affinity column purification-time-resolved fluorescence (TRF) detection. The average recovery was 77% and the LOQ was 0.5 g kg^{-1}. The entire analytical process was completed in 30 min. Rhouati et al. optimized the OTA purification procedure in beer by using aptamer columns with an average recovery of 96% [177].

Table 2. Aptamer sequences of mycotoxins

	Type of aptamer	Aptamer sequences (5'-3')	Dissociation constant (nM)	Ref
Aflatoxin B1 (AFB1)	DNA	TCGTCCATGTGTTGGGCACGTGTTGTCTCTCT GTGTCTCGTGCCCTTCGCTAGGCCCACACGA TGCGTAG	12	
		GTTGGGCACGTGTTGTCTCTCTGTGTCTCGTG CCCTTCGCTAGGCCCACA	30	[170]
Afratoxin B2 (AFB2)	DNA	TCGTCCATGTGTTGGGCACGTGTTGTCTCTCT GTGTCTCGTGCCCTTCGCTAGGCCCACACGA TGCGTAG	30	
		GATCGGGTGTGGGTGGCGTAAAGGGAGCAT CGGACA	200	[164]
Ochratoxin A (OTA)	DNA	CTGGGAGGGAGGGAGGGATCGGGTGTGGGT GGCGTAAAGGGAGCATCGGACACCCGATCC C	-	[171]
		ATACCAGCTTATTCAATTAATCGCATTACCT TATACCAGCTT	100±30	[168]
Ochratoxin A (OTA)	DNA	CTGGGTGTGGGGTGATCAAGGGAGTAGACT		[172]
Ochratoxin A (OTA)	DNA	CGGGTGTGGGTGCCTTGATCCAGGGAGTCT		
Ochratoxin A (OTA)	DNA	GATCGGGTGTGGGTGGCGTAAAGGGAGCAT CGGACAACG		[164]
Fumonisins B1(FB1)	DNA	ATTCAATTACGTCTGCACATACCAGCTTATT CAATTAGATA GTAAGTGCAATCT		
Zearalenone (ZEN)	DNA	TCATCTATCTATGGTACATTACTATCTGTA ATGTGATATG		[173]

3.4.3. Aptamers for Rapid Single Mycotoxin Detection

The current analytical approaches for mycotoxin detection primarily utilize thin layer chromatography (TLC), liquid chromatography-mass spectrometry (LC-MS), (ultra) high performance liquid chromatography (HPLC), gas chromatography (GC), gas chromatography-mass spectrometry (GC-MS), and capillary electrophoresis (CE) [178-181].

Some of these techniques require expensive analytical instruments and professional operators, some require sample pre-treatment, and others maynot meet the sensitivity and specificity requirements. Immunosorbent assays [182, 183], particularly enzyme-linked immunosorbent assay and colloidal gold immunochromatography, can decrease analysis time and satisfy the specificity and sensitivity requirements. However, their application has been limited due to the defects in antibodies. Aptamer recognition can be combined with novel nano-materials in mycotoxin analysis in various electrochemical and optical analytical methods. These methods meet the high specificity, sensitivity and accuracy requirements. As they do not require complex sample pre-treatment, use of aptamers can save resources and analysis time.

(1) Optical Aptamer Analysis for Rapid Single Mycotoxin Detection

Various optical materials have been optimized for use in aptamer modifications. When aptamers interact with their targets, several optical phenomena could occur. Thus, the optical signal detection would allow for qualitative and quantitative target analysis. Fluorescence spectrophotometry, fluorescence polarization and UV spectrophotometry were recently used in optical analysis for mycotoxin detection.

UV spectrophotometry has many advantages in the optical analysis for mycotoxin detection, including simple techniques and time-saving time because it does not require sample pre-treatment. However, its applications have been limited due to their low sensitivity and accuracy. As shown in Figure 3A, Yang et al. [171] reported an OTA aptasensor using aptamer-DNAzyme hairpins as biorecognition elements. After introducing horseradish peroxidase (HRP), G-quadruplex-hemin-DNAzyme was used in the catalytic chromogenic reaction of 3,3',5,5'-tetramethylbenzidine (TMB) (including H_2O_2). This aptasensor was used to analyze OTA in 96 wines within one hour. Yang et al. [184] developed colorimetric OTA detection using the OTA aptamer and unmodified AuNPs (Figure 3B). Without OTA, aptamers were absorbed by the AuNP surface. Red AuNPs began to aggregate upon introduction of the salt solution. When OTA appeared, the OTA aptamer conformation changed from a random coil structure to the G-quadruplex structure, which could not protect AuNP from salt-induced aggregation. This caused the color of AuNP to change from red to blue. It was a rapid, simple, effective quantitative method. Compared with traditional UV spectrophotometry, this approach was more sensitive and accurate. Compared with the antibody-based fluorescence polarization system, the aptamer-based fluorescence polarization detection system has many advantages, such as abroad detection range and high sensitivity. Cruz-Aguado et al. [165] designed a fluorescence polarization method by introducing the OTA aptamer, which rapidly and sensitively detected OTA with an LOD of 5 nM.

Because the aptamer probe has the advantages of sensitivity and background, aptamer-based fluorescence detection has become popular for molecular recognition. Nano-materials, such as carbon nanotubes and graphenes, are natural fluorescence quenchers [185].

Nano-material surfaces can highly delocalize the π bond, allowing for their adsorbtion by ssDNA through π-π stacking interactions. The complicated label steps could be eliminated by using nano-materials to design a molecular beacon. Guo et al. [186] developed a label-free competitive molecular beacon fluorescent aptasensor based upon the self-assembly of single-walled carbon nanotubes (SWNTs). The fluorescence signal is quenched upon SWNT adsorption of the FAM-labeled aptamer. When OTA appeared, it would compete with SWNTs for combinating with the aptamer, allowing for some of the aptamers to be desorbed from the SWNT surface and fluorescence recovery. OTA can be detected in beer upon fluorescent resonance energy transfer (FRET) between FAM and SWNTs (Figure 3C). It is similar to the quenching effect of PVP-coated graphene oxide (Figure 3D) [187]. It has been reported that single-stranded oligonucleotides can greatly enhance Tb^{3+} emission (one important rare earth element) in solution, but duplexes cannot. This feature of Tb^{3+} provides a useful approach to distinguish between single- and double-stranded DNA. Zhang et al. reported [188] a signal-on fluorescent aptasensor for label-free OTA detection in wheat on the basis of structure-switching aptamer, Tb^{3+} and magnetic beads (MBs) (Figure 4).

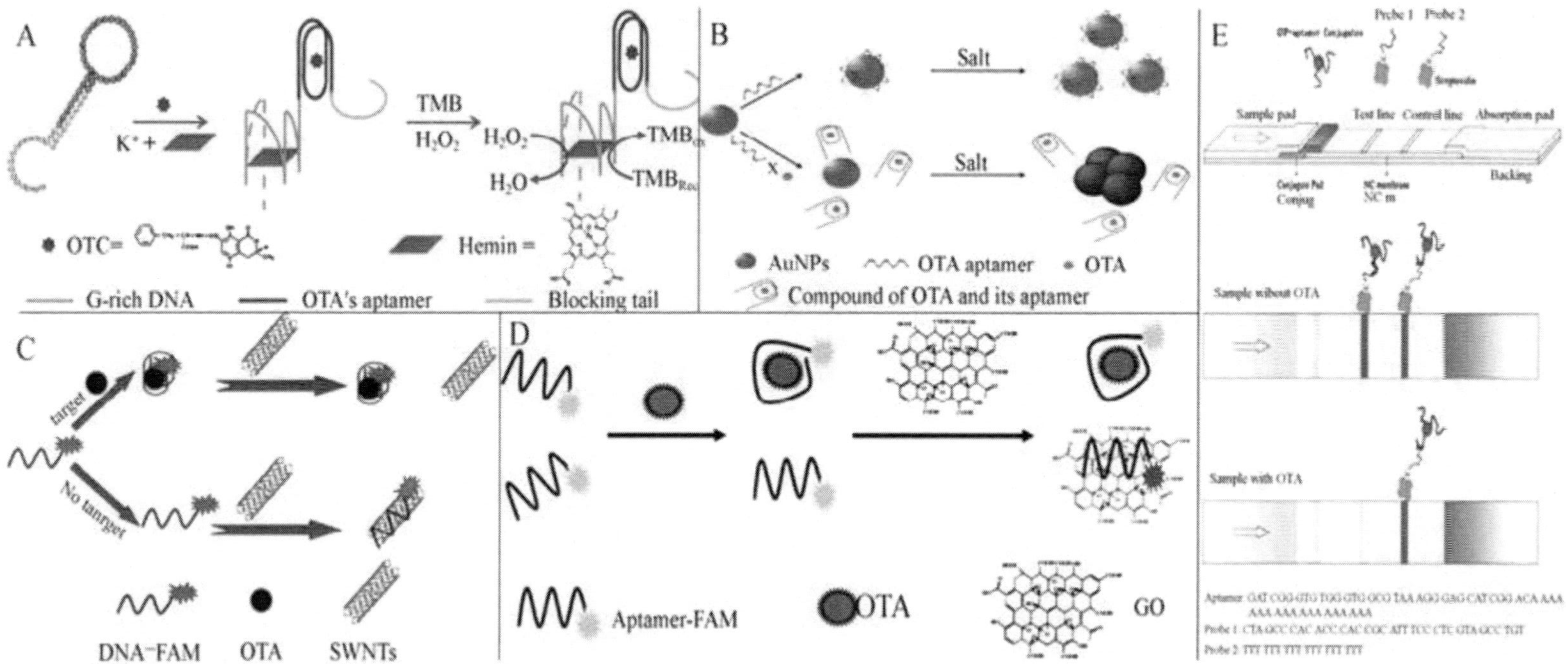

Figure 3. Schematic illustrations of aptamer-based optical OTA detection A. Colorimetric detection by the G-quadruplex-hemin-DNAzyme to catalyze the TMB color reaction; B. Colorimetric detection by unmodified AuNPs; C. Fluorescent detection utilizing SWNTs as quenchers; D. Fluorescent detection utilizing PVP-GO as a quencher; E. Detection by aptamer-based strip.

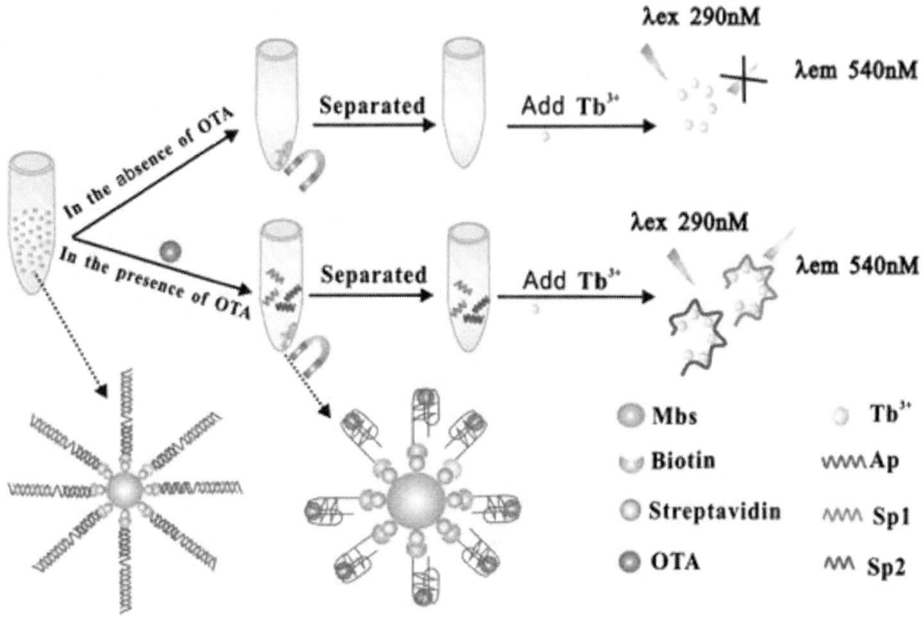

Figure 4. Principle of A signal-on fluorescent aptasensor based on $Tb^{3+}$and structure-switchingaptamer aptasensor.

The anti-OTA aptamer was initially labeled with a biotin group and attached to streptavidin-modified MBs. Two single-stranded signal probes were added and hybridized to the anti-OTA aptamer to form the duplex structure in solution. Because single-stranded DNA can significantly enhance Tb^{3+}emission in solution but duplexes cannot, magnetic bead separation was performed to isolate a supernatant liquid containing no single-stranded DNA (ssDNA) so that it cannot increase Tb^{3+}emission. While in the presence of OTA, it can bind with aptamers to form an OTA-aptamer complex while releasing two single-stranded signal probes. Through magnetic beadseparation, thereleased single-stranded signal probes in the supernatant liquid can largely increase Tb^{3+}fluorescent intensity. By employing the above analysis strategy, this aptasensor can detect OTA at levels as low as 20 pgmL^{-1}with high specificity.

Wang et al. [189] developed an aptamer-based chromatographic strip assay by using the competition between OTA and DNA probes. By intruding the gold nanoparticles labeled aptamers as probes (Apt-GNP), the researchers added two types of OTA complementary sequences DNAs on the test line and the control line. When OTA was present in red wine, Apt-GNP was red on the test line. Upon OTA addition, the red color of the test line is fainted (Figure 3E). This detection system could also be used in the quantum dots detection method [190]. Instead of antibodies, aptamers in the rapid detection test strip overcame thethermolability ofthe antibody, but was also moreconvenient and time-saving.

(2) Aptamer-Based Electrochemical Analytical Techniques for Rapid Single Mycotoxin Detection

Aptamers applied in electrochemical analytical methods include label and label-free modes. Quantum dots (CdS, PbS), enzyme (glucose dehydrogenase, horseradish peroxidase), small molecules (methylene blue, ferrocene) and nanoparticles (Au, Ag) are used in label modes, while electrochemical impedance and quartz crystal microbalance (QCM) are used in the label-free mode.

Methylene blue (MB) is a DNA insertion widely used in electrochemical analysis. Kuang et al. developed a signal-off electrochemical aptasensor to detect OTA in red wine by inserting MB. The aptamer was immobilized on the electrode surface through hybridization with linker DNA (ssDNA1). A gold nanoparticle (GNP)-modified DNA3 was used to amplify the sensing signal. In the presence of MB and OTA, the aptamer was peeled from the electrodes due to competition between OTA and ssDNA1 and ssDNA3 and aptamer. The dissociation decreased the redox current, which was proportional to the OTA concentration in red wine (Figure 5A). Tong et al. [191] designed a signal-on aptasensor for the ultrasensitive detection of OTA in wheat starch. This sensor is based on exonuclease-catalyzed target recycling. First, a ferrocene (Fc)-labeled ssDNA1probe was immobilized on a gold electrode (GE) through Au–S bonding. Fc-labeled ssDNA1 was then hybridized with the complementary aptamer away from the GE surface. The differential pulse voltammetry (DPV) signal was very low when Fc was far away from the electrode surface. Upon OTA addition, aptamer-OTA conjugate formation resulted in aptamer dissociation from the double-strand DNA and the transformation of the probe DNA into a hairpin structure. Because the labeled Fc was close to the electrode surface, the DPV signal significantly improved. This ultrasensitive signal-on aptasensor decreased the high background current of signal-off aptasensors, reduced the detection limit (1 pg mL^{-1}) and enhanced the sensitivity (Figure 5B).

Barthelmebs et al. [192] (Figure 5C) established two competitive modes for OTA detection. In the indirect competitive mode, OTA was immobilized on magnetic beads (OTA-MBs) by a streptavidin-biotin bridging system, and then aptamer-horseradish peroxidase (Apt-HRP) was applied to compete with OTA-MBs and dissociate OTA. The changes in Apt-HRP catalytic activity were used to indirectly detect OTA because dissociative OTA hindered the bond between Apt-HRP and OTA-MBs, leading to a reduction in HRP catalytic reactivity. In the direct competitive assay, aptamer-magnetic beads (Apt-MBs) were prepared using the amino-carboxyl bridging system. In the presence of OTA, OTA was dissociated through competitive binding to the Apt-MBs with alkaline phosphatase-OTA (OTA-ALP). After magnetic separation, the amount of ALP immobilized on MBs decreased, leading to a reduction in DPV.

Analysis of the performance between electrochemical and optical detection showed that the direct assay mode was better than the indirect assay mode (Electrochemical method in wine: LOD direct= 0.15 ng mL^{-1}< LOD indirect= 1.4 ng mL^{-1};UV method: LOD indirect= 3.5 ng mL^{-1}> LOD direct= 0.6 ng mL^{-1}) (Figure 5C).

Electrochemical impedance spectroscopy (EIS) is another electrochemical assay technique used to represent the biological reaction on the electrode interface. This analytical technique has a good interface characterization response and can be used in combination with other various sensors or by itself. Prabhakaret al. [167] developed a highly robust and sensitive impedimetric aptasensor for OTA detection by immobilizing anti-OTA aptamer onto a mixed Langmuir-Blodgett (LB) monolayer comprised of PANI-SA deposited onto indium tin-oxide (ITO)-coated glass plates. When aptamers were immobilized on the electrode surface, a negative charge layer was generated on the electrode surface. Real impedance increased due to electrostatic repulsion in the oxidation-reduction electrode ($Fe(CN)6^{3-/4-}$).

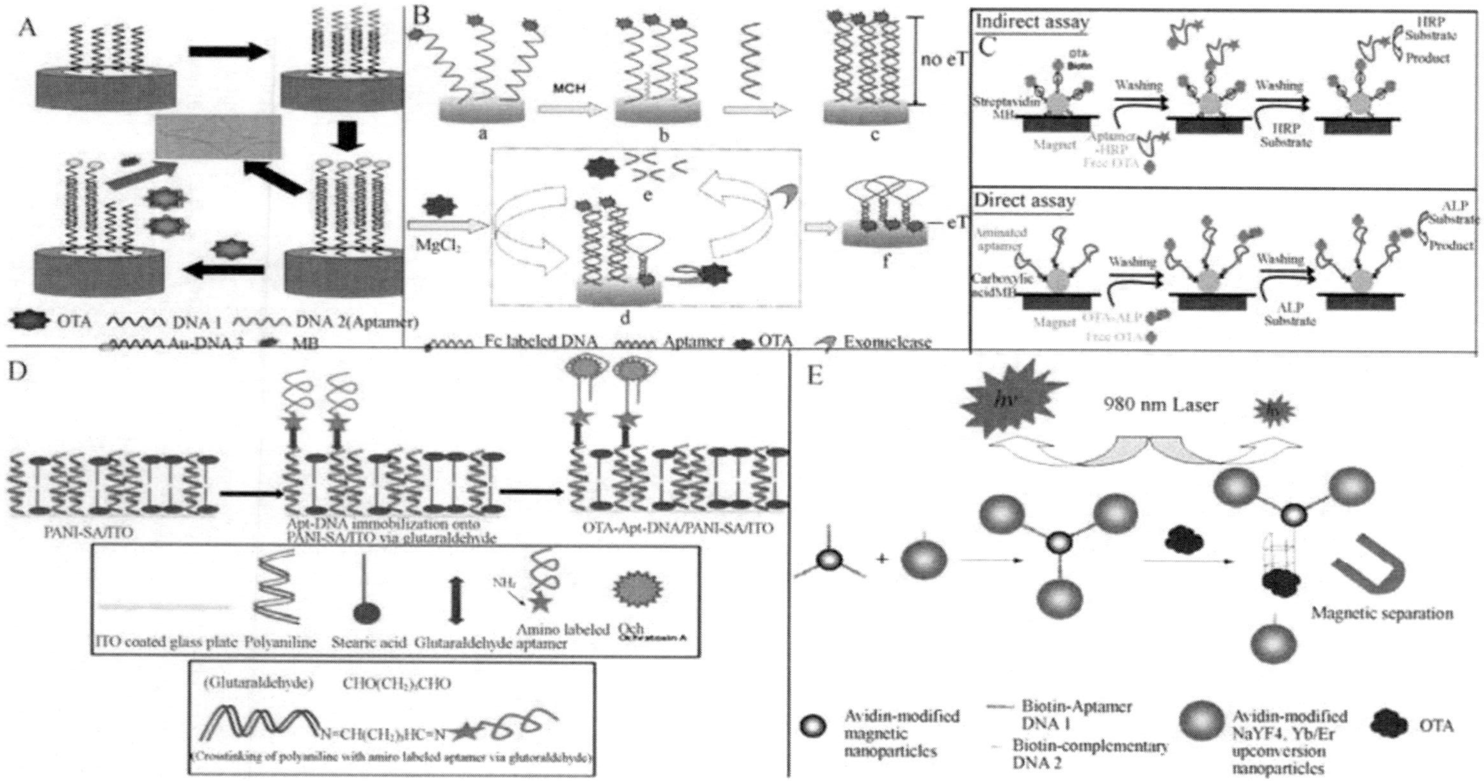

Figure 5. Schematic illustrations for aptamer-based electrochemical OTA detection. A. Signal-off aptasenor (MB inserted probe); B. Signal-on aptasenor (Fc labeled probe); C. Electrochemical aptasenor (Indirect competitive assay and direct competitive assay); D. Electrochemical impedance spectroscopy by Apt-DNA/PANI-SA/ITO aptasensor; E. Electrochemiluminescent detection based on combining magnetic separation of Apt-MNPs and UCNPs luminescent labeling.

As the aptamer dissociated from the electrode in the presence of OTA, the impedance decreased. The Apt-DNA/PANI-SA/ITO aptasensor produced reproducible data and could be reused 13 times after regeneration with NaOH (Figure 5D).

ECL combines chemiluminescence and electrochemistry. It has the advantages of chemiluminescence analysis with a wide linear range, high sensitivity, automation and easy operation, and electrochemical analysis with highly active illuminant, controllability and fewer reagents. Wang et al. designed an electrochemiluminescent aptasensor for detecting OTA in corn, which was assessed by immobilizing aptamer complementary ssDNA onto the surface of a AuNP-functionalized gold electrode. ABEI-labeled aptamer hybridized with the complementary ssDNA and was used as an electro-chemiluminescent probe. In the presence of OTA, the ECL signal decreased due to aptamer recognition of OTA, leading to aptamer dissociation from the ssDNA. Aptamer dissociation resulted in ABEI moving away from the electrode surface. As shown in Figure 5E, Wu et al. [193] designed an assay that immobilized aptamers on the surface of Fe3O4-functionalized MNPs, which were then used to capture and concentrate OTA from bulk samples. The aptamer hybridized with ssDNA labeled with upconversion fluorescent nanoparticles (UCNPs). In the presence of OTA in corn, the distance between Fe3O4-MNPS and UCNPS increased, and the intensity of the electron transfer signal between Fe3O4-MNPS and UCNPS decreased. This method showed similarity and good correlation with ELISA.

3.4.4. Aptamers for Simultaneous Detection of Multiple Mycotoxins

More than four hundred mycotoxins have been found, and most of them have strong toxicity towards humans and animals. These mycotoxins usually contaminate the same types of foods or crops. The contamination status of various mycotoxins in food or crops was tough to detect in reality. If the mycotoxins were inspected individually, it would be time-consuming. Thus, it was necessary to develop an analytical method for the simultaneous detection of multiplemycotoxins, saving time as well as reflecting the extent of real-time contamination. However, because of the difference of the chemical and physical properties of mycotoxins, there are still some puzzles (extraction rates and recovery) in the process of simultaneous detection. Aptamers with high selectivity and sensitivity can be applied to distinguish differences between the analogues and effectively avoid cross-reactions. Anew structure-switching aptasensor to simultaneously detect OTA and FB1 was developed, which was on account of multiplexed fluorescence resonance energy transfer (FRET).In this assay, the multicolor UCNPs are donors and the graphene oxide (GO) is the effective and entire acceptor. This structure-switching aptasensor provided a linear range of 0.1–500 ng mL^{-1} for FB1 and0.05–100 ng mL^{-1} for OTA; LOD was 0.1 ng mL^{-1} for FB1and 0.02 ng mL^{-1} for OTA. This method showed good performance and saved significant analysis time. It was also the first time being applied to aptamers for the simultaneous determination of various mycotoxins.

CONCLUSION

In Over the past few decades, with the development of automatic SELEX, the efficiency as well as selection cycles have been markedly improved. The screened aptamers are more appropriate to detect mycotoxins with various analysis modes. Owing to different analysis techniques (e.g., electrochemical, fluorescence, colorimetric and electrochemiluminescence), aptamer-based inspection strategies continually decrease the detection limits of mycotoxins, reaching ng, pg, and even fg levels, and also introduce various academic subjects (e.g., nanotechnology, intelligent chip and biological sensors), saving detection time and improving selectivity. However, with regard to rapid detection of mycotoxins, the methods described above are still at an initial stage. Thus, it is necessary to study them further to broaden their application fields. Additionally, the development of tendencies of aptasensors in this field may include the following three aspects:

First, more species of mycotoxins, including OTA, should be detected. As far as we know, there are hundreds of mycotoxins possessing toxicity in nature. AFB1 and OTA have been classified as possible Group 1 and 2B carcinogens by The International Agency for Research on Cancer (IARC), respectively. Mycotoxins could generate different toxic metabolites, leading to possible conflicts in detecting these mycotoxins. With the progress of SELEX selection technologies, various mycotoxin aptamers with high sensitivity and selectivity have been selected prosperously. Thus, the applications of aptamer identification technology to detect mycotoxins will be diversified.

Second, more and more complex matrices will be analyzed for mycotoxins using aptasensors. At the present stage, the aptamer identification technology is mainly employed in some simple matrices, such as agricultural products, foods and others. It has been reported that Chinese herbal medicines, especially some 'medicinal and edible herbs,' such as peach seeds, coix seeds, licorice root, hawthorn fruit and lotus seeds, are extremely vulnerable to contamination by mycotoxins in the process of improper handling (e.g., harvesting, processing, storage and distribution). The extent of contamination may be aggravated by the coexistence of various mycotoxins. Meanwhile, intake of contaminated herbs will lead to serious hazards to human health. Aptasensors could be applied to detect mycotoxins in various kinds of contaminated agricultural products. In the future, the progress of aptasensors for the rapid detection of mycotoxins will become a research hotspot.

Third, aptamers could be fixed on solid columns to clean up mycotoxins. Acting as a molecular recognition element, aptamers can be used in the area of purification and separation of mycotoxins. It is significantly cost-saving to develop commercial aptamer affinity columns, to replace expensive immunoaffinity columns. Thus, it will be popular to study aptamer affinity columns in the future.

For rapid detection of various mycotoxins, it is necessary to establish rapid, simple, effective, highly sensitive economic analysis methods, as well as to construct aptasensors via electrochemical, optical and immunology technologies.

References

[1] Wood, C. M., Farrell, A. P., Brauner, C.J. (2011) *Homeostasis and toxicology of essential metals*. Academic Press.

[2] Sambrook, J., Russell David, W. (1989) *Molecular cloning: a laboratory manual. Vol. 3*. Cold spring harbor laboratory press.

[3] Válóczi, A., Hornyik, C., Varga, N., Burgyán, J., Kauppinen, S., Havelda, Z., (2004). Sensitive and specific detection of microRNAs by northern blot analysis using LNA-modified oligonucleotide probes. *Nucleic Acids Res.* 32, e175-e175.

[4] Takamizawa, J., Konishi, H., Yanagisawa, K., Tomida, S., Osada, H., Endoh, H., Harano, T., Yatabe, Y., Nagino, M., Nimura, Y., (2004). Reduced expression of the let-7 microRNAs in human lung cancers in association with shortened postoperative survival. *Cancer Res.* 64, 3753-3756.

[5] Pfeffer, S., Zavolan, M., Grässer, F.A., Chien, M., Russo, J.J., Ju, J., John, B., Enright, A.J., Marks, D., Sander, C., (2004). Identification of virus-encoded microRNAs. *Science*. 304, 734-736.

[6] Dorak, M.T. (2007) *Real-time PCR*. Taylor & Francis.

[7] Chen, C., Ridzon, D.A., Broomer, A.J., Zhou, Z., Lee, D.H., Nguyen, J.T., Barbisin, M., Xu, N.L., Mahuvakar, V.R., Andersen, M.R., (2005). Real-time quantification of microRNAs by stem–loop RT–PCR. *Nucleic Acids Res.* 33, e179-e179.

[8] Rychlik, W., Spencer, W., Rhoads, R., (1990). Optimization of the annealing temperature for DNA amplification *in vitro*. *Nucleic Acids Res.* 18, 6409-6412.

[9] Liang, R.-Q., Li, W., Li, Y., Tan, C.-y., Li, J.-X., Jin, Y.-X., Ruan, K.-C., (2005). An oligonucleotide microarray for microRNA expression analysis based on labeling RNA with quantum dot and nanogold probe. *Nucleic Acids Res.* 33, e17-e17.

[10] Barad, O., Meiri, E., Avniel, A., Aharonov, R., Barzilai, A., Bentwich, I., Einav, U., Gilad, S., Hurban, P., Karov, Y., (2004). MicroRNA expression detected by oligonucleotide microarrays: system establishment and expression profiling in human tissues. *Genome Res.* 14, 2486-2494.

[11] Schmittgen, T.D., Jiang, J., Liu, Q., Yang, L., (2004). A high - throughput method to monitor the expression of microRNA precursors. *Nucleic Acids Res.* 32, e43-e43.

[12] Jiang, J., Lee, E.J., Gusev, Y., Schmittgen, T.D., (2005). Real-time expression profiling of microRNA precursors in human cancer cell lines. *Nucleic Acids Res.* 33, 5394-5403.

[13] O'connor, C., (2008). Fluorescence in situ hybridization (FISH). *Nature Education.* 1, 171.

[14] Gall, J.G., Pardue, M.L., (1969). Formation and detection of RNA-DNA hybrid molecules in cytological preparations. *Proceedings of the National Academy of Sciences*. 63, 378-383.

[15] Kidner, C., Timmermans, M. (2006), *MicroRNA Protocols*. Springer, pp. 159-179.

[16] Ryan, D.G., Oliveira-Fernandes, M., Lavker, R.M., (2006). MicroRNAs of the mammalian eye display distinct and overlapping tissue specificity. *Mol. Vis.* 12, 1175-1184.

[17] Darnell, D., Kaur, S., Stanislaw, S., Davey, S., Konieczka, J., Yatskievych, T., Antin, P., (2007). GEISHA: an in situ hybridization gene expression resource for the chicken embryo. *Cytogenet. Genome Res.* 117, 30-35.

[18] Habbe, N., Koorstra, J.-B.M., Mendell, J.T., Offerhaus, G.J., Ryu, J.K., Feldmann, G., Mullendore, M.E., Goggins, M.G., Hong, S.-M., Maitra, A., (2009). MicroRNA miR-155 is a biomarker of early pancreatic neoplasia. *Cancer Biol. Ther.* 8, 340-346.

[19] Nelson, P. T., Baldwin, D. A., Kloosterman, W. P., Kauppinen, S., Plasterk, R. H., Mourelatos, Z., (2006). RAKE and LNA-ISH reveal microRNA expression and localization in archival human brain. *RNA*. 12, 187-191.

[20] Vettorazzi, A., van Delft, J., de Cerain, A.L., (2013). A review on ochratoxin A transcriptomic studies. *Food Chem. Toxicol.* 59, 766-783.

[21] Manderville, R.A., (2005). A case for the genotoxicity of ochratoxin A by bioactivation and covalent DNA adduction. *Chem. Res. Toxicol.* 18, 1091-1097.

[22] Jennen, D.G., Magkoufopoulou, C., Ketelslegers, H.B., van Herwijnen, M.H., Kleinjans, J.C., van Delft, J.H., (2010). Comparison of HepG2 and HepaRG by whole genome gene expression analysis for the purpose of chemical hazard identification. *Toxicol. Sci.*, kfq026.

[23] Doktorova, T.Y., Ellinger-Ziegelbauer, H., Vinken, M., Vanhaecke, T., van Delft, J., Kleinjans, J., Ahr, H.-J., Rogiers, V., (2012). Comparison of hepatocarcinogen-induced gene expression profiles in conventional primary rat hepatocytes with *in vivo* rat liver. *Arch. Toxicol.* 86, 1399-1411.

[24] Hundhausen, C., Boesch-Saadatmandi, C., Matzner, N., Lang, F., Blank, R., Wolffram, S., Blaschek, W., Rimbach, G., (2008). Ochratoxin a lowers mRNA levels of genes encoding for key proteins of liver cell metabolism. *Cancer Genomics-Proteomics.* 5, 319-332.

[25] Matsuzawa-Nagata, N., Takamura, T., Ando, H., Nakamura, S., Kurita, S., Misu, H., Ota, T., Yokoyama, M., Honda, M., Miyamoto, K.-i., (2008). Increased oxidative stress precedes the onset of high-fat diet–induced insulin resistance and obesity. *Metabolism.* 57, 1071-1077.

[26] Choi, Y. D., Grabowski, P. J., Sharp, P. A., Dreyfuss, G., (1986). Heterogeneous nuclear ribonucleoproteins: role in RNA splicing. *Science.* 231, 1534-1539.

[27] De Maio, A., (1999). Heat shock proteins: facts, thoughts, and dreams. *Shock.* 11, 1-12.

[28] Sergent, O., Pereira, M., Belhomme, C., Chevanne, M., Huc, L., Lagadic-Gossmann, D., (2005). Role for membrane fluidity in ethanol-induced oxidative stress of primary rat hepatocytes. *J. Pharmacol. Exp. Ther.* 313, 104-111.

[29] Boesch - Saadatmandi, C., Wagner, A., Graeser, A., Hundhausen, C., Wolffram, S., Rimbach, G., (2009). Ochratoxin A impairs Nrf2-dependent gene expression in porcine kidney tubulus cells. *J. Anim. Physiol. Anim. Nutr. (Berl.).* 93, 547-554.

[30] Ferrante, M., Raso, G.M., Bilancione, M., Esposito, E., Iacono, A., Meli, R., (2008). Differential modification of inflammatory enzymes in J774A. 1 macrophages by ochratoxin A alone or in combination with lipopolysaccharide. *Toxicol. Lett.* 181, 40-46.

[31] Boesch-Saadatmandi, C., Loboda, A., Jozkowicz, A., Huebbe, P., Blank, R., Wolffram, S., Dulak, J., Rimbach, G., (2008). Effect of ochratoxin A on redox-regulated transcription factors, antioxidant enzymes and glutathione-S-transferase in cultured kidney tubulus cells. *Food Chem. Toxicol.* 46, 2665-2671.

[32] Abdel-Wahhab, M.A., Abdel-Azim, S.H., El-Nekeety, A.A., (2008). Inula crithmoides extract protects against ochratoxin A-induced oxidative stress, clastogenic and mutagenic alterations in male rats. *Toxicon.* 52, 566-573.

[33] Arbillaga, L., Azqueta, A., van Delft, J.H., de Cerain, A.L., (2007). *In vitro* gene expression data supporting a DNA non-reactive genotoxic mechanism for ochratoxin A. *Toxicol. Appl. Pharmacol.* 220, 216-224.

[34] Hassen, W., Ayed-Boussema, I., Bouslimi, A., Bacha, H., (2007). Heat shock proteins (Hsp 70) response is not systematic to cell stress: case of the mycotoxin ochratoxin A. *Toxicology*. 242, 63-70.

[35] Domijan, A.M., Peraica, M., Vrdoljak, A.L., Radić, B., Žlender, V., Fuchs, R., (2007). The involvement of oxidative stress in ochratoxin A and fumonisin B1 toxicity in rats. *Mol. Nutr. Food Res.* 51, 1147-1151.

[36] Palma, N., Cinelli, S., Sapora, O., Wilson, S.H., Dogliotti, E., (2007). Ochratoxin A-induced mutagenesis in mammalian cells is consistent with the production of oxidative stress. *Chem. Res. Toxicol.* 20, 1031-1037.

[37] Giacomo, C.D., Acquaviva, R., Piva, A., Sorrenti, V., Vanella, L., Piva, G., Casadei, G., Fauci, L.L., Ritieni, A., Bognanno, M., (2007). Protective effect of cyanidin 3-O-β-d-glucoside on ochratoxin A-mediated damage in the rat. *Br. J. Nutr.* 98, 937-943.

[38] Costa, S., Utan, A., Cervellati, R., Speroni, E., Guerra, M., (2007). Catechins: natural free-radical scavengers against ochratoxin A-induced cell damage in a pig kidney cell line (LLC-PK1). *Food Chem. Toxicol.* 45, 1910-1917.

[39] Lambert, D., Padfield, P.J., McLaughlin, J., Cannell, S., O'Neill, C.A., (2007). Ochratoxin A displaces claudins from detergent resistant membrane microdomains. *Biochem. Biophys. Res. Commun.* 358, 632-636.

[40] Sava, V., Velasquez, A., Song, S., Sanchez-Ramos, J., (2007). Adult hippocampal neural stem/progenitor cells *in vitro* are vulnerable to the mycotoxin ochratoxin-A. *Toxicol. Sci.* 98, 187-197.

[41] Sutken, E., Aral, E., Ozdemir, F., Uslu, S., Alatas, O., Colak, O., (2007). Protective role of melatonin and coenzyme Q10 in ochratoxin A toxicity in rat liver and kidney. *Int. J. Toxicol.* 26, 81-87.

[42] Arbillaga, L., Azqueta, A., Ezpeleta, O., de Cerain, A.L., (2006). Oxidative DNA damage induced by ochratoxin A in the HK-2 human kidney cell line: evidence of the relationship with cytotoxicity. *Mutagenesis*. 22, 35-42.

[43] Cavin, C., Delatour, T., Marin-Kuan, M., Holzhäuser, D., Higgins, L., Bezencon, C., Guignard, G., Junod, S., Richoz-Payot, J., Gremaud, E., (2007). Reduction in antioxidant defenses may contribute to ochratoxin A toxicity and carcinogenicity. *Toxicol. Sci.* 96, 30-39.

[44] Stemmer, K., Ellinger-Ziegelbauer, H., Ahr, H.-J., Dietrich, D.R., (2007). Carcinogen-specific gene expression profiles in short-term treated Eker and wild-type rats indicative of pathways involved in renal tumorigenesis. *Cancer Res.* 67, 4052-4068.

[45] Gagliano, N., Dalle Donne, I., Torri, C., Migliori, M., Grizzi, F., Milzani, A., Filippi, C., Annoni, G., Colombo, P., Costa, F., (2006). Early cytotoxic effects of ochratoxin A in rat liver: a morphological, biochemical and molecular study. *Toxicology*. 225, 214-224.

[46] Sava, V., Reunova, O., Velasquez, A., Harbison, R., Sanchez-Ramos, J., (2006). Acute neurotoxic effects of the fungal metabolite ochratoxin-A. *Neurotoxicology*. 27, 82-92.

[47] Sava, V., Reunova, O., Velasquez, A., Sanchez-Ramos, J., (2006). Can low level exposure to ochratoxin-A cause parkinsonism? *J. Neurol. Sci.* 249, 68-75.

[48] Marin-Kuan, M., Nestler, S., Verguet, C., Bezencon, C., Piguet, D., Mansourian, R., Holzwarth, J., Grigorov, M., Delatour, T., Mantle, P., (2006). A toxicogenomics approach to identify new plausible epigenetic mechanisms of ochratoxin a carcinogenicity in rat. *Toxicol. Sci.* 89, 120-134.

[49] Ferrante, M.C., Bilancione, M., Raso, G.M., Esposito, E., Iacono, A., Zaccaroni, A., Meli, R., (2006). Expression of COX-2 and hsp72 in peritoneal macrophages after an acute ochratoxin A treatment in mice. *Life Sci.* 79, 1242-1247.
[50] Zeljezic, D., Domijan, A.-M., Peraica, M., (2006). DNA damage by ochratoxin A in rat kidney assessed by the alkaline comet assay. *Braz. J. Med. Biol. Res.* 39, 1563-1568.
[51] Domijan, A.-M., Želježić, D., Kopjar, N., Peraica, M., (2006). Standard and Fpg-modified comet assay in kidney cells of ochratoxin A-and fumonisin B 1-treated rats. *Toxicology*. 222, 53-59.
[52] Tozlovanu, M., Faucet-Marquis, V., Pfohl-Leszkowicz, A., Manderville, R.A., (2006). Genotoxicity of the hydroquinone metabolite of ochratoxin A: structure-activity relationships for covalent DNA adduction. *Chem. Res. Toxicol.* 19, 1241-1247.
[53] Domijan, A.-M., Rudes, K., Peraica, M., (2005). The effect of ochratoxin A on the concentration of protein carbonyls in rats. *Arh. Hig. Rada Toksikol./Arch. Ind. Hyg. Toxicol.* 56, 311-315.
[54] Kamp, H.G., Eisenbrand, G., Janzowski, C., Kiossev, J., Latendresse, J.R., Schlatter, J., Turesky, R.J., (2005). Ochratoxin A induces oxidative DNA damage in liver and kidney after oral dosing to rats. *Mol. Nutr. Food Res.* 49, 1160-1167.
[55] Kamp, H.G., Eisenbrand, G., Schlatter, J., Würth, K., Janzowski, C., (2005). Ochratoxin A: induction of (oxidative) DNA damage, cytotoxicity and apoptosis in mammalian cell lines and primary cells. *Toxicology*. 206, 413-425.
[56] Alvarez-Erviti, L., Leache, C., Gonzalez-Penas, E., De Cerain, A.L., (2005). Alterations induced *in vitro* by ochratoxin A in rat lymphoid cells. *Hum. Exp. Toxicol.* 24, 459-466.
[57] Guerra, M., Galvano, F., Bonsi, L., Speroni, E., Costa, S., Renzulli, C., Cervellati, R., (2005). Cyanidin-3-O-β-glucopyranoside, a natural free-radical scavenger against aflatoxin B1-and ochratoxin A-induced cell damage in a human hepatoma cell line (Hep G2) and a human colonic adenocarcinoma cell line (CaCo-2). *Br. J. Nutr.* 94, 211-220.
[58] Abramson, D., (1987). Measurement of ochratoxin A in barley extracts by liquid chromatography-mass spectrometry. *J. Chromatogr. A*. 391, 315-320.
[59] Mally, A., Pepe, G., Ravoori, S., Fiore, M., Gupta, R.C., Dekant, W., Mosesso, P., (2005). Ochratoxin A causes DNA damage and cytogenetic effects but no DNA adducts in rats. *Chem. Res. Toxicol.* 18, 1253-1261.
[60] Hundhausen, C., Bösch-Saadatmandi, C., Augustin, K., Blank, R., Wolffram, S., Rimbach, G., (2005). Effect of vitamin E and polyphenols on ochratoxin A-induced cytotoxicity in liver (HepG2) cells. *J. Plant Physiol.* 162, 818-822.
[61] Abdel - Wahhab, M.A., Abdel - Galil, M.M., El - Lithey, M., (2005). Melatonin counteracts oxidative stress in rats fed an ochratoxin A contaminated diet. *J. Pineal Res.* 38, 130-135.
[62] Russo, A., La Fauci, L., Acquaviva, R., Campisi, A., Raciti, G., Scifo, C., Renis, M., Galvano, G., Vanella, A., Galvano, F., (2005). Ochratoxin A-induced DNA damage in human fibroblast: protective effect of cyanidin 3-O-β-D-glucoside. *The Journal of nutritional biochemistry*. 16, 31-37.
[63] Zurich, M.-G., Lengacher, S., Braissant, O., Monnet-Tschudi, F., Pellerin, L., Honegger, P., (2005). Unusual astrocyte reactivity caused by the food mycotoxin ochratoxin A in aggregating rat brain cell cultures. *Neuroscience*. 134, 771-782.

[64] Knasmuller, S., Cavin, C., Chakraborty, A., Darroudi, F., Majer, B.J., Huber, W.W., Ehrlich, V.A., (2004). Structurally related mycotoxins ochratoxin A, ochratoxin B, and citrinin differ in their genotoxic activities and in their mode of action in human-derived liver (HepG2) cells: implications for risk assessment. *Nutr. Cancer*. 50, 190-197.

[65] Schaaf, G., Nijmeijer, S., Maas, R., Roestenberg, P., De Groene, E., Fink-Gremmels, J., (2002). The role of oxidative stress in the ochratoxin A-mediated toxicity in proximal tubular cells. *Biochimica et Biophysica Acta (BBA)-Molecular Basis of Disease*. 1588, 149-158.

[66] Petrik, J., Žanić-Grubišić, T., Barišić, K., Pepeljnjak, S., Radić, B., Ferenčić, Ž., Čepelak, I., (2003). Apoptosis and oxidative stress induced by ochratoxin A in rat kidney. *Arch. Toxicol.* 77, 685-693.

[67] Ehrlich, V., Darroudi, F., Uhl, M., Steinkellner, H., Gann, M., Majer, B., Eisenbauer, M., Knasmüller, S., (2002). Genotoxic effects of ochratoxin A in human-derived hepatoma (HepG2) cells. *Food Chem. Toxicol.* 40, 1085-1090.

[68] Lühe, A., Hildebrand, H., Bach, U., Dingermann, T., Ahr, H.-J., (2003). A new approach to studying ochratoxin A (OTA)-induced nephrotoxicity: expression profiling *in vivo* and *in vitro* employing cDNA microarrays. *Toxicol. Sci.* 73, 315-328.

[69] Son, W.-C., Kamino, K., Lee, Y.-S., Kang, K.-S., (2003). Lack of effects of sodium 2-mercaptoethane sulfonate (mesna) on Ochratoxin A induced renal tumorigenicity following life-time oral administration of Ochratoxin A in DA and Lewis rats. *Toxicol. Lett.* 142, 19-27.

[70] Pfohl-Leszkowicz, A., Bartsch, H., Azémar, B., Mohr, U., Estève, J., Castegnaro, M., (2002). MESNA protects rats against nephrotoxicity but not carcinogenicity induced by ochratoxin A, implicating two separate pathways. *Facta Univ. Ser. Med. Biol.* 9, 57-63.

[71] Stemmer, K., Ellinger-Ziegelbauer, H., Ahr, H.-J., Dietrich, D.R., (2009). Molecular characterization of preneoplastic lesions provides insight on the development of renal tumors. *The American journal of pathology*. 175, 1686-1698.

[72] De Saeger, S., Dumoulin, F., Van Peteghem, C., (2004). Quantitative determination of ochratoxin A in kidneys by liquid chromatography/mass spectrometry. *Rapid Commun. Mass Spectrom.* 18, 2661-2668.

[73] Dall'Asta, C., Galaverna, G., Dossena, A., Marchelli, R., (2004). Reversed-phase liquid chromatographic method for the determination of ochratoxin A in wine. *J. Chromatogr. A*. 1024, 275-279.

[74] Zheng, M.Z., Richard, J.L., Binder, J., (2006). A review of rapid methods for the analysis of mycotoxins. *Mycopathologia*. 161, 261-273.

[75] Visconti, A., De Girolamo, A., (2005). Fitness for purpose–Ochratoxin A analytical developments. *Food Addit. Contam.* 22, 37-44.

[76] Leitner, A., Zöllner, P., Paolillo, A., Stroka, J., Papadopoulou-Bouraoui, A., Jaborek, S., Anklam, E., Lindner, W., (2002). Comparison of methods for the determination of ochratoxin A in wine. *Anal. Chim. Acta*. 453, 33-41.

[77] Lindenmeier, M., Schieberle, P., Rychlik, M., (2004). Quantification of ochratoxin A in foods by a stable isotope dilution assay using high-performance liquid chromatography–tandem mass spectrometry. *J. Chromatogr. A*. 1023, 57-66.

[78] Degelmann, P., Becker, M., Herderich, M., Humpf, H.-U., (1999). Determination of ochratoxin A in beer by high-performance liquid chromatography. *Chromatographia*. 49, 543-546.

[79] Gilbert, J., (1999). Overview of mycotoxin methods, present status and future needs. *Nat. Toxins*. 7, 347-352.

[80] Zhang, A., Ma, Y., Feng, L., Wang, Y., He, C., Wang, X., Zhang, H., (2011). Development of a sensitive competitive indirect ELISA method for determination of ochratoxin A levels in cereals originating from Nanjing, China. *Food Control*. 22, 1723-1728.

[81] Liang, Y., Liu, X. J., Liu, Y., Yu, X. Y., Fan, M. T., (2008). Synthesis of three haptens for the class-specific immunoassay of O, O-dimethyl organophosphorus pesticides and effect of hapten heterology on immunoassay sensitivity. *Anal. Chim. Acta*. 615, 174-183.

[82] Candlish, A., Stimson, W.H., Smith, J.E., (1987). Determination of ochratoxin A by monoclonal antibody-based enzyme immunoassay. *Journal-Association of Official Analytical Chemists*. 71, 961-964.

[83] Yu, F.-y., Chi, T.-f., Liu, B.-h., Su, C.-c., (2005). Development of a sensitive enzyme-linked immunosorbent assay for the determination of ochratoxin A. *J. Agric. Food Chem*. 53, 6947-6953.

[84] Clarke, J., Marquardt, R., Oosterveld, A., Frohlich, A., Madrid, F., Dawood, M., (1993). Development of a quantitative and sensitive enzyme-linked immunosorbent assay for ochratoxin A using antibodies from the yolk of the laying hen. *J. Agric. Food Chem*. 41, 1784-1789.

[85] Gendloff, E., Casale, W., Ram, B., Tai, J., Pestka, J., Hart, L., (1986). Hapten-protein conjugates prepared by the mixed anhydride method: Cross-reactive antibodies in heterologous antisera. *J. Immunol. Methods*. 92, 15-20.

[86] Köhler, G., Milstein, C., (1975). Continuous cultures of fused cells secreting antibody of predefined specificity. *Nature*. 256, 495-497.

[87] Fodey, T., Leonard, P., O'Mahony, J., O'Kennedy, R., Danaher, M., (2011). Developments in the production of biological and synthetic binders for immunoassay and sensor-based detection of small molecules. *TrAC, Trends Anal. Chem*. 30, 254-269.

[88] Chung, Y.-J., Jarvis, B.B., Tak, H., Pestka, J.J., (2003). Immunochemical assay for satratoxin G and other macrocyclic trichothecenes associated with indoor air contamination by Stachybotrys chartarum. *Toxicol. Mech. Methods*. 13, 247-252.

[89] Blake, C., Al-Bassam, M., Gould, B., Marks, V., Bridges, J., Riley, C., (1982). Simultaneous enzyme immunoassay of two thyroid hormones. *Clin. Chem*. 28, 1469-1473.

[90] Maragos, C., Busman, M., Sugita-Konishi, Y., (2006). Production and characterization of a monoclonal antibody that cross-reacts with the mycotoxins nivalenol and 4-deoxynivalenol. *Food Addit. Contam*. 23, 816-825.

[91] Tothill, I. E., (2001). Biosensors developments and potential applications in the agricultural diagnosis sector. *Comput. Electron. Agric*. 30, 205-218.

[92] Marco, M.-P., Gee, S., Hammock, B. D., (1995). Immunochemical techniques for environmental analysis I. Immunosensors. *TrAC, Trends Anal. Chem*. 14, 341-350.

[93] González-Martínez, M.Á., Puchades, R., Maquieira, Á., (2001). Comparison of multianalyte immunosensor formats for on-line determination of organic compounds. *Anal. Chem*. 73, 4326-4332.

[94] Maragos, C. M., Thompson, V. S., (1999). Fiber-optic immunosensor for mycotoxins. *Nat. Toxins*. 7, 371-376.

[95] Urraca, J.L., Benito-Peña, E., Pérez-Conde, C., Moreno-Bondi, M.C., Pestka, J.J., (2005). Analysis of zearalenone in cereal and swine feed samples using an automated flow-through immunosensor. *J. Agric. Food Chem.* 53, 3338-3344.

[96] Strachan, N., John, P., Millar, I., (1997). Application of an automated particle - based immunosensor for the detection of aflatoxin B1 in foods. *Food Agric. Immunol.* 9, 177-183.

[97] Carter, R. M., Blake, R. C., Mayer, H. P., Echevarria, A. A., Nguyen, T. D., Bostanian, L.A., (2000). A fluorescent biosensor for detection of zearalenone.

[98] Sapsford, K.E., Ngundi, M.M., Moore, M.H., Lassman, M.E., Shriver-Lake, L.C., Taitt, C. R., Ligler, F.S., (2006). Rapid detection of foodborne contaminants using an Array Biosensor. *Sensors Actuators B: Chem.* 113, 599-607.

[99] Ligler, F. S., Taitt, C.R., Shriver-Lake, L.C., Sapsford, K.E., Shubin, Y., Golden, J.P., (2003). Array biosensor for detection of toxins. *Anal. Bioanal. Chem.* 377, 469-477.

[100] Carter, R., Jacobs, M., Lubrano, G., Guilbault, G., (1997). Rapid detection of aflatoxin B1 with immunochemical optrodes. *Anal. Lett.* 30, 1465-1482.

[101] Weller, M. G., Schuetz, A. J., Winklmair, M., Niessner, R., (1999). Highly parallel affinity sensor for the detection of environmental contaminants in water. *Anal. Chim. Acta*. 393, 29-41.

[102] Homola, J., Yee, S.S., Gauglitz, G., (1999). Surface plasmon resonance sensors: review. *Sensors Actuators B: Chem.* 54, 3-15.

[103] Jorn, C., Lai, E.P., (2005). Interaction of ochratoxin A with molecularly imprinted polypyrrole film on surface plasmon resonance sensor. *React. Funct. Polym.* 63, 171-176.

[104] Dunne, L., Daly, S., Baxter, A., Haughey, S., O'Kennedy, R., (2005). Surface Plasmon Resonance-Based Immunoassay for the Detection of Aflatoxin B1 Using Single-Chain Antibody Fragments. *Spectrosc. Lett.* 38, 229-245.

[105] Daly, S., Dillon, P., Manning, B., Dunne, L., Killard, A., O'KENNEDY, R., (2002). Production and characterization of murine single chain Fv antibodies to aflatoxin B 1 derived from a pre-immunized antibody phage display library system. *Food Agric. Immunol.* 14, 255-274.

[106] Maragos, C., Jolley, M., Nasir, M., (2002). Fluorescence polarization as a tool for the determination of deoxynivalenol in wheat. *Food Addit. Contam.* 19, 400-407.

[107] Lippolis, V., Pascale, M., Visconti, A., (2006). Optimization of a Fluorescence Polarization Immunoassay for Rapid Quantification of Deoxynivalenol in Durum Wheat–Based Products. *Journal of Food Protection®*. 69, 2712-2719.

[108] Shim, W. B., Kolosova, A. Y., Kim, Y. J., Yang, Z. Y., Park, S. J., Eremin, S. A., Lee, I.S., Chung, D. H., (2004). Fluorescence polarization immunoassay based on a monoclonal antibody for the detection of ochratoxin A. *Int. J. Food Sci. Tech.* 39, 829-837.

[109] Colbert, D., Childerstone, M., (1987). Multiple drugs of abuse in urine detected with a single reagent and fluorescence polarization. *Clin. Chem.* 33, 1921-1923.

[110] Bao, J. J., (1997). Capillary electrophoretic immunoassays. *Journal of Chromatography B: Biomedical Sciences and Applications*. 699, 463-480.

[111] Yeung, W., Luo, G., Wang, Q., Ou, J., (2003). Capillary electrophoresis-based immunoassay. *J. Chromatogr. B*. 797, 217-228.

[112] Cho, Y.-J., Lee, D.-H., Kim, D.-O., Min, W.-K., Bong, K.-T., Lee, G.-G., Seo, J.-H., (2005). Production of a monoclonal antibody against ochratoxin A and its application to immunochromatographic assay. *J. Agric. Food Chem.* 53, 8447-8451.

[113] Wang, S., Quan, Y., Lee, N., Kennedy, I.R., (2006). Rapid determination of fumonisin B1 in food samples by enzyme-linked immunosorbent assay and colloidal gold immunoassay. *J. Agric. Food Chem.* 54, 2491-2495.

[114] Collin, R., Schneider, E., Briggs, L., Towers, N., (1998). Development of immunodiagnostic field tests for the detection of the mycotoxin, sporidesmin A. *Food Agric. Immunol.* 10, 91-104.

[115] Zhang, C., Zhang, Y., Wang, S., (2006). Development of multianalyte flow-through and lateral-flow assays using gold particles and horseradish peroxidase as tracers for the rapid determination of carbaryl and endosulfan in agricultural products. *J. Agric. Food Chem.* 54, 2502-2507.

[116] Usleber, E., Schneider, E., Maertlbauer, E., Terplan, G., (1993). Two formats of enzyme immunoassay for 15-acetyldeoxynivalenol applied to wheat. *J. Agric. Food Chem.* 41, 2019-2023.

[117] De Saeger, S., Van Peteghem, C., (1996). Dipstick enzyme immunoassay to detect Fusarium T-2 toxin in wheat. *Appl. Environ. Microbiol.* 62, 1880-1884.

[118] Schneider, E., Usleber, E., Maertlbauer, E., (1995). Rapid detection of fumonisin B1 in corn-based food by competitive direct dipstick enzyme immunoassay/enzyme-linked immunofiltration assay with integrated negative control reaction. *J. Agric. Food Chem.* 43, 2548-2552.

[119] Schneider, E., Usleber, E., Märtlbauer, E., Dietrich, R., Terplan, G., (1995). Multimycotoxin dipstick enzyme immunoassay applied to wheat. *Food Addit. Contam.* 12, 387-393.

[120] Sibanda, L., De Saeger, S., Barna-Vetro, I., Van Peteghem, C., (2002). Development of a solid-phase cleanup and portable rapid flow-through enzyme immunoassay for the detection of ochratoxin A in roasted coffee. *J. Agric. Food Chem.* 50, 6964-6967.

[121] De Saeger, S., Van Peteghem, C., (1999). Flow-through membrane-based enzyme immunoassay for rapid detection of ochratoxin A in wheat. *Journal of Food Protection®*. 62, 65-69.

[122] De Saeger, S., Sibanda, L., Desmet, A., Van Peteghem, C., (2002). A collaborative study to validate novel field immunoassay kits for rapid mycotoxin detection. *Int. J. Food Microbiol.* 75, 135-142.

[123] Pal, A., Dhar, T.K., (2004). An analytical device for on-site immunoassay. Demonstration of its applicability in semiquantitative detection of aflatoxin B1 in a batch of samples with ultrahigh sensitivity. *Anal. Chem.* 76, 98-104.

[124] Schneider, E., Curtui, V., Seidler, C., Dietrich, R., Usleber, E., Märtlbauer, E., (2004). Rapid methods for deoxynivalenol and other trichothecenes. *Toxicol. Lett.* 153, 113-121.

[125] Farré, M., Kantiani, L., Pérez, S., Barceló, D., (2009). Sensors and biosensors in support of EU Directives. *TrAC, Trends Anal. Chem.* 28, 170-185.

[126] Palchetti, I., Mascini, M., (2008). Electroanalytical biosensors and their potential for food pathogen and toxin detection. *Anal. Bioanal. Chem.* 391, 455-471.

[127] Laschi, S., Centi, S., Mascini, M., (2011). Electrochemical arrays coupled with magnetic separators for immunochemistry. *Bioanalytical Reviews*. 3, 11-25.

[128] Grieshaber, D., MacKenzie, R., Voeroes, J., Reimhult, E., (2008). Electrochemical biosensors-sensor principles and architectures. *Sensors*. 8, 1400-1458.
[129] Ricci, F., Volpe, G., Micheli, L., Palleschi, G., (2007). A review on novel developments and applications of immunosensors in food analysis. *Anal. Chim. Acta*. 605, 111-129.
[130] Radi, A.-E., Muñoz-Berbel, X., Cortina-Puig, M., Marty, J.-L., (2009). An electrochemical immunosensor for ochratoxin A based on immobilization of antibodies on diazonium-functionalized gold electrode. *Electrochimica Acta*. 54, 2180-2184.
[131] Rameil, S., Schubert, P., Grundmann, P., Dietrich, R., Märtlbauer, E., (2010). Use of 3-(4-hydroxyphenyl) propionic acid as electron donating compound in a potentiometric aflatoxin M 1-immunosensor. *Anal. Chim. Acta*. 661, 122-127.
[132] Kwon, I., Lee, H.-H., Choi, J., Shin, J.-K., Seo, S.-H., Choi, S.-W., Chun, H.S., (2011). Extended-Gate Metal Oxide Semiconductor Field Effect Transistor-Based Biosensor for Detection of Deoxynivalenol. *Jpn. J. Appl. Phys*. 50, 06GL08.
[133] Romanazzo, D., Ricci, F., Volpe, G., Elliott, C.T., Vesco, S., Kroeger, K., Moscone, D., Stroka, J., Van Egmond, H., Vehniäinen, M., (2010). Development of a recombinant Fab-fragment based electrochemical immunosensor for deoxynivalenol detection in food samples. *Biosens. Bioelectron*. 25, 2615-2621.
[134] Arduini, F., Errico, I., Amine, A., Micheli, L., Palleschi, G., Moscone, D., (2007). Enzymatic spectrophotometric method for aflatoxin B detection based on acetylcholinesterase inhibition. *Anal. Chem*. 79, 3409-3415.
[135] Moressi, M., Zon, A., Fernandez, H., Rivas, G., Solis, V., (1999). Amperometric quantification of Alternaria mycotoxins with a mushroom tyrosinase modified carbon paste electrode. *Electrochem. Commun*. 1, 472-476.
[136] Pohanka, M., Musilek, K., Kuca, K., (2010). Evaluation of aflatoxin B1-acetylcholinesterase dissociation kinetic using the amperometric biosensor technology: prospect for toxicity mechanism. *Protein and peptide letters*. 17, 340-342.
[137] Alonso-Lomillo, M.A., Domínguez-Renedo, O., Ferreira-Gonçalves, L., Arcos-Martínez, M.J., (2010). Sensitive enzyme-biosensor based on screen-printed electrodes for Ochratoxin A. *Biosens. Bioelectron*. 25, 1333-1337.
[138] Vidal, J.C., Bonel, L., Castillo, J.R., (2008). A Modulated Tyrosinase Enzyme-Based Biosensor for Application to the Detection of Dichlorvos and Atrazine Pesticides. *Electroanalysis*. 20, 865-873.
[139] Vidal, J.C., Bonel, L., Duato, P., Castillo, J.R., (2011). Improved electrochemical competitive immunosensor for ochratoxin A with a biotinylated monoclonal antibody capture probe and colloidal gold nanostructuring. *Analytical Methods*. 3, 977-984.
[140] Maragos, C.M., (2009). Recent advances in the development of novel materials for mycotoxin analysis. *Anal. Bioanal. Chem*. 395, 1205-1213.
[141] Li, P., Zhang, Q., Zhang, W., (2009). Immunoassays for aflatoxins. *TrAC, Trends Anal. Chem*. 28, 1115-1126.
[142] Mascini, M., Palchetti, I., Marrazza, G., (2001). DNA electrochemical biosensors. *Fresenius J. Anal. Chem*. 369, 15-22.
[143] Siontorou, C. G., Nikolelis, D. P., Miernik, A., Krull, U. J., (1998). Rapid methods for detection of Aflatoxin M 1 based on electrochemical transduction by self-assembled metal-supported bilayer lipid membranes (s-BLMs) and on interferences with transduction of DNA hybridization. *Electrochimica Acta*. 43, 3611-3617.

[144] Oliveira, S., Diculescu, V., Palleschi, G., Compagnone, D., Oliveira-Brett, A., (2007). Electrochemical oxidation of ochratoxin A at a glassy carbon electrode and in situ evaluation of the interaction with deoxyribonucleic acid using an electrochemical deoxyribonucleic acid-biosensor. *Anal. Chim. Acta.* 588, 283-291.

[145] Rai, M., Varma, A. (2010) *Mycotoxins in food, feed and bioweapons.* Springer.

[146] Blanco-López, M., Lobo-Castanon, M., Miranda-Ordieres, A., Tunon-Blanco, P., (2004). Electrochemical sensors based on molecularly imprinted polymers. *TrAC, Trends Anal. Chem.* 23, 36-48.

[147] Baggiani, C., Anfossi, L., Giovannoli, C., (2008). Molecular imprinted polymers as synthetic receptors for the analysis of myco-and phyco-toxins. *Analyst.* 133, 719-730.

[148] Weiss, R., Freudenschuss, M., Krska, R., Mizaikoff, B., (2003). Improving methods of analysis for mycotoxins: molecularly imprinted polymers for deoxynivalenol and zearalenone. *Food Addit. Contam.* 20, 386-395.

[149] Vidal, J.C., Bonel, L., Ezquerra, A., Hernández, S., Bertolín, J.R., Cubel, C., Castillo, J.R., (2013). Electrochemical affinity biosensors for detection of mycotoxins: A review. *Biosens. Bioelectron.* 49, 146-158.

[150] Shephard, G., Berthiller, F., Burdaspal, P., Crews, C., Jonker, M., Krska, R., Lattanzio, V., MacDonald, S., Malone, R., Maragos, C., (2012). Developments in mycotoxin analysis: an update for 2011-2012. *World Mycotoxin Journal.* 6, 3-30.

[151] Vidal, J.C., Bonel, L., Ezquerra, A., Duato, P., Castillo, J.R., (2012). An electrochemical immunosensor for ochratoxin A determination in wines based on a monoclonal antibody and paramagnetic microbeads. *Anal. Bioanal. Chem.* 403, 1585-1593.

[152] Fernández-Baldo, M.A., Bertolino, F.A., Fernández, G., Messina, G.A., Sanz, M.I., Raba, J., (2011). Determination of Ochratoxin A in apples contaminated with Aspergillus ochraceus by using a microfluidic competitive immunosensor with magnetic nanoparticles. *Analyst.* 136, 2756-2762.

[153] Prieto-Simón, B., Campàs, M., Marty, J.-L., Noguer, T., (2008). Novel highly-performing immunosensor-based strategy for ochratoxin A detection in wine samples. *Biosens. Bioelectron.* 23, 995-1002.

[154] Kaushik, A., Solanki, P. R., Sood, K., Ahmad, S., Malhotra, B. D., (2009). Fumed silica nanoparticles–chitosan nanobiocomposite for ochratoxin-A detection. *Electrochem. Commun.* 11, 1919-1923.

[155] Kaushik, A., Solanki, P. R., Ansari, A. A., Ahmad, S., Malhotra, B. D., (2009). A nanostructured cerium oxide film-based immunosensor for mycotoxin detection. *Nanotechnology.* 20, 055105.

[156] Solanki, P.R., Kaushik, A., Manaka, T., Pandey, M.K., Iwamoto, M., Agrawal, V.V., Malhotra, B.D., (2010). Self-assembled monolayer based impedimetric platform for food borne mycotoxin detection. *Nanoscale.* 2, 2811-2817.

[157] Liu, X.-P., Deng, Y.-J., Jin, X.-Y., Chen, L.-G., Jiang, J.-H., Shen, G.-L., Yu, R.-Q., (2009). Ultrasensitive electrochemical immunosensor for ochratoxin A using gold colloid-mediated hapten immobilization. *Anal. Biochem.* 389, 63-68.

[158] Palchetti, I., Mascini, M., (2012). Electrochemical nanomaterial-based nucleic acid aptasensors. *Anal. Bioanal. Chem.* 402, 3103-3114.

[159] Tothill, I., (2011). Biosensors and nanomaterials and their application for mycotoxin determination. *World Mycotoxin Journal.* 4, 361-374.

[160] Campàs, M., Garibo, D., Prieto-Simón, B., (2012). Novel nanobiotechnological concepts in electrochemical biosensors for the analysis of toxins. *Analyst*. 137, 1055-1067.

[161] Vig, A., Radoi, A., Muñoz-Berbel, X., Gyemant, G., Marty, J.-L., (2009). Impedimetric aflatoxin M 1 immunosensor based on colloidal gold and silver electrodeposition. *Sensors Actuators B: Chem*. 138, 214-220.

[162] Ellington, A.D., Szostak, J.W., (1990). *In vitro* selection of RNA molecules that bind specific ligands. *Nature*. 346, 818-822.

[163] Mairal, T., Özalp, V.C., Sánchez, P.L., Mir, M., Katakis, I., O'Sullivan, C.K., (2008). Aptamers: molecular tools for analytical applications. *Anal. Bioanal. Chem*. 390, 989-1007.

[164] Cruz-Aguado, J.A., Penner, G., (2008). Determination of ochratoxin A with a DNA aptamer. *J. Agric. Food Chem*. 56, 10456-10461.

[165] Cruz-Aguado, J.A., Penner, G., (2008). Fluorescence polarization based displacement assay for the determination of small molecules with aptamers. *Anal. Chem*. 80, 8853-8855.

[166] Bonel, L., Vidal, J.C., Duato, P., Castillo, J.R., (2011). An electrochemical competitive biosensor for ochratoxin A based on a DNA biotinylated aptamer. *Biosens. Bioelectron*. 26, 3254-3259.

[167] Prabhakar, N., Matharu, Z., Malhotra, B., (2011). Polyaniline Langmuir–Blodgett film based aptasensor for ochratoxin A detection. *Biosens. Bioelectron*. 26, 4006-4011.

[168] McKeague, M., Bradley, C.R., Girolamo, A.D., Visconti, A., Miller, J.D., DeRosa, M.C., (2010). Screening and initial binding assessment of fumonisin B1 aptamers. *International journal of molecular sciences*. 11, 4864-4881.

[169] Le, L. C., Cruz-Aguado, J. A., Penner, G. A. (2010). Google Patents.

[170] Shim, W.-B., Mun, H., Joung, H.-A., Ofori, J.A., Chung, D.-H., Kim, M.-G., (2014). Chemiluminescence competitive aptamer assay for the detection of aflatoxin B1 in corn samples. *Food Control*. 36, 30-35.

[171] Yang, C., Lates, V., Prieto-Simón, B., Marty, J.-L., Yang, X., (2012). Aptamer-DNAzyme hairpins for biosensing of Ochratoxin A. *Biosens. Bioelectron*. 32, 208-212.

[172] Barthelmebs, L., Jonca, J., Hayat, A., Prieto-Simon, B., Marty, J.-L., (2011). Enzyme-linked aptamer assays (ELAAs), based on a competition format for a rapid and sensitive detection of ochratoxin A in wine. *Food Control*. 22, 737-743.

[173] Chen, X., Huang, Y., Duan, N., Wu, S., Ma, X., Xia, Y., Zhu, C., Jiang, Y., Wang, Z., (2013). Selection and identification of ssDNA aptamers recognizing zearalenone. *Anal. Bioanal. Chem*. 405, 6573-6581.

[174] Veršilovskis, A., Huybrecht, B., Tangni, E., Pussemier, L., De Saeger, S., Callebaut, A., (2011). Cross-reactivity of some commercially available deoxynivalenol (DON) and zearalenone (ZEN) immunoaffinity columns to DON-and ZEN-conjugated forms and metabolites. *Food Additives & Contaminants: Part A*. 28, 1687-1693.

[175] Chapuis-Hugon, F., du Boisbaudry, A., Madru, B., Pichon, V., (2011). New extraction sorbent based on aptamers for the determination of ochratoxin A in red wine. *Anal. Bioanal. Chem*. 400, 1199-1207.

[176] De Girolamo, A., McKeague, M., Miller, J. D., DeRosa, M. C., Visconti, A., (2011). Determination of ochratoxin A in wheat after clean-up through a DNA aptamer-based solid phase extraction column. *Food Chem*. 127, 1378-1384.

[177] Rhouati, A., Paniel, N., Meraihi, Z., Marty, J.-L., (2011). Development of an oligosorbent for detection of ochratoxin A. *Food Control*. 22, 1790-1796.
[178] Cunha, S., Faria, M., Fernandes, J., (2009). Determination of patulin in apple and quince products by GC–MS using 13 C 5–7 patulin as internal standard. *Food Chem.* 115, 352-359.
[179] Turner, N. W., Subrahmanyam, S., Piletsky, S. A., (2009). Analytical methods for determination of mycotoxins: a review. *Anal. Chim. Acta*. 632, 168-180.
[180] Rasmussen, R. R., Storm, I. M. L. D., Rasmussen, P. H., Smedsgaard, J., Nielsen, K. F., (2010). Multi-mycotoxin analysis of maize silage by LC-MS/MS. *Anal. Bioanal. Chem.* 397, 765-776.
[181] Omurtag, G., Yazicioglu, D., (2001). Occurrence of T-2 toxin in processed cereals and pulses in Turkey determined by HPLC and TLC. *Food Addit. Contam.* 18, 844-849.
[182] Ghali, R., Ghorbel, H., Hedilli, A., (2009). Fumonisin determination in tunisian foods and feeds. ELISA and HPLC methods comparison. *J. Agric. Food Chem.* 57, 3955-3960.
[183] Xu, Y., Huang, Z.-B., He, Q.-H., Deng, S.-Z., Li, L.-S., Li, Y.-P., (2010). Development of an immunochromatographic strip test for the rapid detection of deoxynivalenol in wheat and maize. *Food Chem.* 119, 834-839.
[184] Yang, C., Wang, Y., Marty, J.-L., Yang, X., (2011). Aptamer-based colorimetric biosensing of Ochratoxin A using unmodified gold nanoparticles indicator. *Biosens. Bioelectron.* 26, 2724-2727.
[185] Daniel, W.L., Han, M.S., Lee, J.-S., Mirkin, C.A., (2009). Colorimetric nitrite and nitrate detection with gold nanoparticle probes and kinetic end points. *J. Am. Chem. Soc.* 131, 6362-6363.
[186] Guo, Z., Ren, J., Wang, J., Wang, E., (2011). Single-walled carbon nanotubes based quenching of free FAM-aptamer for selective determination of ochratoxin A. *Talanta*. 85, 2517-2521.
[187] Sheng, L., Ren, J., Miao, Y., Wang, J., Wang, E., (2011). PVP-coated graphene oxide for selective determination of ochratoxin A via quenching fluorescence of free aptamer. *Biosens. Bioelectron.* 26, 3494-3499.
[188] Zhang, J., Zhang, X., Yang, G., Chen, J., Wang, S., (2013). A signal-on fluorescent aptasensor based on Tb 3+ and structure-switching aptamer for label-free detection of ochratoxin A in wheat. *Biosens. Bioelectron.* 41, 704-709.
[189] Wang, L., Ma, W., Chen, W., Liu, L., Ma, W., Zhu, Y., Xu, L., Kuang, H., Xu, C., (2011). An aptamer-based chromatographic strip assay for sensitive toxin semi-quantitative detection. *Biosens. Bioelectron.* 26, 3059-3062.
[190] Wang, L., Chen, W., Ma, W., Liu, L., Ma, W., Zhao, Y., Zhu, Y., Xu, L., Kuang, H., Xu, C., (2011). Fluorescent strip sensor for rapid determination of toxins. *Chem. Commun.* 47, 1574-1576.
[191] Tong, P., Zhang, L., Xu, J.-J., Chen, H.-Y., (2011). Simply amplified electrochemical aptasensor of ochratoxin A based on exonuclease-catalyzed target recycling. *Biosens. Bioelectron.* 29, 97-101.
[192] Barthelmebs, L., Hayat, A., Limiadi, A.W., Marty, J.-L., Noguer, T., (2011). Electrochemical DNA aptamer-based biosensor for OTA detection, using superparamagnetic nanoparticles. *Sensors Actuators B: Chem.* 156, 932-937.

[193] Wu, S., Duan, N., Wang, Z., Wang, H., (2011). Aptamer-functionalized magnetic nanoparticle-based bioassay for the detection of ochratoxin a using upconversion nanoparticles as labels. *Analyst*. 136, 2306-2314.

Chapter 4

OCHRATOXIN A -INDUCED NEPHROTOXICITY

ABSTRACT

Ochratoxin A (OTA) is a mycotoxin that commonly occurs in food products. A large number of mechanistic studies on OTA-induced nephrotoxicity and renal tumor formation are available. The association between the consumption of OTA-contaminated agricultural products and renal carcinoma occurrence has received widespread attention. Nephrotoxicity is the most prominent of the various toxicities associated with OTA. Recently, a number of investigations noted that OTA disrupts mitosis, which results in blocked or asymmetric cell division. This cell division disruption could increase the risk of aneuploidy and exacerbate OTA-induced tumor formation.

Several lines of evidence derived from animal experiments suggest that OTA induces renal system damage and even renal carcinogenesis. In different models, including humans and several animals, such as rats, mice, pigs, dogs, and broiler chickens, OTA-induced renal system damage, including chronic kidney disease, kidney cancer, acute phase response, inflammation, and urothelial carcinoma has been detected. Meanwhile, the mechanisms of OTA-induced renal toxicity have been inferred from pathological changes in the kidney, metabolomic changes in the plasma and urine, global changes in body fluid, protein synthesis inhibition, mitochondrial dysfunction, oxidative stress, DNA damage, calcium homeostasis, the disruption of cell–cell signaling pathways, cell death pathways, cell survival pathways, DNA-OTA adducts, miRNA regulation, DNA methylation, nephrotoxicity amelioration, etc. Here, we review the studies of OTA-induced nephrotoxicity in recent years and summarize how OTA leads to nephrotoxicity.

Keywords: Ochratoxin A, nephrotoxicity, kidney diseases, cancer, cell signaling pathways

1. OCHRATOXIN A-INDUCED RENAL SYSTEM DAMAGE

OTA exerts extensive toxicological effects in animal models, including nephrotoxicity, nephrocarcinogenicity, teratogenicity, neurotoxicity, and immunotoxicity. Mycotoxin-contaminated feed may be ingested and threaten human health. Mycotoxins have been shown to induce a range of adverse effects in the animals' pathological processes, including acute toxicity, developmental toxicity, reproductive toxicity, immunosuppression, and carcinogenicity.

The kidney is an important target of a number of mycotoxins because the high renal blood flow delivers high levels of xenobiotics to the kidneys, and the presence of a variety of transporters enable the active uptake and intracellular accumulation of toxins in the kidney [1]. OTA has been classified as having "clear evidence of carcinogenic activity" based on the increased incidence of renal tumors in male and female rats following chronic exposure [2]. The purpose of this section is to summarize the current research progress on OTA-induced renal system damage.

1.1. Mechanisms of Action for Renal Tumor Formation by Ochratoxin A

The occurrence and toxicity of OTA has been intensively studied. The kidney is the main target organ of OTA toxicity in all animal species tested. Dogs and pigs (LD50 0.2 and 1 mg/kg bw, respectively) have been shown to be more sensitive to the acute toxicity of OTA than rats (LD50 20–30 mg/kg bw) and mice (LD50 46–58 mg/kg bw) [3]. The exposure of pigs to an OTA concentration of 8 mg/kg bw for 90 days was reported to cause impaired kidney function [3]. The chronic feeding of pigs with an OTA-contaminated diet resulted in progressive nephropathy characterized by the degeneration and atrophy of proximal tubules and interstitial fibrosis [4, 5].

In short-term studies in rats, OTA-induced nephrotoxicity is characterized by polyuria, the disorganization of the tubule arrangement, single cell degeneration, and abnormally enlarged mitotic figures, which are usually restricted to the straight segment of the proximal tubule [1]. In short-term exposure, the most striking features of OTA-induced kidney pathology are prominent karyomegaly, polyploidy of proximal tubule cells, and nuclear division without cytokinesis [6].

OTA is also a potent renal carcinogen in rodents. Studies have shown that OTA has the strongest correlation with renal tumor formation in males. A 2-year carcinogenicity bioassay shows that OTA can increase the incidence of renal adenomas and carcinomas. The treatment of male F344 rats with OTA at doses of 70 and 210 μg/kg bw was associated with a high incidence of renal adenomas and carcinomas, whereas no increased tumor rates were observed following exposure to a low dose of 21 μg/kg bw. This research indicates a nonlinear dose-response for renal tumor formation by OTA [2] (Table). In another 2-year bioassay in B6C3F1 mice, benign and malignant renal tumors were significantly generated in male mice fed OTA-contaminated diets at a concentration of 40 μg/kg (combined incidence 63%) [7].

In recent decades, dietary human exposure to OTA has been suspected of involvement in Balkan endemic nephropathy (BEN). BEN is a progressive tubulointerstitial kidney disease that occurs in some countryside areas of Bosnia, Bulgaria, Croatia, Romania, and Serbia and is associated with an increased risk for the development of urothelial cancers. The epidemiological data were insufficient to establish a causal link between OTA and BEN, whereas more recent evidence suggests that chronic exposure to aristolochic acid via flour contaminated with the seeds of *Aristolochia clematitis* may play a role in the etiology of BEN [8, 9]. An exposure assessment in Sri Lanka measured mycotoxin levels in the urine of patients with kidney disease: OTA was present in 93.5 percent of these patients and was also found in individuals without kidney disease [10]. The highest levels of OTA in human blood were observed in sampled populations that included persons with kidney disorders, in which the serum levels were as high as 35–100 ng/mL (ppb). In contrast, the mean serum OTA level in Europe was 0.1–2 ng/mL (ppb) [11]. Another study on Tunisians divided people into those with and without renal disease and showed that OTA levels were elevated in the serum of both populations. Moreover, higher levels of OTA were found in people with renal disease. In one report, the mean value of OTA for the healthy control population was 3.3 ± 1.5 ng/mL (ppb) compared to a mean value of 18 ± 7 ng/mL (ppb) in persons with chronic interstitial nephropathy of unknown origin [12].

Table. Incidence of Selected Non-neoplastic and Neoplastic Renal Lesions in F344 Rats Exposed to OTA at NTP Bioassay Doses [13-16]

	♂				♀			
	Control	21 μg /kg b.w.	70 μg /kg b.w.	210 μg /kg b.w.	Control	21 μg /kg b.w.	70 μg /kg b.w.	210 μg /kg b.w.
14 days								
Karyomegaly	0/5	0/5	0/5	5/5				
Degeneration	0/5	0/5	1/5	5/5[a]				
Proliferation	0/5	0/5	0/5	5/5				
28 days								
Karyomegaly	0/5	0/5	5/5	5/5				
Degeneration	0/5	0/5	5/5	5/5				
Proliferation	0/5	0/5	5/5	5/5				
90 days								
Karyomegaly	0/5	0/5	5/5	5/5				
Degeneration	0/5	0/5	5/5	5/5				
Proliferation	0/5	0/5	5/5	5/5				
9 months								
Karyomegaly	0/15	0/15	15/15	15/15	0/15	0/15	15/15	15/15
Degeneration	0/15	0/15	0/15	0/15	0/15	0/15	0/15	0/15
15 months								
Karyomegaly	0/15	0/14	14/14	15/15	0/15	0/14	15/15	15/15
Degeneration	0/15	0/14	0/14	15/15	0/15	0/14	0/15	0/15
2 years								
Karyomegaly	0/50	1/51	51/51	50/50	0/50	8/51	50/50	50/50
Degeneration	0/50	0/51	50/51	49/50	0/50	0/51	49/50	49/50
Proliferation	0/50	0/51	10/51	26/50	0/50	0/51	3/50	16/50
Adenoma	1/50	1/51	6/51	10/50	0/50	0/51	1/50	5/50
Carcinoma	0/50	0/51	16/51	30/50	0/50	0/51	1/50	3/50
Combined	1/50	1/51	20/51	36/50	0/50	0/51	2/50	8/50

[a]Qualitative changes in BrdU-labeling pattern without significant increases in the number of labeled cells.

1.2. Chronic Kidney Disease and Kidney Cancer

Renal tumor formation has been demonstrated with more than 100 chemicals, and remarkably, nearly all types of renal neoplasms are described in human pathology and in small rodents. Several renal carcinogens are thought to induce tumor formation through the direct interaction of the parent compound or a reactive intermediate formed in the kidney with DNA to produce pre-mutagenic lesions.

Chronic kidney disease (CKD) is a serious worldwide public health problem. As defined by the American National Kidney Foundation, the percentage of patients with stage 3 or greater CKD is approximately 5% [17]. Patients with CKD are at increased overall risk for cancer [18]. Increasingly, studies have uncovered the adverse relationship between long-term CKD and cancer. Renal function is thought to be a crucial factor to consider in future comprehensive therapeutic strategies [19].

Cancer, a malignant tumor or malignant neoplasm, is a group of diseases involving abnormal cell growth with the potential to invade or spread to other parts of the body. Kidney cancer can be divided into two types, renal cell carcinoma (RCC) and transitional cell carcinoma (TCC) of the renal pelvis. In the United States, kidney cancer represents the sixth most common malignant neoplasia, with more than 60,000 estimated new cases for 2014 [20].

Although tumor generation is secondary to chronic renal toxicity and compensatory cell proliferation, it has been extensively researched. Toxins may induce a high rate of cell proliferation, decreasing the time available for DNA damage repair, and thus increasing the probability of mutations. These mutations will then lead to the spontaneous transformation of renal epithelial cells and promote the clonal expansion of these cells, thereby increasing the risk of neoplasia [21]. Rached et al. [14] reported tissue regeneration and cell proliferation as key steps of tumor formation that changed after OTA exposure. According to these findings, Mally [16] believes that the OTA inhibition of histone acetyltransferase (HAT) activity causes the disruption of mitosis and cell proliferation and genetic instability, which could be a mechanism of action for renal tumor formation.

1.3. Acute Phase Response and Inflammation

In kidney cells, the acute phase response and inflammation are important to the immune response and cause tissue destruction. Cells receive this signal-mediated response to cellular insults by extracellular factors such as OTA or other toxins and from physical stresses. The kidney tissues of male Wistar rats treated with a low dose (1 mg/kg) of OTA showed acute toxicity [22]. Meanwhile, with regard to acute phase response and inflammation, several proteins and genes were associated with OTA-induced nephrotoxicity.

In this model, OTA induces the up-regulation of ceruloplasmin, which is a copper-dependent acute phase protein that carries more than 95% of the total copper in healthy human plasma and expressed in the kidney during inflammation [23, 24]. A contrapsin-like protease inhibitor, known as a negative acute phase protein, is down-regulated by OTA, suggesting potential support of the acute phase response and cancer development [25]. Ceruloplasmin and contrapsin-like protease inhibitor are likely to be useful as indexes for evaluating the OTA-induced acute phase response [22]. Zinc fingers are small, compact protein subunits that are particularly important in gene regulation. The up-regulated expression of zinc finger proteins by OTA is believed to be an immediate-early response and has also been suggested to play an important role in the repair of tissue lesions [22, 26].

With regard to nephrotoxicity, severe tissue lesions lead to the activation of several mechanisms of blood clotting after OTA treatment. Alpha-fibrinogen and the tissue factor genes play important roles in coagulation after tissue lesion formation. By binding to PAR-1, thrombin mediates the conversion of fibrinogen into fibrin fibers to form blood clots [27]. Transcription factors (TFs) are known to initiate the intrinsic pathway of blood clotting in traumatized tissue and are also involved in the PAR-1-mediated blood clotting pathway [28]. The up-regulated response of alpha-fibrinogen and the tissue factor are associated with the increasing occurrence of necrosis in the OTA-induced histopathological examinations [22]. Tumor necrosis factor-α (TNF-α) is a potent pro-inflammatory factor. It participates in a large variety of biological activities and also may play a role in the context of OTA-induced tubulointerstitial nephritis. OTA potentiated the proapoptotic activity of TNF-α [29].

1.4. Urothelial Carcinomas

OTA is a potent nephrotoxin that is known to damage the proximal tubule [30]. It is worth mentioning that there has been a special concern for toxins that induce urothelial carcinoma. The primary effects in humans related to the exposure to OTA were nephropathies and the development of tumors in the urinary tract [31]. Urothelial carcinoma is a malignant tumor differentiated from the urothelial epithelium that may be involved in pathologies of the kidneys, ureters, bladder, and urethra. In a study of OTA and its metabolite ochratoxin α (OTα), these compounds were detected in 60 urine samples of inhabitants from Coimbra city (Portugal); OTA conjugation with glucuronic acid in human urine occurs, and the presence of OTA was found in 42 samples at concentrations above the LOQ and ranging between 0.021 and 0.105 ng/mL [32]. Another study assessed exposure to OTA in a group of 72 adult inhabitants of the city of Lleida (Spain). The presence of OTA occurrence was greater than that of OTα. The range of concentrations was 0.057–0.562 ng/mL and 0.056–2.894 ng/mL for OTA and OTα, respectively [31]. In upper urinary tract carcinoma, the tumor would become an obstacle in the urinary tract, and the increasing tract pressure could extend to all nephrons and ultimately affect kidney function.

1.5. Biotransformation and Toxicokinetics of Ochratoxin A

Although orally administered OTA is absorbed predominantly from the small intestine, OTA has a very high affinity to plasma proteins, and 99.9% of the circulating OTA is bound to plasma proteins. Thus, OTA can be transported to many organs, especially the kidneys. Maximal plasma concentrations were reached within 24-48 h after the application of a single oral dose of 0.5 mg/kg bw to male and female F344 rats [33]. It is worth noting that due to its minimal metabolism and slow excretion, OTA can maintain a plasma half-life of 230 h [33]. To kidney tissue in rodents and humans, this is a disaster because the kidney has abundant blood vessels, and the tubular reabsorption system keeps OTA in this tissue for a long time. The chemical reactivity of OTA has been proposed to affect metabolic pathways because of its slow biotransformation. There is abundant evidence to suggest that the sex- and organ-specific toxicity of OTA are partly attributable to the toxicokinetics of OTA. Zepnik et al. [33] detected OTA concentrations in the kidneys and livers of male and female F344 rats following a single oral dose of OTA (0.5 mg/kg bw) over a period of 96 days. OTA accumulated in the kidneys to a concentration of 480 pmol/g tissue in males 24 h after OTA administration. In contrast, in the livers of both male and female rats, OTA concentrations were less than 12 pmol/g tissue with a maximum 24 h after administration. It is important to emphasize that OTA levels in the kidneys of female rats were significantly lower (200 pmol/g tissue) than those observed in males, which may be related to the differential expression of organic anion transporters increasing the renal absorption of OTA in male rats [14, 34].

The electrochemical and photochemical oxidation of OTA results in the formation of a quinone derivative of OTA (OTQ)/hydroquinone derivative of OTA (OTQ/OTHQ) redox couple. Through redox cycling, these execrable metabolites lead to reactive oxygen species generation and associated oxidative DNA lesions [35,36]. In addition, the electrophilic OTQ reacts with tissue nucleophiles, as shown by the formation of a glutathione conjugate following the autoxidation of OTHQ in the presence of glutathione [37].

1.6. Uptake into the Proximal Tubule Epithelium

OTA can be specifically transported into the kidney epithelial cells of the straight segment of the proximal tubule. OTA is widely distributed in the kidneys compared with the livers or other tissues in a range of animal species including rats and pigs [33, 38-40]. In a report of male and female F344 rats that were administered a single oral dose of OTA (500 μg/kg bw), the concentrations of OTA in the kidney were significantly higher compared with the liver (>20-fold). Moreover, higher OTA concentrations were found in male rats compared with females [33], consistent with the known gender disparity in tumor formation with chronic OTA exposure.

The basolateral uptake of OTA into proximal tubule cells occurs primarily via the kidney-specific rat organic anion transporter Oat1 (and presumably Oat3), which causes the site-specific accumulation of OTA [41]. Both Oat1 and Oat3 are expressed more abundantly in the kidneys of male versus female rats due to the stimulatory effect of androgens (and progesterone) and the inhibitory effect of estrogens [42, 43].

This androgen-induced expression may be a good explanation for the increased accumulation of OTA in male rat kidneys and the associated susceptibility to OTA toxicity. In contrast, the expression of rat renal Oat2 is localized to the brush border of proximal tubule cells and participates in apical efflux and the tubular secretion of many organic anions, which is higher in female rats [42, 44, 45], suggesting that OTA may be more efficiently secreted in females. In addition, the rat nephron can extensively reabsorb OTA, and the proximal straight tubules are the major site of OTA reabsorption [46, 47]. In proximal tubule cells, several organic anion carriers expressed at the apical membrane contribute to OTA reabsorption in rats, such as the organic anion transporting polypeptide (Oatp1), the kidney-specific organic anion transporter (Oat-k2), organic anion transporter (Oat5), and oligopeptide transporter (Pept2) [46-49]. Again, the endocrine regulation of the expression of apical transporters such as Oatp1 [50], which promotes the reabsorption of organic anions [51, 52], appears to contribute to renal OTA accumulation in male rats. However, the extent of accumulation in the kidneys of female rats appears to be limited by higher tubular secretion combined with lower reabsorption. In one report, a greater than two-fold greater urinary excretion of OTA was found in female F344 rats compared with males after a single oral dose (500 μg/kg bw) [33]. Similarly, interspecies differences in the expression of renal transporters mediating basolateral uptake, secretion, and the reabsorption of OTA may lead to the differences in tumor response between rats and mice. Based on these reports, sufficient evidence exists to indicate that the preferential uptake of OTA into proximal tubule cells via apical and basolateral carriers expressed in a sex-hormone–dependent manner accounts for the site-specific accumulation of OTA, particularly in the male rat kidney [42-44, 53].

2. Pathology of Ochratoxin A-Induced Renal Toxicity

In recent years, OTA-induced nephrotoxicity has been studied more by studying the physiological state of cells. In this part, we summarize some of these OTA-induced pathological changes.

2.1. Pathological Mechanisms

In the kidneys, OTA-induced pathology was present at all dose levels administered with a clear dose response. After the administration of 2 mg/kg of body weight to rats, the entire outer stripe of the outer medulla (OSOM), involving the straight segments of the proximal tubule, was basophilic with some disorganization of the tubule arrangement (Figure 1) [54].

The pictures reveal abundant apoptotic cells. OTA makes the cells degenerate or detach into the tubule lumen. There were also abnormally enlarged and misshapen nuclei scattered through the S3 tubules of the OSOM, representing nuclear division without increasing polyploidy and cytokinesis. Mitotic figures were substantially increased in the affected tubules. The occasional abnormally enlarged nuclei in the figures indicate that the polyploid nuclei have the capacity to continue nuclear division [54].

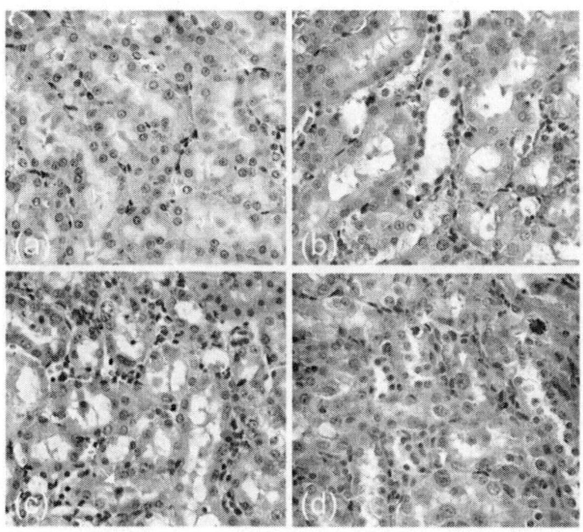

Figure 1. H & E stained kidney sections of a control rat (a) and of rats treated with OTA at 0.25 (b), 1 (c), and 2 (d) mg/kg body weight for 2 weeks. Apoptosis and detachment of cells into the tubulus lumen is frequently observed in S3 tubules of the OSOM in response to OTA treatment (arrow with small arrowhead). Abnormally enlarged and sometimes misshapen nuclei (arrow with large arrowhead) are scattered through the S3 segment, with a clear dose response. Particularly at the highest dose, mitotic figures (arrow with curved arrowhead) are substantially increased compared to controls, and sometimes aberrant mitosis is present [54].

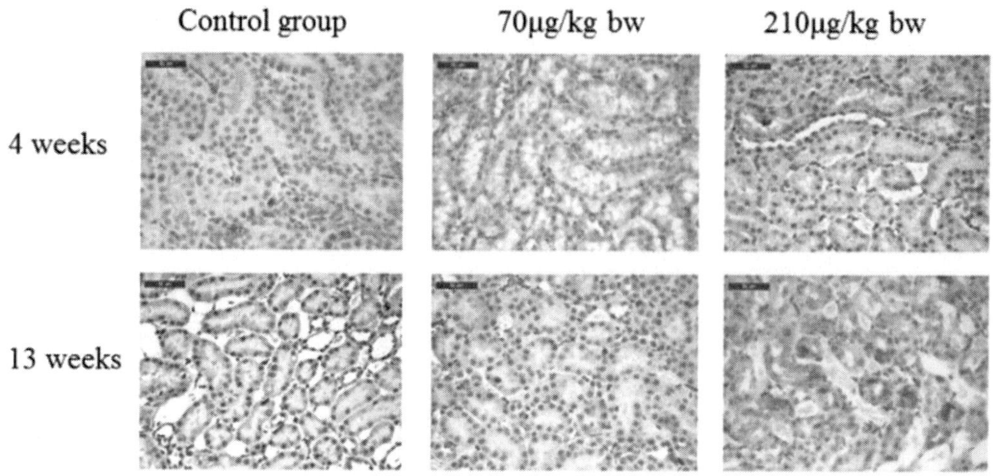

Figure 2. Expression of kidney injury molecule-1 in the kidneys of male F344/N rats treated with OTA at 0, 70, or 210 μg/kg bw for 4 weeks or 13 weeks. Kim-1 was detected on the membrane of renal tubular epithelial cells. Original magnification = 400× [57].

Kidney injury molecule-1 (Kim-1) is a transmembrane glycoprotein that is not expressed in the mature kidney but is up-regulated and shed into the urine following nephrotoxic injury [55]. For this reason, Kim-1 is considered a specific and sensitive biomarker of kidney injury [56]. Kim-1 was located primarily in the tubular epithelial cell membranes with low expression in the cytoplasm. After rats were treated with 70 or 210 μg/kg bw OTA for 4 or 13 weeks, Kim-1 was immunostained (Figure 2). Kim-1 was detectable in the outer stripe of the outer medulla, consistent with the location of damage induced by OTA toxicity observed in the microscopic anatomic pathology [57]. Kidney damage and reduced kidney function are believed to be potent risk factors for heart failure [58], but there are few reports that show the relationship between OTA and heart failure.

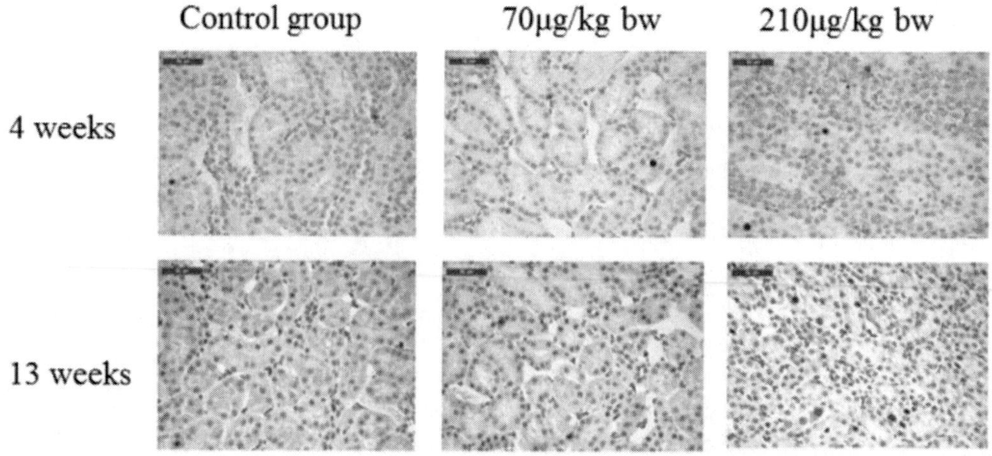

Figure 3. Expression of proliferating cell nuclear antigen (PCNA) in the kidneys of male F344/N rats treated with OTA at 0, 70, or 210 μg/kg bw for 4 weeks or 13 weeks. Original magnification = 400× [57].

3.1. Protein Synthesis Inhibition

Protein synthesis inhibitors stop or slow cell growth or proliferation by disrupting the processes that lead directly to the generation of new proteins, and these inhibitors then interfere with normal metabolism. Toxins such as OTA also function via protein synthesis inhibition both *in vitro* and *in vivo* through the down-regulation of RNA and protein expression levels [59].

Protein synthesis inhibitors work at different stages of mRNA translation into proteins. Aminoacyl-tRNA synthetases play a key role in the faithful translation of the genetic code because they catalyze the attachment of each amino acid to its cognate tRNA. The human proteome includes two sets of aaRSs that are encoded by nuclear genes and are involved in either cytosolic or mitochondrial protein synthesis [77]. OTA has been demonstrated to inhibit protein synthesis, including the inhibition of aminoacyl-tRNA synthetases, valyl-tRNA synthetase, and phenylalanyl-tRNA synthetase expression [78-80]. Moreover, there are other proteins that are regulated in the initiation, elongation, and termination stages. Marin-Kuan et al. [81] observed that the expression of prostaglandin F2 receptor negative regulator and the eukaryotic translation initiation factor 4E binding protein 1 were up-regulated; elongation factor 1-alpha 1, which promotes the GTP-dependent binding of aminoacyl-tRNA to the A-site of ribosomes during protein biosynthesis, was down-regulated [82]. This is direct evidence indicating OTA-induced protein inhibition in the rat kidney.

3.2. Mitochondrial Dysfunction

Because mitochondria represent a significant point at which most metabolic and signaling pathways converge, these cellular organelles play an important role in many types of metabolic reprogramming, such as in the context of stress, disease, and cancer [83]. It is common for many toxins such as OTA to induce mitochondrial dysfunction. Cuezva et al. [84] believe that one of the cancer cell phenotypes is the ineluctable metabolic reprogramming resulting from the onset of cellular proliferation. During cell proliferation, the synchronous fluctuation of increased glycolysis concurrent with restrained oxidative phosphorylation is indispensably required. Thanks to the key role of mitochondria in energy metabolism and their unquestionable tumor suppressor function, scientists have studied the mitochondria as new therapeutic targets for diseases and cancer. In addition, they are the central hinge for the biosynthesis of nucleic acids, amino acids, and lipids in cells and the main providers of NADH and NADPH; thus, they are significant regulators of the redox balance [85, 86].

Mitochondria are particularly susceptible to oxidative damage [87]. The major endogenous source of cellular ROS is the mitochondrial electron transport chain; OTA has the ability to disturb the mitochondrial electron transport chain [88]. Additionally, OTA increased the mitochondrial membrane potential and led to giant mitochondria and the formation of straight extensions on a porcine renal proximal tubular epithelial cell line [30]. Our team for the first time used a mitoproteomics approach based on the iTRAQ technique to globally investigate the potential mechanisms of OTA-induced renal toxicity, utilizing the mitochondria of human embryonic kidney 293 (HEK 293) cells [89]. In the study, the changes of cell viability, intracellular ROS production, and $\Delta\Psi m$ caused by OTA and/or

NAC were assessed. Pre-treatment with NAC could reverse the OTA-induced ROS increase and ΔΨm decrease. According to the iTRAQ-based mitoproteomics, the expression of sixty-six proteins was significantly changed as a result of OTA treatment, and the dynamically changed proteins were primarily involved in the following pathways: the mitochondrial electron transport chain (mETC), protein synthesis, and stress response and cell death. Interestingly, the adverse effects induced by OTA at the protein level could be almost entirely reversed by NAC.

First, the electron transport chain in the mitochondria (mETC) was disturbed by OTA. In the study, OTA suppressed the expression of complex II (SDHA and SDHB), complex III (UQCRFS1 and UQCRC2), Complex V (ATP5B), TIMM44, and TOMM70A. In fact, the majority of the respiratory chain subunits in human (including complexes I, III, IV, and V) are not encoded directly by mitochondrial DNA. After the transcription by nuclear genes and translation on cytosolic ribosomes, the subunit proteins are finally imported into mitochondria, during which the translocase of the outer membrane (TOM) and translocase of the inner membrane (TIM) act as import machinery [90]. Therefore, the down-regulation of TIMM44 and TOMM70A as a result of OTA treatment may most likely affect the import of complexes III and V. OTA also inhibited the associations between Complex II (SDHA and SDHB) and ACO2, implying that the TCA cycle would be conspicuously disrupted. Additionally, SDHA and SDHB were reported to be important tumor suppressor genes [91, 92], and thus repressing the expression of SDHA and SDHB may be one mechanism of the carcinogenicity of OTA. However, the pretreatment with NAC successfully reversed the down-regulation of SDHA protein caused by OTA and the down-regulation of SDHB at both the gene transcription and protein levels.

A strong relationship among complex V (ATP5B), VDAC1, and HK1 was reported. In the study, VDAC1 showed a down-regulation after OTA treatment. VDAC1 is an essential component of MPTP [93], and it participates in the limited exchange of ADP, ATP, and small hydrophilic molecules across the outer membrane. Cells with a deficiency of VDAC1 have a 4-fold reduced capacity for ATP synthesis and low levels of both ATP and ADP. Because there is a strong correlation between cell growth and ATP accumulation, this result suggested that limited metabolites are exchanged between mitochondria and the cytosol [94]. OTA induced the down-regulation of both VDAC1 and ATP5B, forecasting a decrease of the ATP level, which is a signal of energy metabolism consistent with a disrupted TCA cycle.

The secondary ATP binding site in HK1 is a key factor in the interaction between HK1 and VDAC1. A large accumulation of ATP always appears in cancer cells, and they may protect the cells from apoptosis in a synergistic way because the tertiary structure of VDAC1 could be stabilized by high levels of ATP and therefore avoid pro-apoptotic factor binding [95]. According to Abu-Hamad's group, HK1 binding to VDAC1 could inhibit cytochrome c release, thus promoting the survival of tumor cells [96]. In the present study, the results implied that the inhibition of HK1 induced by OTA could promote the release of cytochrome c from mitochondria into the cytosol. These results also suggested another mechanism by which OTA might accelerate apoptosis by inducing a decrease of ATP levels.

In addition, mETC is the major endogenous source of cellular ROS [88]. The increase of ROS was observed in this study, and it may be a consequence of the disruption of complex III (UQCRFS1 and UQCRC2) caused by OTA treatment. Moreover, OTA also decreased ΔΨm, likely as a side effect of the perturbation of both complex III and V. Reasonably, pre-treatment with the antioxidant NAC reversed the increase of ROS and the decrease of ΔΨm induced by OTA. Furthermore, OTA-induced adverse effects at the protein level could also be reversed by NAC, demonstrating that NAC can protect HEK 293 cells from OTA-induced mitochondrial damage.

Second, OTA can inhibit protein synthesis. TSFM, IARS2, EPRS, and LRPPRC were down-regulated upon OTA treatment. TSFM, IARS2, and EPRS are mitochondrially encoded mRNAs, and their reductions may be a result of the down-regulation of LRPPRC because it was reported that decreased LRPPRC expression could lead to lower steady-state levels of most of the mitochondrially encoded mRNAs [97]. Thereby, OTA achieved its inhibition of protein translation and biosynthesis.

Third, OTA can induce a stress response and cell death. TRAP1, a representative mitochondrial HSP, was down-regulated at both the transcriptional and protein levels by OTA, consequently indicating a decrease of the conversion of ATP to AD+ Q. In the meantime, the down-regulation of ATP5B could lead to a reduction of the conversion of ADP to ATP. Additionally, a study from Masuda et al. revealed that the suppression of TRAP1 expression might play a significant role in the formation of ROS, leading to the induction of apoptosis [98]. NAC pre-treatment reversed the decrease of TRAP1 expression in addition to the increase of ROS.

LONP1 is another protein that was down-regulated by OTA at both the transcriptional and protein levels. Mitochondrial ACO2 plays an important role in the TCA cycle and is very sensitive to oxidative inactivation. Because LONP1 can selectively recognize and degrade the oxidized, hydrophobic form of ACO2, its down-regulation may lead to a decrease of the degradation of oxidized ACO2, thus perturbing the TCA cycle. ATP-stimulated LONP1 was believed to be a crucial defense against oxidative stress [99]. Consequently, the defenses against ROS should be reduced by the down-regulation of ATP5B and LONP1 induced by OTA.

OTA also induced the down-regulation of RPS3, DIABLO, AIFM1, and OPA1, among which, RPS3 has a role in the defense against genotoxicity. It is reported that RPS3-deficient cells were more vulnerable to genotoxic stress than the control cells, and the sensitivity could be reversed by the overexpression of RPS3 [100]. Therefore, the suppression of RPS3 might be a contributor to the genotoxicity of OTA.

DIABLO activates caspases in the cytochrome c/Apaf-1/caspase-9 pathway [101]. The release of cytochrome c and DIABLO coincides with $\Delta\Psi m$ depolarization [102]. It should be noted that the loss of $\Delta\Psi m$ and the release of AIF from mitochondria are key steps in apoptosis [103]. Therefore, this is one of the OTA-induced apoptosis pathways mediated by the decrease in $\Delta\Psi m$ and the release of DIABLO and AIFM1 from mitochondria into the cytosol.

OPA1 plays an important role in maintaining normal cristae structure and function, thus preserving inner membrane structure and protecting cells from apoptosis [104]. Consistent with the down-regulation of OPA1, OTA induced the down-regulation of AIFM1, DIABLO, and ATP5B and decreased $\Delta\Psi m$. These changes may contribute to OTA-induced apoptosis and have been validated by Western blot analysis of the expression and transfer of CYCS.

In summary, Figure 5 presents a hypothetical model of the toxicity mechanism of OTA-induced mitochondrial damage.

Mitochondrial electron transport complex II (SDHA and SDHB), complex III (UQCRFS1 and UQCRC2), and Complex V (ATP5B) are down-regulated upon OTA treatment. The repression of VDAC1 and ATP5B led to a reduction of ATP levels, which might be responsible for OTA-induced apoptosis via the decrease of OPA1 and responsible for TCA cycle dysfunction via the suppression of LONP1 expression and the interaction between VDAC1 and HK1. Additionally, the anti-oxidative capacity of HEK 293 cells decreased in line with the suppression of LONP1. Moreover, the suppression of TRAP1 may be partly understood as a response to the decrease of ATP levels, and its deficiency and OTA-induced changes in UQCRC2 resulted in the increase of ROS and the decrease of ΔΨm, which was validated at the molecular level by evidence for the perturbation of complexes III and V. In the end, the decrease in ΔΨm and the release of AIFM1 and DIABLO from mitochondria to the cytosol contributed to OTA-induced apoptosis.

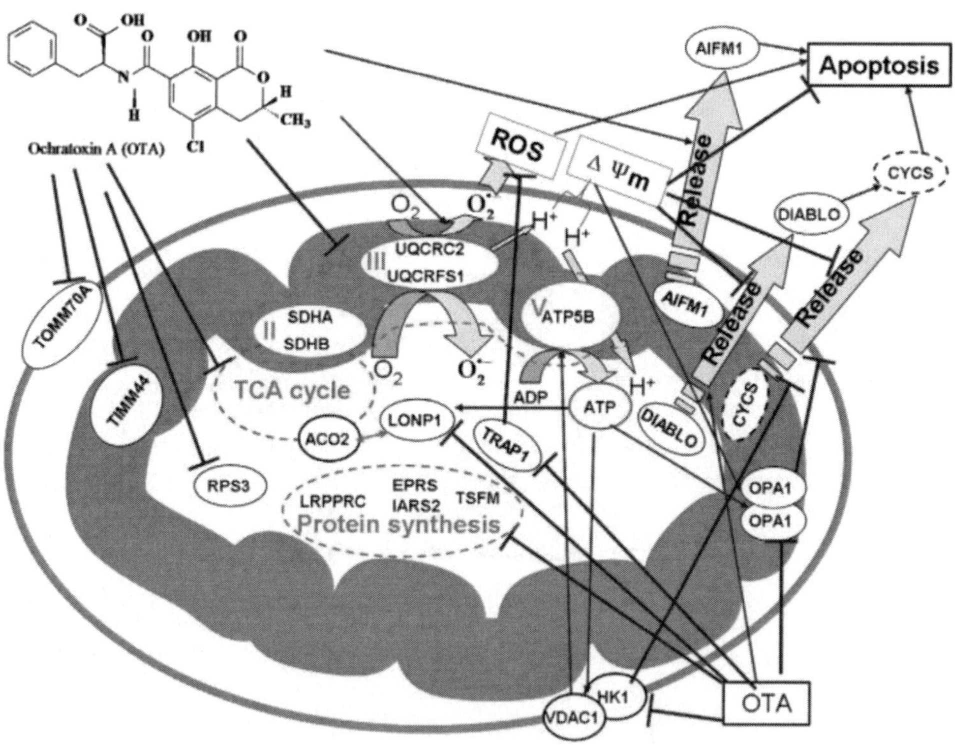

Figure 5. A hypothetical model of the toxicity mechanism for OTA-induced mitochondrial damage. Inhibition (⊥), promotion (→), and interactions (↔) in response to OTA [105].

Next, we investigated the role of multifunctional protein LONP1, which was found down-regulated by OTA in the previously mentioned mitoproteomics, in OTA-induced renal cytotoxicity [106].

Using RNA interference, we first constructed LONP1-deficient HEK 293 cells which were more sensitive to OTA according to cell viability assays. Then, we applied iTRAQ-based proteomics to identify significant variations of protein expression between LONP1-deficient cells and control cells in the absence and presence of OTA. Thirty-four differentially expressed proteins were found between LONP1-deficient cells and control cells, and 44 differentially expressed proteins were found as a result of both LONP1 knockdown and OTA exposure. Function and pathway analysis revealed that LONP1 played a protective role in resistance to OTA through 4 processes: defending against OTA-induced oxidative stress in the mitochondria; regulating the synthesis, modification, and repair of protein; maintaining the balance of carbohydrate metabolism; and assisting in mtDNA maintenance.

First, the down-regulation of LONP1 increases oxidative stress. OTA can induce an ROS increase and even cause oxidative damage in cells, which is a major mechanism for its toxicity. Proteomic analysis showed that several proteins contributing to defense against oxidative stress were up-regulated as a result of LONP1 deficiency. SODC is a well-known antioxidant enzyme that defends against ROS, particularly superoxide anion radicals [107]. GSTP1 is a type of glutathione S-transferase capable of using GSH as an electron donor in peroxidative reactions, providing protection from oxidative byproducts [108]. It plays an important role in the detoxification of xenobiotics and endogenous toxicants [109]. PRDX6 is involved in the redox regulation of cells and plays a role in the regulation of phospholipid turnover and protection against oxidative injury [110]. In addition, the up-regulation of heat shock proteins (HSP60 and HS90A) was also demonstrated as protection against oxidative stress [111]. The up-regulation of these antioxidants was most likely a sign of the increase of ROS and oxidative stress caused by LONP1 deficiency because LONP1 was reported to participate in the defense against free radical damage in mitochondria [112].

We presumed that one reason why LONP1-deficient cells were more sensitive to OTA was their inability to efficiently respond to the OTA-induced oxidative stress in the mitochondria. The inability of LONP1 to defend against oxidative stress led to the up-regulation of proteins with similar functions to allow cells to survive in the presence of OTA.

Second, low concentrations of OTA disrupts protein metabolism and induces resistance to apoptosis in LONP1-deficient cells. LONP1 plays an important role in mitochondrial protein quality control and can degrade oxidatively damaged proteins in the mitochondria [113]. Its down-regulation presumably has a negative effect on mitochondrial protein metabolism. The Hsp family of molecular chaperones has a similar function as LONP1 in protein quality control [114]. Some members of the Hsp family, such as HSP90A and HSP60, were up-regulated as a result of LONP1 deficiency, and the up-regulation of HSC70 was remarkable in response to OTA according to the proteome results.

OTA not only disrupted mitochondrial protein metabolism; several ER proteins expressed at different levels and involved in protein metabolism were also detected in the proteome. CALR is a calcium-binding chaperone that promotes protein folding, oligomeric assembly, and quality control in the ER through the calreticulin/calnexin cycle [115]. PDIA1 is a multifunctional protein catalyzing the formation, breakage, and rearrangement of disulfide bonds [116]. GLU2B is related to protein folding and post-translational protein modification [117]. Because LONP1 localizes in the mitochondria, it is more likely that the up-regulation of ER proteins is not a direct result of LONP1 deficiency. They are most likely recruited for maintaining normal protein synthesis and modification in LONP1-deficient cells when threatened by OTA. Thus, OTA is found to disturb protein synthesis and modification and may induce ER stress-induced apoptosis.

LONP1-deficient cells are more vulnerable to OTA. However, low concentrations of OTA presumably induce anti-apoptotic action instead of eliciting excess cell death because the proteome results showed a trend of resistance to apoptosis. For example, LEG7, which is upstream of JNK activation and cytochrome c release, was inhibited. In addition, LMNA and LMNB1, components of the nuclear envelope involved in the execution phase of apoptosis as substrates for caspases, were down-regulated [118]. The degradation of these proteins was a sign of nuclear degradation during apoptosis. They were not degraded but up-regulated in this study, which was evidence against apoptosis.

Third, carbohydrate metabolism was accelerated as a consequence of LONP1 deficiency and OTA treatment. Some crucial enzymes in the TCA cycle and glycolysis were up-regulated as a result of the down-regulation of LONP1. IDHP and NNTM have roles in the production of NADPH, and MDHM is a key enzyme in the TCA cycle [119-121]. Their up-regulation was a sign of TCA cycle acceleration or energy production. One more indicator of energy production increase was the up-regulation of PGK1, an ATP-generating enzyme of the glycolytic pathway [122]. In response to OTA, two other glycolytic enzymes, ENOA and KPYM, were up-regulated. Coincidentally, KPYM also catalyzes an ATP-generating process [123]. The enhancement of carbohydrate metabolism and energy production induced by OTA was also observed in Wang's research [124]. They detected many proteins involved in carbohydrate metabolism enhanced in Arabidopsis thaliana seedlings treated with OTA.

One possible explanation is that both LONP1 deficiency and OTA treatment brought the cell to a high-energy-demand state to adapt itself to the intracellular changes or to maintain regular metabolic activities under certain stresses.

Finally, LONP1 assists in mtDNA maintenance. LONP1 is a protein component of mitochondrial nucleoids with the ability to bind mtDNA [125]. It is capable of degrading and regulating proteins during the replication and expression of mtDNA, and it can remodel mitochondrial nucleoids. Matsushima's group showed that Lon protease regulated mitochondrial DNA copy numbers and transcription by selectively degrading mitochondrial transcription factor A [126]. The results showed that the down-regulation of LONP1 expression caused Histone H4 to have a sharp reduction, which has a negative effect on mtDNA replication, repair, and transcription regulation. OTA down-regulated the expression of several proteins involved in DNA and RNA metabolic processes. DDX52 is related to RNA transport and degradation and has role in transcription and translation, and SSRP1 participates in DNA replication and DNA damage repair. This most likely is an indication that OTA can induce damage to both DNA and RNA metabolism in LONP1-deficient cells. Other significant proteins that may participate in OTA-induced renal toxicity include OPA1 and Smac.

OPA1, a dynamin GTPase located in the mitochondria, is a key component in the outer and inner mitochondrial membrane dynamics. Because OPA1 is able to form a diffusion barrier for proteins that are stored in mitochondrial cristae, it is essential for mitochondrial fusion and the regulation of apoptosis. Proteolytic processing, as a response to apoptotic signals, may contribute to the disassembly of OPA1 oligomers and the release of the caspase activator CYC into the mitochondrial intermembrane space [127]. It has been reported that the suppression of the expression of OPA1 via shRNA led to increased mitochondrial fragmentation, cytochrome c release from the mitochondria, and apoptosis as a consequence [128]. In contrast, overexpressing OPA1 can reverse this phenomenon [129]. Additionally, according to Baricault et al. the apoptosis induction, $\Delta\Psi m$ dissipation, or the inhibition of ATP synthase could lead to a decrease of mitochondrial ATP levels, which would also have an impact on OPA1 processing [130].

Smac is a pro-apoptotic molecule that is released from mitochondria to the cytosol during apoptosis [131]. It was reported that the release of CYC and SMAC coincides with the loss of $\Delta\Psi m$ [102]. Moreover, the loss of $\Delta\Psi m$ is key step in apoptosis [103]. Therefore, the mitochondrial depolarization and the release of Smac from mitochondria to the cytosol may regulate OTA-induced apoptosis. The OTA-induced decrease of Smac in the mitochondria might imply the execution of apoptosis because Smac is being released to the cytosol [105].

3.3. Oxidative Stress and DNA Damage

Oxidative stress is essentially an imbalance between the systemic generation of reactive oxygen species and a biological system's ability to detoxify the reactive intermediates or to repair the resulting damage. At this point, oxidative stress can cause disruptions in the normal mechanisms of cell signal transduction.

Previous studies demonstrated that OTA induced nephrotoxicity through oxidative DNA damage *in vitro* [68,132]. OTA-induced nephrotoxicity resulting from oxidative stress was also reported *in vivo* [133,134]. OTA-induced oxidative stress was also considered a source of some cancers in both *in vivo* and *in vitro* studies [135-137]. Male F344 rats (6-7 weeks old) were gavaged with OTA at different doses (0, 70, or 210 μg/kg bw, denoted as C, L, and H groups, respectively) for 4 or 13 weeks (5 days per week). After the administration of OTA, the kidney weights were decreased in the high-dose group compared to the control group at 13 weeks.

Clusterin is induced in a wide variety of acute and chronic experimental renal diseases [138]. The expression of clusterin was significantly induced in low- and high-dose OTA groups at 13 weeks. Vimentin expression is a predictor of renal dysfunction, and it was increased significantly in the kidneys of OTA-treated groups at 13 weeks. Vimentin was a sensitive and specific marker for conventional renal cell carcinoma. Vimentin contributes to cancer cell migration, invasion, and metastasis [139]. The JAK2/STAT3/vimentin signaling pathway participates in the regulation of the proliferation and migration of human colon cancer cells [140].

SOD is an important indicator of oxidative stress. However, the effects of OTA on SOD activity in the kidneys is not conclusive [141-145]. Qi et al. [57] showed that no significant change in SOD was observed in the kidney of OTA-treated groups. MDA is a lipid peroxidation product formed by ROS interaction with polyunsaturated lipids. The change of MDA induced by OTA is controversial [142, 146, 147]. In the present study of rats, the renal lipid peroxide changes were not significantly different between the OTA-treated and control groups [57].

The production of reactive oxygen species (including free radicals and peroxides) is a particularly destructive aspect of oxidative stress. Some of the less reactive of these species (such as superoxide) can be converted by oxidoreduction reactions with transition metals or other redox cycling compounds into more aggressive radical species [148]. This progression can cause extensive cellular damage. In the kidney, oxidative stress is regarded as a deregulated pathway after OTA administration to rats. One study found that OTA induced a dose-dependent increase in reactive oxygen species in primary rat proximal tubule cells and in the LLC-PK1 cell line, which led to a depletion of intracellular glutathione and cell death [149]. Another study found that low doses of OTA could activate apoptotic processes and induce oxidative damage in both proximal and distal tubule cells in Wistar rat kidneys [150].

ROS can function as "redox messengers" in intracellular signaling. Accumulating evidence has indicated that ROS exert several effects on essential biological processes, including normal cell growth, programmed cell death, and cellular senescence [151]. The molecular probe DCFH-DA can be used to elevate the steady state levels of ROS. Under treatment with 12.5, 25, and 50 μM OTA for 24 h, the production of ROS was increased 1.19-, 1.85-, and 4.24-fold, respectively, compared with control cells [152].

The majority of genes related to oxidative stress share the nuclear factor-erythroid 2-related factor 2 (Nrf2) and the ARE motifs (antioxidant regulatory element) in their promoter region. Marin-Kuan et al. [81] observed a reduction in Nrf2-dependent gene expression in the kidney of rats fed chronically with a diet containing OTA. Nrf2 is a redox-sensitive transcription factor that regulates the expression of antioxidant and cytoprotective genes. Nrf2 deficiency enhances susceptibility to both ischemic and nephrotoxic acute kidney injury and identifies this transcription factor [153]. Nrf2 can be negatively regulated by cytoplasmic Kelch-like ECH associated protein 1 (Keap1), thereby providing an inducible antioxidant defense. The antioxidant properties of Nrf2 are thought to be exerted primarily by stimulating the transcription of antioxidant proteins [154]. The Keap1-Nrf2 pathway regulates both mitochondrial and cytosolic ROS production through NADPH oxidase. OTA has been reported to inhibit Nrf2 activation and Nrf2 gene transcription; together with OTA-induced Nrf2 protein depletion, this would render the cell defenseless against physiological and compound-induced oxidative stress [155].

NF-κB controls the transcription of DNA, cytokine production, and cell survival. In many studies of toxin-induced stress effects, NF-κB has been studied because it belongs to the category of "rapid-acting" primary transcription factors. These factors are present in cells in an inactive state and do not require new protein synthesis to become activated; this class includes c-Jun, STATs, and nuclear hormone receptors. NF-κB activation is highly variable and can be effected by reactive oxygen species (ROS), tumor necrosis factor alpha (TNFα), interleukin 1-beta (IL-1β), bacterial lipopolysaccharides (LPS), isoproterenol, cocaine, and ionizing radiation [156, 157]. OTA significantly increased NF-κB activity in cell cultures [158]. It is well documented that this activation by OTA is associated with cell growth regulatory pathways, cancer progress, or even cancer development.

The overproduction of ROS can induce irreversible oxidative damage to DNA or other transient or permanent cellular alterations, thus causing genomic instability and consequent malignant transformation [159]. Cellular DNA damage is involved in mutagenesis, external (exogenous) and internal metabolic (endogenous) processes, and the development of cancer. Toxins can change the cellular genome, generating errors in the transcription of RNA and the ensuing translation into proteins necessary for signaling and cellular function.

Increased intracellular ROS could directly attack DNA and generate various types of DNA damage. Single-cell gel electrophoresis, also known as the alkaline comet assay, is a sensitive fluorescent microscope-based method that has been widely used to detect DNA damage including DNA strand breaks, DNA adduct formations, and DNA-DNA cross links [160]. OTA treatment caused obvious DNA breaks in a dose-dependent manner. A dose-dependent decrease in the mRNA levels of cyclin A2, cyclin E1, and CDK2 in HEK 293 cells has been detected with various concentrations of OTA for 24 h. The results show the OTA-induced S phase arrest and suggest that OTA-induced S phase arrest might be due to OTA-elicited DNA damage and changes in S phase-related proteins [152].

The permanent exposure to endogenous and exogenous agents leads to DNA damage in organs. DNA damage can be divided into base modifications, strand breaks, crosslinks, and mismatches. However, there are also several DNA repair pathways. Major DNA repair pathways are mismatch repair (MMR), nucleotide excision repair (NER), base excision repair

(BER), homologous recombinational repair (HR), and non-homologous end joining (NHEJ). Each pathway is directed to specific types of damage, and a given type of damage can be targeted by several pathways. [161]. If DNA is not repaired, this damage can cause mutations, diseases, and cell death.

The cellular responses to DNA damage include processes that address tolerance, apoptosis, and the direct correction of the damage by DNA repair mechanisms, which may require the activation of checkpoint pathways. The histone H2AX is a key component in DNA repair that becomes rapidly phosphorylated on a serine four residues from the carboxyl terminus (serine c-4) to form γ-H2AX at nascent DNA double-strand break sites [162]. γ-H2AX is usually used as a marker of DNA double-strand breaks. In OTA-induced nephrotoxicity, γ-H2AX is obviously up-regulated [163].

In the human renal proximal tubular epithelial cell line HK-2, OTA has been reported to produce single-strand DNA breaks and oxidative DNA damage. Moreover, these data also suggest that OTA is not acting as a direct genotoxic carcinogen [164].

3.4. Calcium Homeostasis

Calcium is known to be one of the major players in cell signaling. It has been demonstrated that intracellular Ca^{2+} levels play an important role in cell growth and proliferation regulation [165, 166]. The loss of normal calcium homeostasis is an early and critical event in the development of toxic injury. This reduction of cell viability is independent of Ca^{2+}. A disruption of calcium homeostasis, leading to a sustained increase of calcium levels in the cytosol, has been associated with cytotoxicity in response to a variety of agents in various cell types [167-170]. Therefore, many research specialists have investigated the effects of OTA on cellular calcium homeostasis, cell proliferation, and viability and its interaction with angiotensin II and epidermal growth factor. There is some evidence to suggest that OTA may mediate some of its effects by altering intracellular Ca^{2+} homeostasis [171-174].

In response to OTA treatment, an alteration of Ca^{2+} homeostasis can be detected. OTA induces short spikes in intracellular Ca^{2+} concentrations resembling physiological responses [159,173]. In human kidney epithelial cells, OTA was shown to stimulate cell proliferation by interfering with intercellular Ca^{2+}. The disruption of calcium homeostasis may lead to the disruption of cell cycle control and serve as a trigger for cell proliferation in rat kidney. *In vitro* experiments provided evidence of an effect of OTA on intracellular Ca^{2+} homeostasis [159, 175, 176]. Ca^{2+} homeostasis modulation by OTA was also observed *in vivo* [14]. In immortalized human kidney epithelial cells-1 (IHKE-1), as little as nanomolar concentrations of OTA can lead to rapidly and reversible oscillations in intracellular calcium. This effect was dependent upon extracellular calcium, phospholipase C, and cAMP-dependent protein kinase. Moreover, OTA potentiated the calcium-dependent stimulation of cell proliferation induced by epidermal growth factor and angiotensin II [159]. These effects were observed in the absence of cytotoxicity, suggesting that OTA may interfere with calcium-dependent cell signaling and thereby lead to increased cell proliferation. An increased rate of microsomal Ca^{2+} uptake was observed after OTA administration *in vivo* and *in vitro* [177]. Gene profiling analysis suggested a modulation of genes involved in calcium homeostasis by OTA [81, 178]. In these studies, OTA was also found to alter the expression of genes associated with mitogenic growth factor signaling, such as insulin-like growth factor-1, insulin receptor, and GTP-binding protein. Thus, temporal changes in intracellular Ca^{2+} may occur as a result of growth factor receptor activation and the transduction of extracellular signals that involve Ca^{2+} as a second messenger, or they may occur via the direct interaction of OTA with Ca^{2+} and/or regulators of Ca^{2+} homeostasis [14]. The increase in calcium pump activity was inconsistent with previous observations that OTA enhances lipid peroxidation, a condition known to inhibit the calcium pump [177]. These data indicate that Ca^{2+}-dependent signal transduction pathways may be affected by OTA treatment [179].

Regucalcin (RGN) is involved in calcium homeostasis and implicated in cell survival and proliferation. Apart from oxidative stress, alterations of Ca^{2+} homeostasis and cytoskeleton structure may be the first responses to OTA. One study found that RGN was inhibited strongly at 7 and 21 days [178]. Toxicogenomic data have shown a significant down-regulation of regucalcin [81]. The disruption of Ca^{2+} homeostasis in the rat kidney by OTA is also supported by microarray analysis, which demonstrates changes in the expression of genes involved in calcium signaling, including RGN [22,81].

Although the inhibition of Na^+/H^+-exchange and H^+-ATPase did not reduce the OTA-induced acidification, the stimulation of NHE (maximal flux increases) in a Ca^{2+}-dependent manner, as a function of pHc, was revealed. In addition, OTA induced a hyperpolarization of the mitochondrial membrane potential (Ψm) in a Ca^{2+}-dependent manner. The study found that OTA activates mitochondria and NHE by interfering with cellular Ca^{2+} homeostasis [180]. The pre-incubation of microsomes with NADPH had a profound inhibitory effect on calcium uptake, but OTA was able to overcome the inhibition [175].

OTA interaction with other chemical substances can also effect calcium homeostasis. OTA and citrinin (CTN) lead to a dose-dependent increase in the cytosolic calcium level. However, in combination, the two treatments did not provoke an additional increase in calcium signaling. In conclusion, the disturbance of Ca^{2+} homeostasis caused by OTA and CTN plays a significant role in genotoxicity and death [181].

3.5. Disruption of Cell–Cell Signaling Pathways

Cell signaling processes such as the inhibition of gap junction intercellular communication (GJIC) [174], interference with microtubule dynamics, and mitotic spindle formation [182] or increased histone deacetylase (HDAC) enzymatic activity [183] have been described after exposure to OTA.

The effects on the cytoskeleton and cell organization-related genes were also observed in the transcriptomic studies performed by Arbillaga et al. [178]. In their study, advillin, which has been reported to play a role in actin remodeling, was the most highly induced gene across the entire dataset (*in vitro* and *in vivo* models). Moreover, this gene expression study further indicated that OTA may increase histone deacetylase activity and histone acetyltransferase inhibition [183]. In the study by Lühe et al. [22], some genes related to chromatin unpacking were up-regulated, such as the nucleosome assembly protein, which functions in DNA replication and repair.

The *Wnt* pathway regulates the *β-catenin* cellular concentration, which is the protein that directly participates in gene transcription and cell adhesion [184]. Wnt-induced intracellular signaling is an extensively studied mechanism reported to underpin the development and progression of cancer. Wnt-induced intracellular signaling sequesters *β-catenin* from GSK-3β and promotes its accumulation in the nucleus, where *β-catenin* works synchronously with T-cell factor/lymphocyte enhancer binding factor to stimulate the expression of target genes [185]. It is known that the Wnt signaling pathway is involved in diverse cell-cell interactions and that its alteration leads to tumor formation [186, 187]. In addition, a 13-week exposure to OTA caused hypermethylation in the promoters of the critical cell adhesion-related genes *E-cadherin* and *N-cadherin*, leading to the reduction of corresponding mRNA expression accompanied by the transcriptional activation of the Wnt and PI3K/AKT pathways [188].

In the last review by Mally [16], in which a novel MoA for OTA as a renal carcinogen in rats is proposed, the inhibition of histone acetyltransferases and the disruption of mitosis are presented as key events in the postulated MoA.

3.6. Cell Death and Cell Survival

Programmed cell death (apoptosis) always occurs as a response to changes in the cell microenvironment, including the deficiency of certain survival requirements and stresses from lethal factors. Crucial regulators in cell death and survival include surrounding cells, the extracellular matrix, and soluble mediators [189], Figure 6. OTA can induce apoptosis in human proximal tubule-derived cells (IHKE) [190] and dog renal collecting duct-derived cells (MDCK-C7) [29,190] at nanomolar concentrations. Interestingly, OTA potentiated the apoptotic effect of tumor necrosis factor (TNF) in MDCK-C7 cells [29]. Moreover, larger amounts of OTA also led to apoptosis in hamster kidney (HaK) and human HeLa cells [191].

As reported, OTA can affect various signal transduction pathways which lead to cell death and survival in a cell type- and dosage-dependent manner. Among these pathways, the ERK1/2 and JNK/SAPK MAPK pathways are major targets, whereas knowledge about their involvement in OTA-induced cell death is fragmented. The MAPK/ERK pathway (the Ras-Raf-MEK-ERK pathway) consists of a chain of proteins that convey a signal from a receptor on the cell surface to the nuclear DNA. In transcriptome experiments performed with OTA, MAPK signaling pathways were shown to be significantly affected at both time points (6 h and 24 h) in the *in vitro* study performed [178].

MAPKs are members of the evolutionarily conserved serine/threonine kinase family, which can respond to a variety of physical and chemical stresses and play significant roles in cell adaptation and survival [192]. The increasing incidence of renal carcinoma in the kidneys of chronically OTA-fed male rats was reported to be related to the activation of the MAPK-ERK pathway [183]. The three-tiered cascade of MAPK activation includes MAP kinase kinase kinases (MAPKKKs, MAP3Ks), which phosphorylate and activate MAP kinase kinases (MAPKKs, MAP2Ks), which subsequently phosphorylate and activate MAPKs [193].

Within this cell signaling process, apoptosis signal-regulating kinase 1 (ASK1) is a MAP3K family member that activates both the MKK4/MKK7-JNK and MKK3/MKK6-p38 pathways [194]. ASK1 is known to play a pivotal role in oxidative stress- and endoplasmic reticulum stress-induced cell death [195]. Recently, Liang et al. created ASK1-deficient HEK 293 cells via RNA interference and then globally investigated the regulatory mechanism of ASK1 in OTA-induced renal toxicity using iTRAQ-based quantitative proteomics. Interestingly, ASK1 deficiency can alleviate OTA-induced ROS generation and the mitochondrial membrane depolarization and thereby desensitize the cells to OTA-induced apoptosis. Thirty-three and 24 proteins were identified as differentially expressed upon OTA treatment in control and ASK1-deficient cells, respectively. The conclusion from pathway analysis is that ASK1 plays an essential role in promoting OTA-induced renal cytotoxicity. ASK1 took part in the OTA-induced suppression of mRNA splicing, nucleotide metabolism, the cell cycle, DNA repair, and the activation of lipid metabolism, which will be described in more detail below.

OTA can induce oxidative stress, which is the major trigger of ASK1 activation; thus, whether ASK1 would have a reverse effect on oxidative stress was determined. Interestingly, ASK1 is able to amplify the ROS generation, very likely because of the generation of oxidants and the decrease of certain antioxidants. According to Chambers et al. JNK deficiency mediated the amplification of ROS generation during stress; however, p38 deficiency did not have the same effect [196]. Therefore, the authors speculated that ASK1-mediated ROS production is achieved primarily through JNK activation.

Mitochondrial membrane depolarization is an apoptotic signal that precedes DNA fragmentation in a variety cell types, and consequently it is considered a sign of early apoptosis [197]. ASK1 plays an essential role in the loss of $\Delta\Psi m$ as well as the activation of the mitochondrial-dependent apoptotic pathway.

Existing studies focused on the impact of OTA on metabolic pathways including oxidative stress, apoptosis, DNA damage and repair, cell cycle arrest, and lipid peroxidation [197, 198]. The proteome data indicated novel metabolic pathways influenced by OTA, including mRNA splicing, nucleotide metabolism, and lipid metabolism (especially the biosynthesis and transfer of lipids). For example, RRM1 and RRM2 are enzymes that catalyze the biosynthesis of deoxyribonucleotides, and TYMS participates in the biosynthesis of de novo mitochondrial thymidylate. Unscheduled DNA synthesis and replication may occur when the formation of the four basic DNA substrates are blocked. It was reported that OTA at low concentrations (750 nM–1 mM) could lead to a dose-dependent induction of DNA repair in rat hepatocytes and porcine urinary bladder epithelial cells [199]. Nevertheless, the proteome data suggested that OTA induced the down-regulation of important proteins in nucleotide excision repair (NER), including POLR2A, POLR2B, and PCNA; XAB2 in transcription-coupled repair (TCR); and PRPF19 in double-strand break (DSB) repair, which reveals that a higher concentration (20 mM) of OTA led to the inhibition rather than the induction of DNA repair. On one hand, OTA induces DNA damage; on the other, it can impair DNA repair mechanisms, further aggravating the aftermath of DNA damage.

Apart from the induction of DNA damage, OTA can also induce damage at the RNA level. Several proteins involved in mRNA splicing were down-regulated by OTA. Among the major pre-mRNA processing events, pre-mRNA splicing can occur post-transcriptionally or co-transcriptionally [200]. Moreover, HNRNPL, which plays an important role in the alternative splicing through which different mRNAs encoding distinct protein products are generated, including those involved in apoptosis [201], is down-regulated by OTA. Altogether, the results indicate that OTA suppressed mature mRNA formation and consequently affects the translation and expression of critical proteins.

The fact that the cell cycle was inhibited by OTA was also shown in the study. For example, CDK1 plays an essential role in promoting G1 progression, the G1-S transition, and the G2-M transition. CDC5L is a positive regulator of G2/M progression, and KIF11 is needed to establish a bipolar spindle during M phase; the mentioned proteins were all down-regulated by OTA unfortunately.

The knockdown of ASK1 almost reversed all of the OTA-induced inhibition or activation of the proteins in the previously mentioned pathways. Surprisingly, the alteration of proteins involved in nucleotide and lipid metabolism was completely counteracted by ASK1 deficiency, indicating that the activation of ASK1 is likely a key factor in the OTA-induced inhibition of nucleotide metabolism and the acceleration of lipid metabolism. However, ASK1 was partly involved in the mediation of the OTA-induced repression of mRNA splicing, DNA repair, and the cell cycle. Kuo et al. [202] reported that isoobtusilactone A-induced cell cycle arrest in human breast cancer cells, coupled with the increase of p21 expression and the decrease of cyclin A, cyclin B1, cdc2, and cdc25 expression, resulted from ROS generation and through ASK1 signaling pathways. The observations in this study further validated that ASK1 regulated the OTA-induced cell cycle arrest and was associated with the suppression of CDK1 expression. However, the activating effect of ASK1 on cell cycle progression was also observed because ASK1 knockdown resulted in the down-regulation of LIG1, MKI67, and STAG2. Moreover, p38 was reported to act as a dual regulator of ROS signaling, either activating or suppressing cell cycle progression depending on different activation stimuli [203]. As it is downstream of ASK1, it is reasonable to assume that ASK1 might also have a dual role in the regulation of the cell cycle.

DNA damage induced by genotoxins, such as DNA adducts, DNA double strand breaks, DNA cross-links, and DNA methylation, can always lead to cell death, typically via apoptosis [204]. The activation of cell cycle checkpoints and repair mechanisms is the response of cells to DNA lesions; otherwise, failure of the repair system will cause cell death and even tumorigenesis [205]. The repression of RNA synthesis is another trigger of cell death upon DNA damage that will result in a decline in MKP1 and sustained JNK activation [204]. ASK1 inhibits the DNA repair mechanisms by down-regulating NER (PCNA, POLR2B) and TCR (XAB2), which might further cause prolonged JNK activation and apoptosis initiation. In addition, the interactions between proteins, such as the binding of p53 to the SH3 domains of Bcl2, play a key role in p53-dependent apoptosis after RNA synthesis is inhibited [206]. Taking these points together, the authors speculate that apoptosis was induced through the suppression of certain genes or through interactions between proteins when ASK1 mediated the blockade of RNA synthesis (especially mRNA splicing).

In summary, ASK1 has a role in the amplification of oxidative stress, whereas it is activated by OTA through ROS generation. Then, the activated ASK1 mediates apoptosis initiation through several pathways, such as impairing nucleotide metabolism, causing DNA damage, affecting cell cycle checkpoints and DNA repair mechanisms, and blocking of RNA synthesis. When DNA is damaged, the repair cascade was initiated at the cell cycle checkpoints. Once the repair fails, an apoptotic pathway is activated. Meanwhile, DNA damage can restrain RNA synthesis, resulting in the suppression of certain anti-apoptotic genes.

At higher ochratoxin concentrations (50 μM) on a porcine renal proximal tubular epithelial cell line, the induction of apoptosis and changes of cytoskeleton organization were observed [30].

Genes related to apoptotic processes and DNA damage were down-regulated in two reports (DIABLO, FAS, FADD, TNFRSF1A, MYC, PCNA in [207] and Annexin V and MM-1 in [30]). The research from Lühe et al. [22] also confirmed that OTA might influence apoptotic processes because the expression of Annexin V and clusterin was restrained. In a study of ochratoxin A-induced renal toxicity in male F344 rats, genes involved in tissue regeneration and cell proliferation including c-MYC and PLK2 presented a slight induction at the last time point tested (21 days) [178]. The up-regulation of several genes related to cell proliferation and cell death, such as AKT, PLK, and c-MYC, was also observed in a longer term exposure study [81]. The proto-oncogene c-MYC, up-regulated in the aforementioned *in vivo* studies, was down-regulated *in vitro* [30, 164], and it is also associated with apoptosis [208].

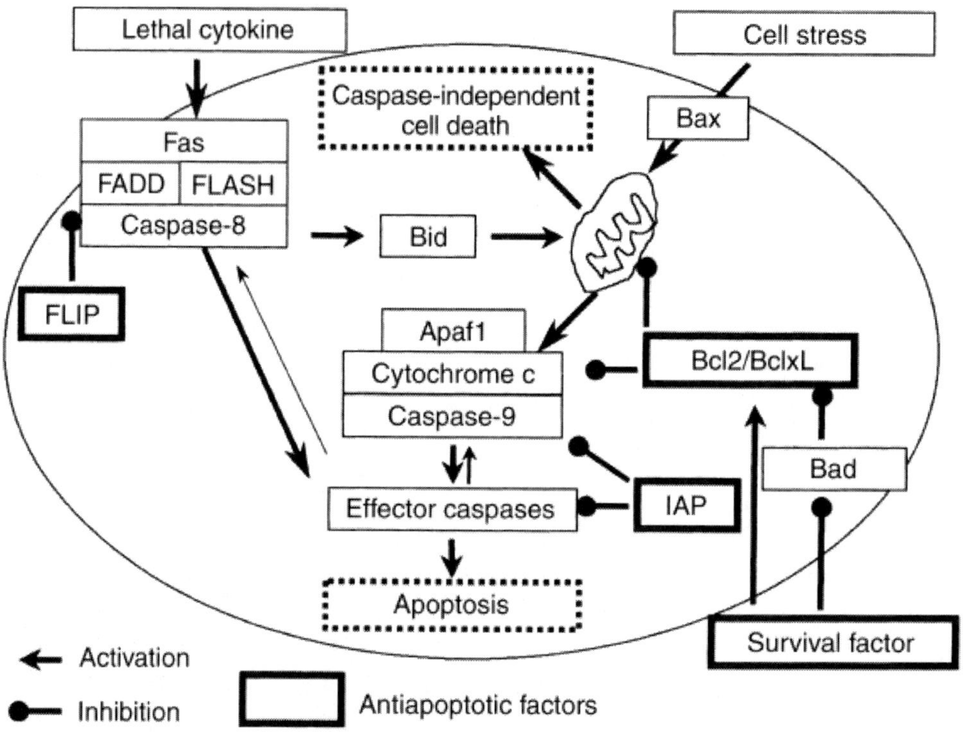

Figure 6. Intracellular pathways for cell death can be activated by death receptors or cell stress and are antagonized by molecules such as Bcl-2/Bcl-xL, FLIP, IAP, and extracellular survival factors [189].

Early histopathological changes induced by OTA in the rat kidney are coupled with remarkable changes in the expression of several key regulators of mitosis, including molecules participating in centrosome function, chromosome condensation and segregation, cytokinesis, spindle checkpoint, and G2/M progression. However, the genes associated with G1 progression, G1–S transition, and growth factor signaling were generally not altered by OTA exposure. Furthermore, the immunohistochemical analysis validated the up-regulation of mitosis-control proteins in S3 proximal tubule cells. This cell type is the major target of OTA carcinogenicity, most likely because of the preferential uptake and intracellular accumulation of OTA at this site [209, 210].

As reported, a commonly considered tumor suppressor, p21WAF1/CIP1, was significantly up-regulated in the kidneys of OTA-treated rats, especially in cells with giant nuclei. It is believed that p21WAF1/CIP1 plays a key role in post-mitotic G1 arrest and aberrant exit from mitosis, which serves to prevent tetraploid cells from re-entering the cell cycle [211]. Nevertheless, accumulating evidence reveals that p21WAF1/CIP1 can interfere with stress-signaling pathways (c-Jun N-terminal kinase and p38) and prevent the activation of pro-caspase-3 by binding to it and therefore inhibiting apoptosis [212]. The overexpression of p21WAF1/CIP1 in tetraploid or polyploid cells might restrain secondary apoptosis after mitotic slippage and facilitate the survival of cells with nuclear aberrations. This implies that the aberrant expression of p21WAF1/CIP1 may be a contributor to genomic instability and even tumor formation [213,214]. In the research of Chang et al. [213], the transient over-expression of p21WAF1/CIP1 caused the inactivation of the Cdk1cdc2–cyclin B1 complex during mitosis, which is likely responsible for the transcriptional suppression and loss of mitosis-control proteins and the resulting growth arrest and abnormal mitosis in recovering cells [215].

Another important mechanism in cell death and survival is autophagy, which is a self-destruction system conserved among eukaryotes. The first step of autophagy is the formation of an autophagosome, a double membrane structure entrapping cytoplasmic components including organelles. Then, the autophagosome is fused with a lysosome, forming an autolysosome. The captured cytoplasmic components are then degraded by lysosomal hydrolases [216]. Thus, autophagy and also mitophagy are important for organisms to adapt to various stresses. The function of autophagy or mitophagy in OTA-induced renal toxicology deserves investigation.

Nix is a selective autophagy receptor for mitochondrial clearance [217]; our team investigated the regulation of Nix upon autophagic response to OTA-induced renal toxicity in 2014. The results revealed that OTA induced autophagic and mitophagic activities. Nix-deficient HEK 293 cells were more vulnerable to OTA-induced cell death than controls. OTA treatment induced significant alteration of the mRNA expression of Bad, SDHA, and AIFM1 and the protein expression of VDAC, LONP1, SDHA, and AIF. Particularly, the up-regulation of pro-apoptotic Bad and AIF induced by OTA was prominent only in Nix-deficient cells, which might be responsible for the higher ratio of cell death. These results imply that Nix plays a central role in autophagy and mitophagy, thus protecting cells against OTA-induced renal toxicity.

The fact that OTA induces autophagy and mitophagy was demonstrated. The degradation of toxic protein aggregates and unneeded organelles can be achieved by autophagy; however, both excess and insufficient autophagy seems to be able to promote cell injury. Therefore, the appropriate regulation of autophagy is essential for cellular homeostasis [218]. Autophagy can suppress cell death by adaptive responses to various forms of stress [219]. Coinciding with this notion, OTA accelerates autophagy in the early stage of its renal toxicity. Unfortunately, massive autophagy may also kill cells [219], which might account for the more severe cell death caused by OTA [89]. Mitophagy refers to mitochondrial degradation by autophagy, during which autophagic delivery to lysosomes is the major degradation pathway. The results reveal that an increase in mitophagic flux activity is induced by OTA treatment, indicating that OTA can induce mitochondrial dysfunction and damage. The ROS generation and mitochondrial membrane potential changes affected by OTA are further evidence [89]. Because Nix functions as a regulated mitophagy receptor, Nix knockdown was used to investigate the role of mitophagy in OTA-induced renal cytotoxicity. Because NIX deficiency can decrease the sensitivity of mitophagy events and increase cell death, the essential role of mitophagy in the response to OTA-induced toxicity was indicated.

Second, OTA disturbs the homeostasis between apoptosis and autophagy. The mitochondrial pathway and the death receptor pathway are two major apoptosis pathways. The former is mainly regulated by members of the *Bcl-2* protein family, including anti-apoptotic proteins (Bcl-2, Bcl-XL, Bcl-w, etc.) and pro-apoptotic proteins (Bax, Bad, Bid, Bak, Bcl-XS, Bim, BNip3, Nix, etc.) [220]. Fas and tumor necrosis factor are the key players in the death receptor pathway. The binding and auto-activation of procaspase-8 is achieve by the adaptor molecule FADD/Mort-1 being recruited to the receptor when Fas ligand binds to its receptor. Once activated, *cas*pase-*8* can process effector caspases, such as *caspase-3, -6*, and *-7*, inducing a cascade of caspase cleavage of their substrates [221].

Interestingly, OTA remarkably restrained the expression of several anti-apoptotic genes (*Bcl-XL and Bcl-2*) and pro-apoptotic genes (*Bid and Bax*), demonstrating a coordinated response to balance apoptotic death. However, the knockdown of Nix does not reverse the effect of OTA on the expression of these 4 apoptosis-related genes. Surprisingly, the mRNA level of Bad was only up-regulated by OTA in combination with Nix deficiency, which may account for the increased sensitivity of Nix-deficient cells to OTA. In the cytotoxic response to OTA, Nix deficiency leads to defective autophagy, eventually further inducing selective apoptosis factors.

Although the mechanisms by which apoptosis and autophagy interact in response to the stress of OTA are not entirely understood, it is highly likely that the suppression of autophagy contributes to a bio-energetic shortage triggering apoptosis [219]. OTA induced the down-regulation of caspase-3, caspase-8, and caspase-9 in the study. In the Fas signaling pathway, the proximal substrate of caspase-8 is Bid. Full-length Bid is localized in the cytosol; however, its truncated form translocates to the mitochondria, inducing the release of CYC from mitochondria [222]. Apart from caspase-8, caspase-3 can also cleave Bid, inducing the release of CYC in response to caspase cleavage. Cytosolic CYC binds to Apaf-1, which then permits the recruitment of procaspase. The auto-activation of procaspase-9 is a result of oligomerization [221]. Activated caspase-9 initiates a caspase cascade involving the downstream executioner's caspase-3, -6, and -7 [223]. Additionally, in the mitochondrial pathway, the apoptosome, the complex of CYC, Apaf-1, and caspase-9, is an essential activator of the effector caspases [221]. Caspases-3 and -7 are momentous regulators of apoptosis in the mitochondrial pathway, as reported by Lakhani et al. [224].

The results revealed that treatment with OTA inhibited the expression of the three apoptotic genes caspase-3, caspase-8, and caspase-9. This phenomenon implies that HEK 293 cells respond to OTA by the induction of cell survival programs through the suppression of apoptotic genes in the early stage of OTA-induced renal cytotoxicity. The inhibitor-of-apoptosis (IAP) family of genes has an evolutionarily conserved role in regulating programmed cell death [225]. Human c-IAP1, c-IAP2, and XIAP are capable of preventing the activation of caspases-3, -6, and -7 by blocking the activation of procaspase-9 induced by cytochrome c, whereas these IAPs cannot prevent the caspase-8-induced proteolytic activation of procaspase-3 [226]. Caspase-9 is a crucial initiator caspase required for apoptosis signaling through the mitochondrial pathway [227]. Nix deficiency remarkably increases the expression of caspase-9, demonstrating that apoptosis via the mitochondrial pathway might be increased when autophagy is inhibited.

Nix deficiency and OTA treatment lead to the up-regulation of AIF, implying a cell death reaction. This also suggests that autophagy is essential for cells to counteract OTA-induced renal cytotoxicity. Moreover, with Nix deficiency, the expression of AIFM1 is restrained by

OTA in renal cells. This implies that the accumulating AIF induces a feedback repression mechanism to inhibit the expression of AIFM1.

Moreover, the TCA cycle and the mETC was disrupted by OTA. Compelling evidence has shown that VDAC1 might function in the crosstalk between the cytoplasm and mitochondria by interacting with enzymes related to energy metabolism and mitochondrial-induced apoptosis [228]. The low levels of VDAC1 have been determined to limit metabolite exchange between mitochondria and cytosol [94], and the up-regulated VDAC1 in Nix shRNA cells increased the synthesis of ATP. LONP1 is capable of selectively recognizing and degrading the oxidized, hydrophobic form of aconitase 2 (ACO2) [99]. SDHA is known to be a member of the mitochondrial electron transport complex II. LONP1 activation promoting the TCA cycle may explain the up-regulation of SDHA induced by OTA. Under normal circumstances, ATP production and homeostasis are regulated and associated with autophagy [229]. Nix knockdown can inhibit autophagic activity, and thus the production of ATP was reduced. To adapt to this circumstance, VDAC1 was accumulated, followed by the up-regulation of LONP1 and SDHA in response to OTA. Thus, the TCA cycle and mitochondrial electron transport chain operation was accelerated. In brief, autophagy has been found to play a critical role in counteracting OTA-induced perturbations to the TCA cycle and mETC.

3.7. miRNA Regulation

MicroRNAs (miRNAs) are found in eukaryotes and function in the regulation and control of endogenous non-coding RNAs. miRNA length is approximately 20-25 nucleotides. Recent studies have shown that miRNAs are involved in the regulation of various processes, including developmental processes, virus defense, hematopoiesis, organ formation, cell proliferation and apoptosis, fat metabolism, etc. [230]. Moreover, miRNAs have been associated with two issues of crucial importance: the development of renal carcinoma and cancer metastasis [231, 232]. miRNA profiling analysis has been an important method to evaluate kidney toxicity.

Dai et al. [233] analyzed miRNA profiles in the kidneys of rats gavaged with OTA. Male rats were divided into different groups and gavaged with OTA in corn oil for 2, 4, 13, and 26 weeks. The dose of OTA was 0, 70, or 210 μg/kg body weight. Mean body weights were slightly reduced after 2, 3, 5, 7, and 10 weeks of OTA administration at 210 μg/kg. Kidney weight was also significantly reduced. Distinguishable but not severe histopathological damage after 13 weeks were found by serum clinical chemistry and histopathology analysis. Thirteen weeks was chosen as the optimal duration of OTA treatment for miRNA profiling because in preliminary experiments, the nephrotoxicity of OTA reached a turning point in the 13-week group.

Hierarchical clustering analysis results showed that CK and CH groups are closer in Euclidean distance. There is a strange phenomenon that hierarchical clustering analysis of the gene expression in rats fed with OTA for less than 12 months indicates that a cluster was noted between the late (12 months) and the early (7 days) responses [234]. These results suggest that the nephrotoxicity induced by OTA may be related to the age of rats. The hierarchical clustering and the small RNA and miRNA length distribution pattern suggest that a series of signaling pathways are affected and orchestrated in the early response to OTA toxicity and that the transition period might be of paramount importance in the regulation of nephrotoxicity.

To further demonstrate this phenomenon, STEM software was used to examine the expression pattern of all known miRNAs. Two profiles (1 and 5) showed 77 important miRNAs. All of these miRNAs were reduced in CM and elevated in CH kidneys in both profiles. Among the 77 miRNAs, miR-141, miR-130a, miR-130b, miR-129, miR-218b, and miR-3588 were considered for further studies and were decreased by at least 2-fold in CM groups compared to CH groups.

"Phosphatidylinositol signaling system", "pancreatic cancer", and "MAPK signaling pathway" were greatly enriched as evidenced by both separative and collective analyses. Many previous studies focused on the relationship between the MAPK pathway and OTA toxicity [29, 183, 234, 235]. Marin-Kuan, M et al. administered 300 μg/kg OTA to male rats for 12 months to study the effect on the MAPK pathway. They detected different aspects of the kidney and showed that the MAPK pathway was significant in the impact of OTA on rat kidneys. The experiment also confirmed that the long-term administration of OTA to rats can lead to progressive repression of the MAPK pathway [233].

It was interesting to find that in the four and thirteen week's groups, the expression of miR-130b was increased, whereas in the twenty-six weeks groups, it was decreased. This may suggest that short- and long-term treatment with OTA would result in opposite effects on miR-130b.

For a more precise knowledge of the effect of OTA on miRNA, we assessed the expression of Drosha, Dicer1, and DRCG8, which are necessary for miRNA processing. Drosha and Dicer1 were significantly reduced, but DRCG8 did not change.

In previous studies, cancer formation has been associated with the decreased expression of Dicer and Drosha. Cellular transformation and tumorigenesis is thought to result from the abrogation of Drosha, DRCG8, or Dicer. After the OTA treatment of rats, Drosha and Dicer1 were both decreased in the kidney, which may indicate that OTA leads to nephrotoxicity and renal carcinogenesis by affecting miRNA processing. From another aspect, OTA can increase the expression of Akt1, c-Myc [236], and PCNA [34], proteins associated with cell survival and proliferation. These genes are associated with the enhanced proliferation and invasion in tumor cells. Therefore, OTA may induce Akt, c-Myc, and PCNA to mediate the carcinogenicity. Dicer also plays an important role in the maturation of short interfering RNAs (siRNAs) [237, 238]. siRNAs are RNAs that can specifically degrade target mRNAs. Therefore, the decrease of Dicer by OTA may affect the siRNA processing as well. Based on the reduction of Dicer in CM and CH groups and Drosha in the CH group, the impacts of OTA on rats may also include the inhibition of miRNA and siRNA formation.

miRNA levels are tightly regulated at the post-transcriptional level by a number of RNA-binding proteins [239]. Microprocessors are composed of Drosha and DGCR8. Drosha is a nuclear protein that recognizes and trims pre-miRNAs, whereas DGCR8 is its cofactor, also known as DiGeorge syndrome critical region gene 8 protein. Furthermore, Drosha can directly regulate the levels of the Microprocessor complex by cleaving hairpin structures in DGCR8 mRNA, thereby decreasing DGCR8 protein levels [240]. In the kidneys of rats with OTA nephrotoxicity, Drosha was significantly suppressed [241]. The upgraded Dicer is also reported, indicating an impairment of miRNA biogenesis in response to OTA [241]. Although there was little knowledge to clearly affirm OTA-induced miRNA changes, in many tumors, miRNA formation can be observed to be impaired. Therefore, there may be a relationship between cancer and miRNA processing.

3.8. DNA Methylation

DNA methylation is one of the most common epigenetic alterations. Its role in several chronic diseases has been a topic of considerable interest in recent years. Notably, some carcinogens alter the levels of DNA methylation, which is part of the underlying biological mechanisms by which carcinogens promote the development of tumors. Many studies have shown that environmental contaminants, such as arsenic, cadmium, chromium (VI), and nickel, are capable of altering DNA methylation status, including hypermethylation and hypomethylation [242-244]. Additionally, aflatoxin B1 (AFB1), a dietary mycotoxin, could cause methylation changes in humans [245, 246]. A previous study conducted in our laboratory showed that OTA exposure induced global DNA hypomethylation in HepG2 cells [247]. The studies of Selvakumar et al. [248] and Wing et al. [249] showed that disrupting methylation modulation may be a key mechanism in kidney disease. Thus, we hypothesized that the potential mechanisms of OTA-mediated nephrotoxicity may involve its ability to disrupt the epigenetic profile through the modification of DNA methylation.

Global DNA hypomethylation has been associated with chromosomal instability and aberrant gene expression, both of which could contribute to the onset and/or progression of renal disease and increased tumor frequency and malignancy [250]. After 4 weeks, high-dose OTA treatment (210 μg/kg diet) significantly decreased the extent of global methylation, whereas low-dose treatment (70 μg/kg diet) resulted in no decrease compared to the control. More remarkably, OTA treatment for 13 weeks caused statistical increases ($p<0.01$) in global DNA methylation in the kidneys of both low-dose and high-dose treatment groups. However, after 26 weeks of OTA exposure, nephritic global DNA methylation showed no significant change. Additionally, no significant changes were found in any of the control groups. The results of global methylation analysis showed that the degree of global methylation in kidney DNA was changed dynamically and that treatment with OTA promoted a decrease and increase of genomic methylation in a dose-dependent manner by 4 and 13 weeks, respectively. In fact, it was not uncommon that dynamic alterations of global DNA methylation existed during the process of OTA-induced nephrotoxicity. A similar phenomenon was found in ACI rats during estrogen-induced breast carcinogenesis [251] and in rat liver epithelial cells exposed to chronic cadmium [252]. It was proposed that hypomethylation was most likely an early event in OTA-induced nephrotoxicity, whereas hypermethylation accumulated in more advanced stages. Furthermore, our results showed that the variations of genomic methylation were dose dependent after 4 and 13 weeks of OTA treatment. However, after 26 weeks of exposure, the levels of global DNA methylation showed no further significant changes. One possible reason was that hyper- and hypo-methylation counteract one another in different anatomical regions or in different genomic loci, resulting in a lack of overall global DNA methylation changes. These data suggested a dynamic and dose-dependent progressive nature of global DNA methylation, indicating its crucial role in OTA-induced nephrotoxicity. However, the degree of global methylation measured in this study did not necessarily reflect the extent of OTA toxicity.

The DNA methyltransferases (DNMTs) were examined because they are positive regulators of DNA methylation [253] and are involved in the accurate maintenance of DNA methylation patterns. Thus, the expression levels of DNMTs could affect the global DNA methylation status as well as the DNA methylation of specific loci. The results showed that the relative mRNA transcripts of DNMT1 and DNMT3b were significantly increased after 13 weeks of OTA exposure. At that time, the level of DNMT3a in the kidney tissue was significantly decreased rather than increased by OTA. It was shown that the initial loss of genomic methylation was accompanied by a down-regulation of DNMT1. The subsequent increase of global DNA methylation was accompanied by an up-regulation of DNMT1 and de novo DNMT3b. In this study, the mRNA levels of DNMT1 generally correlated with the overall extent of DNA methylation. Thus, it was possible that OTA mediated the expression of DNMT1, contributing to the changes in the levels of global DNA methylation in renal tissue of rats exposed to OTA.

The overexpression of DNMTs, especially DNMT3b, may be responsible for aberrant gene-specific de novo methylation events and result in gene silencing [254,255]. In the present study, the progressive accumulation of DNA methylation in the promoters of critical cell adhesion-related genes (E-cadherin and N-cadherin) was consistent with the increasing DNMT1 and DNMT3b mRNA expression levels. These results suggested that DNMTs may have roles in modulating genomic and some gene-specific methylation during long-term OTA exposure.

Cadherins are a family of transmembrane glycoproteins that mediate calcium-dependent cell-cell adhesion in multicellular systems [256]. Moreover, cadherins have important roles in regulating morphogenesis and have been linked with the invasion and metastasis of various human malignancies [257,258], especially nephropathies [259]. The cadherin superfamily is composed of classical cadherins, which include E-cadherin, the predominant constituents of adherens junctions [260], and N-cadherin, the most widely expressed cadherin in the kidneys [261]. Accumulating evidence suggests that E-cadherin promoter hypermethylation is associated with reduced transcription in renal cell carcinoma [262], and methylation-silenced N-cadherin has been found in aged rat kidney tissue [263]. It has been previously reported that OTA decreased the expression of the E-cadherin ($p < 0.05$) and N-cadherin genes in the kidneys of Wistar rats after treatment with 289 μg/kg OTA for 90 days [264,265]. In our study, MSP was used to determine the effect of OTA on DNA methylation in the promoter regions of E-cadherin and N-cadherin. The methylation status in E-cadherin and N-cadherin promoters was unaffected over the 4-week exposure period. However, after 13 weeks of OTA exposure, the promoters of E-cadherin and N-cadherin genes were hypermethylated in the kidneys. RT-PCR analysis showed that OTA exposure significantly decreased E-cadherin and N-cadherin expression after 13 weeks. Therefore, the transcriptional inactivation of the E-cadherin and N-cadherin genes was most likely associated with DNA hypermethylation in the kidney after 13 weeks of OTA exposure.

Hypermethylation of the E-cadherin promoter region has been reported to be correlated with the loss or reduction of expression in several human tumors [266,267] and was often found in renal cancer [262]. Furthermore, there are several mechanisms for abnormal E-cadherin expression in tumors, and E-cadherin appears to have a dual purpose in mammalian cells, not only regulating intercellular adhesion but also modulating a variety of intracellular signal transduction pathways, including Wnt/β-catenin, PI3K/Akt, and NF-κB signaling [268]. In our experiment, genes related to Wnt, PI3K, and NF-κB pathways showed increased expression in individual kidney samples from the control and 13-week OTA exposure groups. The activation of the Wnt signaling pathway has been linked to tumorigenesis [269]. Moreover, an activated PI3K/Akt pathway was one of the most pertinent hallmarks involved in cancer progression in OTA-treated Eker rats [270]. It was previously reported that E-cadherin could negatively regulate Wnt/β-catenin signaling by interacting with β-catenin, suggesting that the loss of E-cadherin expression could contribute to the up-regulation of this pathway in human cancers [271]. Lau et al. [272] observed a down-regulation of the E-cadherin-induced activation of the PI3K/Akt pathway in ovarian cancer cells. Thus, we suggested that the E-cadherin-mediated activation of the Wnt and PI3K/Akt signaling pathway was a possible mechanism of OTA toxicity.

It is worth noting that methylation-silenced N-cadherin occurred in aged rat kidney tissue, and the loss of N-cadherin expression increased progressively with age [263]. In addition, the activation of several signaling pathways, as described above, such as the Wnt pathway, PI3K/Akt, and NF-κB signaling, was related to cellular aging [273-276]. Thus, it was hypothesized that OTA induced the methylation silencing of N-cadherin and the activation of relevant signaling pathways, which might further lead to the senescence of nephrocytes. These findings made it tempting to speculate a possible link between aging and an increased risk for OTA-induced kidney injury.

In summary, our data demonstrated that long-term OTA exposure induced DNA methylation changes and resultant signaling pathway disorders. The kidneys of rats exposed to OTA initially exhibited reduced DNMT1 mRNA expression and genomic hypomethylation, as expected, but upon longer exposures, these tissues showed increases in both global DNA methylation and DNMT1 mRNA expression. Methylation silencing of the E-cadherin and N-cadherin genes may be related to the disruption of several signaling pathways, especially the Wnt and PI3K pathways, which are linked to tumorigenesis. Thus, altered DNA methylation in the kidneys of OTA-exposed rats may represent a mechanism for controlling the expression of important genes in kidney function and carcinogenesis.

3.9. DNA-OTA Adducts

The formation of OTA-DNA adducts seems to be another mechanism of OTA-induced nephrotoxicity and renal carcinogenicity. For decades, many reports of OTA-DNA adducts have been published.

In 1990, Pfohl-Leszkowicz et al. showed that OTA treatment induced the formation of DNA adducts in different organs in mice in a dose- and time-dependent manner. DNA was isolated from kidney, liver, and spleen excised from mice 24, 48, and 72 h after oral treatment with OTA at 0.6, 1.2, and 2.5 mg/kg body weight. After the administration of 2.5 mg/kg body weight, 40 adducts per 109 nucleotides were found in kidney DNA after 72 h [277]. In humans, OTA-mediated DNA adducts were also detected in tumor tissue from the kidney and bladder of Bulgarian patients undergoing surgery for cancer [278]. Similar results concerning OTA-DNA adducts *in vivo* were since observed [279, 280].

In vitro, Obrecht-Pflumio incubated microsomes prepared from mice or rabbit kidney, used as metabolic activators, with commercial salmon testes DNA and OTA, with NADPH or arachidonic acid used as cofactors [281]. In the mouse kidney system, up to 126 DNA adducts for 109 nucleotides were detected using the ^{32}P post-labeling method. Similar results were also obtained in rabbit kidney microsomes [281]. They further demonstrated that OTA was metabolized to genotoxic metabolites that interacted with the guanine residues of DNA because the DNA adducts were found only on dGMP when dAMP, dGMP, dTMP, and dCMP were used as substrates under the same conditions as with DNA.

Dai's group used the photochemistry of OTA to assess its reactivity toward dGMP, and a carbon- and oxygen-bonded C8-OTA dGMP adduct (OTB-dG) was identified [36]. Thereafter, scientists tried to identify the structures of OTA-DNA adducts.

According to Manderville's review, after the bioactivation of OTA, phenoxyl radicals, benzoquinone electrophiles, and carbon-centered radicals are able to react directly with DNA to form covalent DNA adducts, as shown in Figure 7. Two chemical structures of authentic OTA-DNA adducts exist, as shown in Figure 8; OTB-dG is the major OTA-DNA adduct [282].

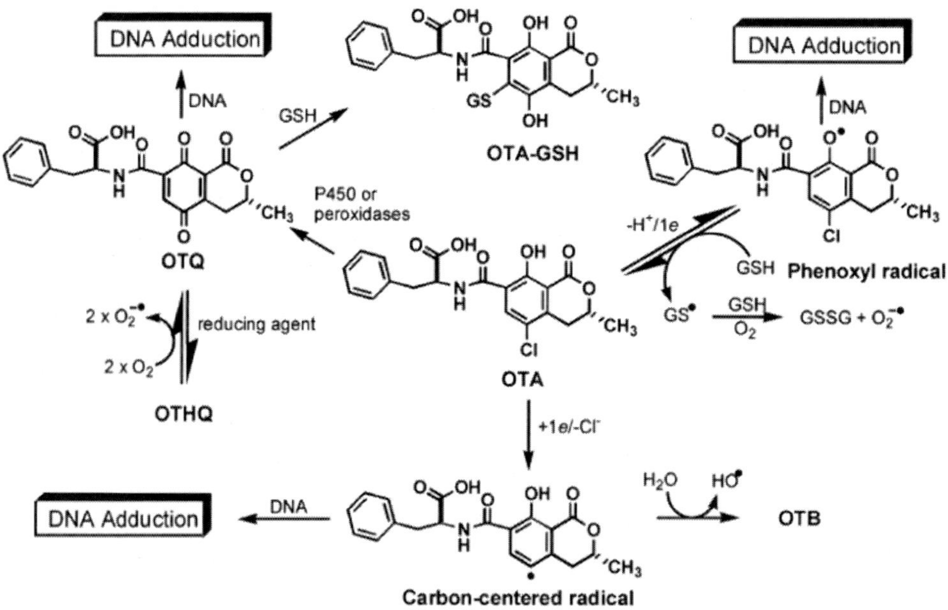

Figure 7. Proposed Pathways for the Bioactivation of OTA [282].

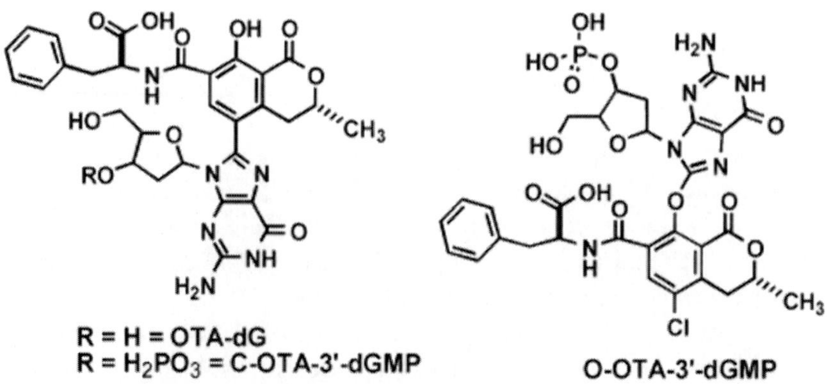

Figure 8 Chemical structures of authentic OTA-DNA adducts [282].

Recently, Sharma et al. used a robust computational approach to investigate the site-specific incorporation of OTB-dG into DNA. They determined the conformational preferences of OTB-dG in three ionization states at three guanine positions in the NarI recognition sequence opposite cytosine. Representative adducted DNA helices were derived from over 2,160 ns of simulation and ranked via free energies. Their results suggest that DNA-OTA adducts likely adopts a mixture of conformations. Nevertheless, the preferred conformation depends on the flanking bases and ionization state because of deviations in discrete local interactions at the lesion site, supporting recent experiments suggesting the induction of double-strand breaks and deletion mutations upon OTA exposure [283].

However, a group of scientists hold an opposite view that there is little chance for OTA to form covalent DNA adducts and lack of proof in their own studies; thus whether OTA can form DNA adducts is debatable.

Gautier's study showed no relationship between OTA metabolism and glutathione conjugate formation with rat, mouse, or human kidney microsomes or postmitochondrial supernatants. There was also no evidence of DNA adduct formation when OTA was incubated with P450 enzyme systems in the presence of calf thymus DNA (<20 adducts/109 DNA bases) [284]. In Mally's *in vivo* study, only traces of OTHQ were found in the urine of male F344 rats treated with high doses of OTA (2 mg/kg bw) for 2 weeks, suggesting that this metabolite of OTA is not formed to a relevant extent. In agreement with the *in vitro* data, OTA-dG was not detected by LC-MS/MS in either liver or kidney DNA extracted from treated animals [285]. Moreover, the DNA binding of OTA and OTB was detected in male rats given a single dose of ^{14}C-OTA or ^{14}C-OTB using accelerator mass spectrometry. The ^{14}C content in treated animals was not significantly different from controls. This report indicates that OTA does not form covalent DNA adducts in high yields [285].

4. Nephrotoxicity Amelioration

Ochratoxin A has been established as one of the most abundant food-contaminating mycotoxins; it is generally present in the food chain. Dietary antioxidants have gained much attention in recent years for their antioxidative and therapeutic properties. Recently, attention has been focused on the antidotal mechanism of natural antioxidant constituents.

Polyphenols constitute the most numerous and widely distributed groups of natural antioxidants in the plant kingdom [286]. The most abundant are phenolic acids and flavonoids accounting for 60% and 30%, respectively, of total polyphenols in the diet—approximately 1 g/day. The derivatives of the most abundant phenolic acids in plants are hydroxybenzoic and hydroxycinnamic acids. Within the hydroxycinnamic acid group are chlorogenic and caffeic acids, which are very abundant in most fruits and coffee [287]. It has been shown that luteolin and chlorogenic and caffeic acids protect against OTA-induced toxicity and protect cells from various forms of oxidant stresses by scavenging reactive oxygen species (ROS) [288].

Flavones such as apigenin and luteolin are the most common phenolics in plant-based foods. Another flavone, quercetin, is a dietary antioxidant that can combat OTA-induced toxicity in the Vero (a normal African green monkey kidney cell line) cell line. Quercetin modulated the antioxidant response through the activation of the Nrf2 pathway to defend against OTA-induced toxicity. Morphological studies including scanning electron microscopy, cell cycle analysis, and the inhibition of a caspase cascade indicated that quercetin has the ability to prevent OTA-induced apoptosis and DNA fragmentation. Quercetin also exhibited antigenotoxic potential by attenuating OTA-induced DNA damage and micronucleus formation [289].

N-acetylcysteine (NAC) is a cysteine donor that maintains or even increases the intracellular levels of glutathione [290]. NAC is a thiol-containing compound and the precursor of GSH; thus, it has a broad range of effects including antioxidant activity and intracellular glutathione level induction [291]. NAC pretreatment significantly prevented OTA-induced G1 phase arrest in human peripheral blood mononuclear cells [292]. The OTA-induced DNA damage and subsequent S-phase arrest may be due to the overproduction of ROS. NAC inhibited OTA-induced cytotoxicity by neutralizing excessive ROS production. Concretely, NAC prevented the OTA-induced oxidative stress and cell damage by two mechanisms: (i) directly scavenging the ROS and (ii) indirectly ameliorating DNA damage. For this reason, NAC should be explored as a protective agent against OTA-induced toxicity [163].

REFERENCES

[1] Mally, A., Dekant, W., (2009). Mycotoxins and the kidney: modes of action for renal tumor formation by ochratoxin A in rodents. *Mol. Nutr. Food Res.* 53, 467-478.

[2] NTP, (1989). Toxicology and carcinogenesis studies of ochratoxin A (cas no. 303-47-9) in F344/N Rats (Gavage Studies). *Natl.Toxicol. Program Tech. Rep. Ser.* . 358, 142.

[3] Krogh, P., Elling, F., Gyrd-Hansen, N., Hald, B., Larsen, A.E., Lillehoj, E.B., Madsen, A., Mortensen, H.P., Ravnskov, U., (1976). Experimental porcine nephropathy: changes of renal function and structure perorally induced by crystalline ochratoxin A. *Acta Pathol. Microbiol. Scand. A.* 84, 429-434.

[4] Krogh, P., Elling, F., Friis, C., Hald, B., Larsen, A.E., Lillehoj, E.B., Madsen, A., Mortensen, H.P., Rasmussen, F., Ravnskov, U., (1979). Porcine nephropathy induced by long-term ingestion of ochratoxin A. *Vet. Pathol.* 16, 466-475.

[5] Stoev, S.D., Paskalev, M., MacDonald, S., Mantle, P.G., (2002). Experimental one year ochratoxin A toxicosis in pigs. *Exp. Toxicol. Pathol.* 53, 481-487.

[6] Maaroufi, K., Zakhama, A., Baudrimont, I., Achour, A., Abid, S., Ellouz, F., Dhouib, S., Creppy, E.E., Bacha, H., (1999). Karyomegaly of tubular cells as early stage marker of the nephrotoxicity induced by ochratoxin A in rats. *Hum. Exp. Toxicol.* 18, 410-415.

[7] Bendele, A.M., Carlton, W.W., Krogh, P., Lillehoj, E., (1985). Ochratoxin A carcinogenesis in the (C57BL/6J× C3H) F1 mouse. *J. Natl. Cancer Inst.* 75, 733-742.

[8] Attaluri, S., Bonala, R.R., Yang, I.Y., Lukin, M.A., Wen, Y.J., Grollman, A.P., Moriya, M., Iden, C.R., Johnson, F., (2010). DNA adducts of aristolochic acid II: total synthesis and site-specific mutagenesis studies in mammalian cells. *Nucleic Acids Res.* 38, 339-352.

[9] Arlt, V.M., Stiborova, M., vom Brocke, J., Simoes, M.L., Lord, G.M., Nortier, J.L., Hollstein, M., Phillips, D.H., Schmeiser, H.H., (2007). Aristolochic acid mutagenesis: molecular clues to the aetiology of Balkan endemic nephropathy-associated urothelial cancer. *Carcinogenesis.* 28, 2253-2261.

[10] Desalegn, B., Nanayakkara, S., Harada, K.H., Hitomi, T., Chandrajith, R., Karunaratne, U., Abeysekera, T., Koizumi, A., (2011). Mycotoxin Detection in Urine Samples from Patients with Chronic Kidney Disease of Uncertain Etiology in Sri Lanka. *Bull. Environ. Contam. Toxicol.* 87, 6-10.

[11] Clark, H.A., Snedeker, S.M., (2006). Ochratoxin A: Its cancer risk and potential for exposure. *J Toxicol Env Heal B*. 9, 265-296.

[12] Zaied, C., Bouaziz, C., Azizi, I., Bensassi, F., Chour, A., Bacha, H., Abid, S., (2011). Presence of ochratoxin A in Tunisian blood nephropathy patients. Exposure level to OTA. *Exp. Toxicol. Pathol.* 63, 613-618.

[13] Boorman, G.A., Mcdonald, M.R., Imoto, S., Persing, R., (1992). Renal lesions induced by ochratoxin A exposure in the F344 rat. *Toxicol. Pathol.* 20, 236-245.

[14] Rached, E., Hard, G.C., Blumbach, K., Weber, K., Draheim, R., Lutz, W.K., Özden, S., Steger, U., Dekant, W., Mally, A., (2007). Ochratoxin A: 13-week oral toxicity and cell proliferation in male F344/N rats. *Toxicol. Sci.* 97, 288-298.

[15] Rached, E., Hoffmann, D., Blumbach, K., Weber, K., Dekant, W., Mally, A., (2008). Evaluation of putative biomarkers of nephrotoxicity after exposure to ochratoxin a *in vivo* and *in vitro*. *Toxicol. Sci.* 103, 371-381.

[16] Mally, A., (2012). Ochratoxin A and mitotic disruption: mode of action analysis of renal tumor formation by ochratoxin A. *Toxicol. Sci.* 127, 315-330.

[17] Coresh, J., Selvin, E., Stevens, L.A., Manzi, J., Kusek, J.W., Eggers, P., Van Lente, F., Levey, A.S., (2007). Prevalence of chronic kidney disease in the United States. *JAMA*. 298, 2038-2047.

[18] Maisonneuve, P., Agodoa, L., Gellert, R., Stewart, J.H., Buccianti, G., Lowenfels, A.B., Wolf, R.A., Jones, E., Dsiney, A.P., Briggs, D., (1999). Cancer in patients on dialysis for end-stage renal disease: an international collaborative study. *The Lancet*. 354, 93-99.

[19] Xiong, G., Chen, X., Li, X., Fang, D., Zhang, L., Yang, L., Zhang, L., Yao, L., He, Z., Zhou, L., (2014). Prevalence and factors associated with baseline chronic kidney disease in China: a 10-year study of 785 upper urinary tract urothelial carcinoma patients. *J. Formos. Med. Assoc.* 113, 521-526.

[20] Siegel, R., Ma, J., Zou, Z., Jemal, A., (2014). Cancer statistics, 2014. *CA Cancer J. Clin.* 64, 9-29.

[21] Dietrich, D.R., Swenberg, J.A., (1991). Preneoplastic lesions in rodent kidney induced spontaneously or by non-genotoxic agents: predictive nature and comparison to lesions induced by genotoxic carcinogens. *Mutation Research/Fundamental and Molecular Mechanisms of Mutagenesis*. 248, 239-260.

[22] Lühe, A., Hildebrand, H., Bach, U., Dingermann, T., Ahr, H.-J., (2003). A new approach to studying ochratoxin A (OTA)-induced nephrotoxicity: expression profiling *in vivo* and *in vitro* employing cDNA microarrays. *Toxicol. Sci.* 73, 315-328.

[23] Kalmovarin, N., Friedrichs, W.E., O'brien, H.V., Linehan, L.A., Bowman, B.H., Yang, F., (1991). Extrahepatic expression of plasma protein genes during inflammation. *Inflammation*. 15, 369-379.

[24] Hellman, N.E., Gitlin, J.D., (2002). Ceruloplasmin metabolism and function. *Annu. Rev. Nutr.* 22, 439-458.

[25] Ohkubo, K., Ogata, S., Misumi, Y., Takami, N., Ikehara, Y., (1991). Molecular cloning and characterization of rat contrapsin-like protease inhibitor and related proteins. *J. Biochem.* 109, 243-250.

[26] Kojima, S., Hayashi, S., Shimokado, K., Suzuki, Y., Shimada, J., Crippa, M.P., Friedman, S.L., (2000). Transcriptional activation of urokinase by the Krüppel-like factor Zf9/COPEB activates latent TGF-β1 in vascular endothelial cells. *Blood.* 95, 1309-1316.

[27] Macfarlane, S.R., Seatter, M.J., Kanke, T., Hunter, G.D., Plevin, R., (2001). Proteinase-activated receptors. *Pharmacol. Rev.* 53, 245-282.

[28] Riewald, M., Ruf, W., (2001). Mechanistic coupling of protease signaling and initiation of coagulation by tissue factor. *Proceedings of the National Academy of Sciences.* 98, 7742-7747.

[29] Gekle, M., Schwerdt, G., Freudinger, R., Mildenberger, S., Wilflingseder, D., Pollack, V., Dander, M., Schramek, H., (2000). Ochratoxin A induces JNK activation and apoptosis in MDCK-C7 cells at nanomolar concentrations. *J. Pharmacol. Exp. Ther.* 293, 837-844.

[30] Gennari, A., Pazos, P., Boveri, M., Callaghan, R., Casado, J., Maurici, D., Corsini, E., Prieto, P., (2004). New insights into the mechanisms involved in renal proximal tubular damage induced *in vitro* by ochratoxin A. *J. Biochem. Mol. Toxicol.* 18, 43-49.

[31] Coronel, M., Marin, S., Tarragó, M., Cano-Sancho, G., Ramos, A., Sanchis, V., (2011). Ochratoxin A and its metabolite ochratoxin alpha in urine and assessment of the exposure of inhabitants of Lleida, Spain. *Food Chem. Toxicol.* 49, 1436-1442.

[32] Pena, A., Seifrtová, M., Lino, C., Silveira, I., Solich, P., (2006). Estimation of ochratoxin A in Portuguese population: new data on the occurrence in human urine by high performance liquid chromatography with fluorescence detection. *Food Chem. Toxicol.* 44, 1449-1454.

[33] Zepnik, H., Volkel, W., Dekant, W., (2003). Toxicokinetics of the mycotoxin ochratoxin A in F 344 rats after oral administration. *Toxicol. Appl. Pharmacol.* 192, 36-44.

[34] Mally, A., Volkel, W., Amberg, A., Kurz, M., Wanek, P., Eder, E., Hard, G., Dekant, W., (2005). Functional, biochemical, and pathological effects of repeated oral administration of ochratoxin A to rats. *Chem. Res. Toxicol.* 18, 1242-1252.

[35] Calcutt, M.W., Gillman, I.G., Noftle, R.E., Manderville, R.A., (2001). Electrochemical oxidation of ochratoxin A: Correlation with 4-chlorophenol. *Chem. Res. Toxicol.* 14, 1266-1272.

[36] Dai, J., Wright, M.W., Manderville, R.A., (2003). Ochratoxin A forms a carbon-bonded C8-deoxyguanosine nucleoside adduct: Implications for C8 reactivity by a phenolic radical. *J. Am. Chem. Soc.* 125, 3716-3717.

[37] Dai, J., Park, G., Wright, M.W., Adams, M., Akman, S.A., Manderville, R.A., (2002). Detection and characterization of a glutathione conjugate of ochratoxin A. *Chem. Res. Toxicol.* 15, 1581-1588.

[38] Aoudia, N., Callu, P., Grosjean, F., Larondelle, Y., (2009). Effectiveness of mycotoxin sequestration activity of micronized wheat fibres on distribution of ochratoxin A in plasma, liver and kidney of piglets fed a naturally contaminated diet. *Food Chem. Toxicol.* 47, 1485-1489.

[39] Corcuera, L.A., Ibanez-Vea, M., Vettorazzi, A., Gonzalez-Penas, E., Cerain, A.L., (2011). Validation of a UHPLC-FLD analytical method for the simultaneous quantification of aflatoxin B1 and ochratoxin a in rat plasma, liver and kidney. *J. Chromatogr. B Analyt. Technol. Biomed. Life Sci.* 879, 2733-2740.

[40] Zhang, B., Shen, X.L., Liang, R., Li, Y., Huang, K., Zhao, C., Luo, Y., Xu, W., (2014). Protective role of the mitochondrial Lon protease 1 in ochratoxin A-induced cytotoxicity in HEK293 cells. *J. Proteomics*. 101, 154-168.

[41] Tsuda, M., Sekine, T., Takeda, M., Cha, S.H., Kanai, Y., Kimura, M., Endou, H., (1999). Transport of ochratoxin A by renal multispecific organic anion transporter 1. *J. Pharmacol. Exp. Ther.* 289, 1301-1305.

[42] Buist, S.C., Klaassen, C.D., (2004). Rat and mouse differences in gender-predominant expression of organic anion transporter (Oat1-3; Slc22a6-8) mRNA levels. *Drug Metab. Dispos.* 32, 620-625.

[43] Ljubojevic, M., Herak-Kramberger, C.M., Hagos, Y., Bahn, A., Endou, H., Burckhardt, G., Sabolic, I., (2004). Rat renal cortical OAT1 and OAT3 exhibit gender differences determined by both androgen stimulation and estrogen inhibition. *Am. J. Physiol. Renal Physiol.* 287, F124-138.

[44] Ljubojevic, M., Balen, D., Breljak, D., Kusan, M., Anzai, N., Bahn, A., Burckhardt, G., Sabolic, I., (2007). Renal expression of organic anion transporter OAT2 in rats and mice is regulated by sex hormones. *Am. J. Physiol. Renal Physiol.* 292, F361-372.

[45] Smith, A.G., Francis, J.E., (1983). Evidence for the Active Renal Secretion Of S-Pentachlorophenyl-N-Acetyl-L-Cysteine by Female Rats. *Biochem. Pharmacol.* 32, 3797-3801.

[46] Dahlmann, A., Dantzler, W.H., Silbernagl, S., Gekle, M., (1998). Detailed mapping of ochratoxin a reabsorption along the rat nephron *in vivo*: The nephrotoxin can be reabsorbed in all nephron segments by different mechanisms. *J. Pharmacol. Exp. Ther.* 286, 157-162.

[47] Zingerle, M., Silbernagl, S., Gekle, M., (1997). Reabsorption of the nephrotoxin ochratoxin A along the rat nephron *in vivo*. *J. Pharmacol. Exp. Ther.* 280, 220-224.

[48] Anzai, N., Jutabha, P., Enomoto, A., Yokoyama, H., Nonoguchi, H., Hirata, T., Shiraya, K., He, X., Cha, S.H., Takeda, M., Miyazaki, H., Sakata, T., Tomita, K., Igarashi, T., Kanai, Y., Anzai, N., (2005). Functional characterization of rat organic anion transporter 5 (Slc22a19) at the apical membrane of renal proximal tubules. *J. Pharmacol. Exp. Ther.* 315, 534-544.

[49] Anzai, N., Jutabha, P., Endou, H., (2010). Molecular Mechanism of Ochratoxin A Transport in the Kidney. *Toxins (Basel)*. 2, 1381-1398.

[50] Lu, R., Kanai, N., Bao, Y., Wolkoff, A.W., Schuster, V.L., (1996). Regulation of renal oatp mRNA expression by testosterone. *Am. J. Physiol.* 270, F332-337.

[51] Gotoh, Y., Kato, Y., Stieger, B., Meier, P.J., Sugiyama, Y., (2002). Gender difference in the Oatp1-mediated tubular reabsorption of estradiol 17beta-D-glucuronide in rats. *Am. J. Physiol. Endocrinol. Metab.* 282, E1245-1254.

[52] Kato, Y., Kuge, K., Kusuhara, H., Meier, P.J., Sugiyama, Y., (2002). Gender difference in the urinary excretion of organic anions in rats. *J. Pharmacol. Exp. Ther.* 302, 483-489.

[53] Lu, R., Kanai, N., Bao, Y., Wolkoff, A.W., Schuster, V.L., (1996). Regulation of renal oatp mRNA expression by testosterone. *Am J Physiol-Renal*. 270, F332-F337.

[54] Mally, A., Voelkel, W., Amberg, A., Kurz, M., Wanek, P., Eder, E., Hard, G., Dekant, W., (2005). Functional, biochemical, and pathological effects of repeated oral administration of ochratoxin A to rats. *Chem. Res. Toxicol.* 18, 1242-1252.

[55] Prozialeck, W., Vaidya, V., Liu, J., Waalkes, M., Edwards, J., Lamar, P., Bernard, A., Dumont, X., Bonventre, J., (2007). Kidney injury molecule-1 is an early biomarker of cadmium nephrotoxicity. *Kidney Int.* 72, 985-993.

[56] Bonventre, J.V., (2008). Kidney Injury Molecule - 1 (KIM - 1): A specific and sensitive biomarker of kidney injury. *Scand. J. Clin. Lab. Invest.* 68, 78-83.

[57] Qi, X., Yu, T., Zhu, L., Gao, J., He, X., Huang, K., Luo, Y., Xu, W., (2014). Ochratoxin A induces rat renal carcinogenicity with limited induction of oxidative stress responses. *Toxicol. Appl. Pharmacol.* 280, 543-549.

[58] Driver, T.H., Katz, R., Ix, J.H., Magnani, J.W., Peralta, C.A., Parikh, C.R., Fried, L., Newman, A.B., Kritchevsky, S.B., Sarnak, M.J., (2014). Urinary kidney injury molecule 1 (KIM-1) and interleukin 18 (IL-18) as risk markers for heart failure in older adults: the Health, Aging, and Body Composition (Health ABC) study. *Am. J. Kidney Dis.* 64, 49-56.

[59] Marin-Kuan, M., Cavin, C., Delatour, T., Schilter, B., (2008). Ochratoxin A carcinogenicity involves a complex network of epigenetic mechanisms. *Toxicon.* 52, 195-202.

[60] Murphy, B., Clark, S., Donald, I., Pinsky, M., Vedady, D., (1974). Conversion of maternal cortisol to cortisone during placental transfer to the human fetus. *Am. J. Obstet. Gynecol.* 118, 538.

[61] Takahashi, T., Caviness Jr, V., (1993). PCNA-binding to DNA at the G1/S transition in proliferating cells of the developing cerebral wall. *J. Neurocytol.* 22, 1096-1102.

[62] Paunesku, T., Mittal, S., Protic, M., Oryhon, J., Korolev, S., Joachimiak, A., Woloschak, G., (2001). Proliferating cell nuclear antigen (PCNA): ringmaster of the genome. *Int. J. Radiat. Biol.* 77, 1007-1021.

[63] Braun, T., Li, S., Moss, T.J., Newnham, J.P., Challis, J.R., Gluckman, P.D., Sloboda, D.M., (2007). Maternal betamethasone administration reduces binucleate cell number and placental lactogen in sheep. *J. Endocrinol.* 194, 337-347.

[64] Stemmer, K., Ellinger-Ziegelbauer, H., Ahr, H.-J., Dietrich, D.R., (2007). Carcinogen-specific gene expression profiles in short-term treated Eker and wild-type rats indicative of pathways involved in renal tumorigenesis. *Cancer Res.* 67, 4052-4068.

[65] Stemmer, K., Dietrich, D., (2008). Gender disparity in early ochratoxin A mediated renal toxicity. *Toxicologist.* 102, 21.

[66] Stemmer, K., Ellinger-Ziegelbauer, H., Ahr, H.-J., Dietrich, D.R., (2009). Molecular characterization of preneoplastic lesions provides insight on the development of renal tumors. *The American journal of pathology.* 175, 1686-1698.

[67] Xia, K., He, X.Y., Dai, Q., Cheng, W.H., Qi, X.Z., Guo, M.Z., Luo, Y.B., Huang, K.L., Zhao, C.H., Xu, W.T., (2014). Discovery of systematic responses and potential biomarkers induced by ochratoxin A using metabolomics. *Food Addit Contam A.* 31, 1904-1913.

[68] Schaaf, G.J., Nijmeijer, S.M., Maas, R.F.M., Roestenberg, P., de Groene, E.M., Fink-Gremmels, J., (2002). The role of oxidative stress in the ochratoxin A-mediated toxicity in proximal tubular cells. *Bba-Mol Basis Dis.* 1588, 149-158.

[69] Pfohl-Leszkowicz, A., Manderville, R.A., (2007). Ochratoxin A: An overview on toxicity and carcinogenicity in animals and humans. *Mol. Nutr. Food Res.* 51, 61-99.

[70] Stoev, S.D., Stoeva, J.K., Anguelov, G., Hald, B., Creppy, E.E., Radic, B., (1998). Haematological, biochemical and toxicological investigations in spontaneous cases with different frequency of porcine nephropathy in Bulgaria. *J Vet Med A.* 45, 229-236.

[71] Jia, L.W., Chen, J., Yin, P.Y., Lu, X., Xu, G.W., (2008). Serum metabonomics study of chronic renal failure by ultra performance liquid chromatography coupled with Q-TOF mass spectrometry. *Metabolomics.* 4, 183-189.

[72] Waters, N.J., Waterfield, C.J., Farrant, R.D., Holmes, E., Nicholson, J.K., (2005). Metabonomic deconvolution of embedded toxicity: Application to thioacetamide hepato- and nephrotoxicity. *Chem. Res. Toxicol.* 18, 639 654.

[73] Dunn, W.B., Broadhurst, D.I., Deepak, S.M., Buch, M.H., McDowell, G., Spasic, I., Ellis, D.I., Brooks, N., Kell, D.B., Neyses, L., (2007). Serum metabolomics reveals many novel metabolic markers of heart failure, including pseudouridine and 2-oxoglutarate. *Metabolomics.* 3, 413-426.

[74] Smith, J.L., Wishnok, J.S., Deen, W.M., (1994). Metabolism And Excretion Of Methylamines In Rats. *Toxicol. Appl. Pharmacol.* 125, 296-308.

[75] Wang, H.F., Bai, J., Chen, G., Li, W., Xiang, R.W., Su, G.Y., Pei, Y.H., (2013). A metabolic profiling analysis of the acute hepatotoxicity and nephrotoxicity of Zhusha Anshen Wan compared with cinnabar in rats using H-1 NMR spectroscopy. *J. Ethnopharmacol.* 146, 572-580.

[76] Boudonck, K.J., Mitchell, M.W., Nemet, L., Keresztes, L., Nyska, A., Shinar, D., Rosenstock, M., (2009). Discovery of Metabolomics Biomarkers for Early Detection of Nephrotoxicity. *Toxicol. Pathol.* 37, 280-292.

[77] Bonnefond, L., Fender, A., Rudinger-Thirion, J., Giegé, R., Florentz, C., Sissler, M., (2005). Toward the full set of human mitochondrial aminoacyl-tRNA synthetases: characterization of AspRS and TyrRS. *Biochemistry.* 44, 4805-4816.

[78] Creppy, E.E., Mayer, M., Kern, D., Schlegel, M., Steyn, P.S., Vleggaar, R., Dirheimer, G., (1981). *In vitro* inhibition of yeast valyl-tRNA synthetase by the valine homologue of ochratoxin A. *Biochimica et Biophysica Acta (BBA)-Nucleic Acids and Protein Synthesis.* 656, 265-268.

[79] Creppy, E., Kern, D., Steyn, P., Vleggaar, R., Röschenthaler, R., Dirheimer, G., (1983). Comparative study of the effect of ochratoxin A analogues on yeast aminoacyl-tRNA synthetases and on the growth and protein synthesis of hepatoma cells. *Toxicol. Lett.* 19, 217-224.

[80] Creppy, E., Størmer, F., Kern, D., Röschenthaler, R., Dirheimer, G., (1983). Effects of ochratoxin A metabolites on yeast phenylalanyl-tRNA synthetase and on the growth and *in vivo* protein synthesis of hepatoma cells. *Chem. Biol. Interact.* 47, 239-247.

[81] Marin-Kuan, M., Nestler, S., Verguet, C., Bezencon, C., Piguet, D., Mansourian, R., Holzwarth, J., Grigorov, M., Delatour, T., Mantle, P., (2006). A toxicogenomics approach to identify new plausible epigenetic mechanisms of ochratoxin a carcinogenicity in rat. *Toxicol. Sci.* 89, 120-134.

[82] Vettorazzi, A., van Delft, J., de Cerain, A.L., (2013). A review on ochratoxin A transcriptomic studies. *Food Chem. Toxicol.* 59, 766-783.

[83] Goldenthal, M.J., Marín-García, J., (2004). Mitochondrial signaling pathways: a receiver/integrator organelle. *Mol. Cell. Biochem.* 262, 1-16.

[84] Cuezva, J.M., Ortega, Á.D., Willers, I., Sánchez-Cenizo, L., Aldea, M., Sánchez-Aragó, M., (2009). The tumor suppressor function of mitochondria: translation into the clinics. *Biochimica et Biophysica Acta (BBA)-Molecular Basis of Disease*. 1792, 1145-1158.

[85] DeBerardinis, R.J., Mancuso, A., Daikhin, E., Nissim, I., Yudkoff, M., Wehrli, S., Thompson, C.B., (2007). Beyond aerobic glycolysis: transformed cells can engage in glutamine metabolism that exceeds the requirement for protein and nucleotide synthesis. *Proceedings of the National Academy of Sciences*. 104, 19345-19350.

[86] Sciacovelli, M., Gaude, E., Hilvo, M., Frezza, C., (2014). The metabolic alterations of cancer cells. *Methods Enzymol.* 542, 1-23.

[87] Musicco, C., Capelli, V., Pesce, V., Timperio, A.M., Calvani, M., Mosconi, L., Cantatore, P., Gadaleta, M.N., (2011). Rat liver mitochondrial proteome: Changes associated with aging and acetyl-L-carnitine treatment. *J. Proteomics*. 74, 2536-2547.

[88] Azad, M.B., Chen, Y.Q., Gibson, S.B., (2009). Regulation of Autophagy by Reactive Oxygen Species (ROS): Implications for Cancer Progression and Treatment. *Antioxid Redox Sign*. 11, 777-790.

[89] Shen, X.L., Zhang, Y., Xu, W.T., Liang, R., Zheng, J.J., Luo, Y.B., Wang, Y., Huang, K.L., (2013). An iTRAQ-based mitoproteomics approach for profiling the nephrotoxicity mechanisms of ochratoxin A in HEK 293 cells. *Journal of proteomics*. 78, 398-415.

[90] Mick, D.U., Fox, T.D., Rehling, P., (2011). Inventory control: cytochrome c oxidase assembly regulates mitochondrial translation. *Nat. Rev. Mol. Cell Biol.* 12, 14-20.

[91] Burnichon, N., Briere, J.J., Libe, R., Vescovo, L., Riviere, J., Tissier, F., Jouanno, E., Jeunemaitre, X., Benit, P., Tzagoloff, A., Rustin, P., Bertherat, J., Favier, J., Gimenez-Roqueplo, A.P., (2010). SDHA is a tumor suppressor gene causing paraganglioma. *Hum. Mol. Genet.* 19, 3011-3020.

[92] Cervera, A.M., Apostolova, N., Crespo, F.L., Mata, M., McCreath, K.J., (2008). Cells silenced for SDHB expression display characteristic features of the tumor phenotype. *Cancer Res.* 68, 4058-4067.

[93] Chen, Y., Gaczynska, M., Osmulski, P., Polci, R., Riley, D.J., (2010). Phosphorylation by Nek1 regulates opening and closing of voltage dependent anion channel 1. *Biochem. Biophys. Res. Commun.* 394, 798-803.

[94] Abu-Hamad, S., Sivan, S., Shoshan-Barmatz, V., (2006). The expression level of the voltage-dependent anion channel controls life and death of the cell. *Proc. Natl. Acad. Sci. U. S. A.* 103, 5787-5792.

[95] Rosano, C., (2011). Molecular model of hexokinase binding to the outer mitochondrial membrane porin (VDAC1): Implication for the design of new cancer therapies. *Mitochondrion*. 11, 513-519.

[96] Abu-Hamad, S., Zaid, H., Israelson, A., Nahon, E., Shoshan-Barmatz, V., (2008). Hexokinase-I protection against apoptotic cell death is mediated via interaction with the voltage-dependent anion channel-1 - Mapping the site of binding. *J. Biol. Chem.* 283, 13482-13490.

[97] Mick, D.U., Fox, T.D., Rehling, P., (2011). Inventory control: cytochrome c oxidase assembly regulates mitochondrial translation. *Nat Rev Mol Cell Bio*. 12, 14-20.

[98] Masuda, Y., Shima, G., Aiuchi, T., Horie, M., Hori, K., Nakajo, S., Kajimoto, S., Shibayama-Imazu, T., Nakaya, K., (2004). Involvement of tumor necrosis factor

receptor-associated protein 1 (TRAP1) in apoptosis induced by beta-hydroxyisovalerylshikonin. *J. Biol. Chem.* 279, 42503-42515.

[99] Bota, D.A., Davies, K.J.A., (2002). Lon protease preferentially degrades oxidized mitochondrial aconitase by an ATP-stimulated mechanism. *Nat Cell Biol.* 4, 674-680.

[100] Kim, T.S., Kim, H.D., Kim, J., (2009). PKC delta-dependent functional switch of rpS3 between translation and DNA repair. *Bba-Mol Cell Res.* 1793, 395-405.

[101] Du, C.Y., Fang, M., Li, Y.C., Li, L., Wang, X.D., (2000). Smac, a mitochondrial protein that promotes cytochrome c-dependent caspase activation by eliminating IAP inhibition. *Cell.* 102, 33-42.

[102] Rehm, M., Dussmann, H., Prehn, J.H.M., (2003). Real-time single cell analysis of Smac/DIABLO release during apoptosis. *J Cell Biol.* 162, 1031-1043.

[103] Lui, J.C.K., Kong, S.K., (2007). Heat shock protein 70 inhibits the nuclear import of apoptosis-inducing factor to avoid DNA fragmentation in TF-1 cells during erythropoiesis. *Febs Lett.* 581, 109-117.

[104] Olichon, A., Baricault, L., Gas, N., Guillou, E., Valette, A., Belenguer, P., Lenaers, G., (2003). Loss of OPA1 perturbates the mitochondrial inner membrane structure and integrity, leading to cytochrome c release and apoptosis. *J. Biol. Chem.* 278, 7743-7746.

[105] Shen, X.L., Zhang, Y., Xu, W., Liang, R., Zheng, J., Luo, Y., Wang, Y., Huang, K., (2013). An iTRAQ-based mitoproteomics approach for profiling the nephrotoxicity mechanisms of ochratoxin A in HEK 293 cells. *J. Proteomics.* 78, 398-415.

[106] Zhang, B.Y., Shen, X.L., Liang, R., Li, Y.Z., Huang, K.L., Zhao, C.H., Luo, Y.B., Xu, W.T., (2014). Protective role of the mitochondrial Lon protease 1 in ochratoxin A-induced cytotoxicity in HEK293 cells. *J. Proteomics.* 101, 154-168.

[107] Lin, Y.L., Lai, Z.X., (2013). Superoxide dismutase multigene family in longan somatic embryos: a comparison of CuZn-SOD, Fe-SOD, and Mn-SOD gene structure, splicing, phylogeny, and expression. *Mol. Breed.* 32, 595-615.

[108] Gallagher, E.P., Gardner, J.L., Barber, D.S., (2006). Several glutathione S-transferase isozymes that protect against oxidative injury are expressed in human liver mitochondria. *Biochem. Pharmacol.* 71, 1619-1628.

[109] Hayes, J.D., Flanagan, J.U., Jowsey, I.R., (2005). Glutathione transferases. *Annu. Rev. Pharmacol. Toxicol.* 45, 51-88.

[110] Chen, J.W., Dodia, C., Feinstein, S.I., Jain, M.K., Fisher, A.B., (2000). 1-Cys peroxiredoxin, a bifunctional enzyme with glutathione peroxidase and phospholipase A2 activities. *J. Biol. Chem.* 275, 28421-28427.

[111] Kalmar, B., Greensmith, L., (2009). Induction of heat shock proteins for protection against oxidative stress. *Adv. Drug Del. Rev.* 61, 310-318.

[112] Lionaki, E., Tavernarakis, N., (2013). Oxidative stress and mitochondrial protein quality control in aging. *J. Proteomics.* 92, 181-194.

[113] Ngo, J.K., Davies, K.J.A., (2007). Importance of the Lon protease in mitochondrial maintenance and the significance of declining Lon in aging. *Molecular Mechanisms And Models Of Aging.* 1119, 78-87.

[114] Feder, M.E., Hofmann, G.E., (1999). Heat-shock proteins, molecular chaperones, and the stress response: Evolutionary and ecological physiology. *Annu. Rev. Physiol.* 61, 243-282.

[115] Gelebart, P., Opas, M., Michalak, M., (2005). Calreticulin, a Ca2+-binding chaperone of the endoplasmic reticulum. *Int. J. Biochem. Cell Biol.* 37, 260-266.

[116] Wilkinson, B., Gilbert, H.F., (2004). Protein disulfide isomerase. *Bba-Proteins Proteom.* 1699, 35-44.
[117] Pelletier, M.F., Marcil, A., Sevigny, G., Jakob, C.A., Tessier, D.C., Chevet, E., Menard, R., Bergeron, J.J.M., Thomas, D.Y., (2000). The heterodimeric structure of glucosidase II is required for its activity, solubility, and localization *in vivo*. *Glycobiology.* 10, 815-827.
[118] Broers, J.L.V., Ramaekers, F.C.S., Bonne, G., Ben Yaou, R., Hutchison, C.J., (2006). Nuclear lamins: Laminopathies and their role in premature ageing. *Physiol. Rev.* 86, 967-1008.
[119] Jo, S.H., Son, M.K., Koh, H.J., Lee, S.M., Song, I.H., Kim, Y.O., Lee, Y.S., Jeong, K.S., Kim, W.B., Park, J.W., Song, B.J., Huh, T.L., (2001). Control of mitochondrial redox balance and cellular defense against oxidative damage by mitochondrial NADP(+)-dependent isocitrate dehydrogenase. (vol 276, pg 16168, 2001). *J. Biol. Chem.* 276, 26732-26732.
[120] Gameiro, P.A., Laviolette, L.A., Kelleher, J.K., Iliopoulos, O., Stephanopoulos, G., (2013). Cofactor Balance by Nicotinamide Nucleotide Transhydrogenase (NNT) Coordinates Reductive Carboxylation and Glucose Catabolism in the Tricarboxylic Acid (TCA) Cycle. *J. Biol. Chem.* 288, 12967-12977.
[121] Matsuda, T., Takahashi-Yanaga, F., Yoshihara, T., Maenaka, K., Watanabe, Y., Miwa, Y., Morimoto, S., Kubohara, Y., Hirata, M., Sasaguri, T., (2010). Dictyostelium Differentiation-Inducing Factor-1 Binds to Mitochondrial Malate Dehydrogenase and Inhibits Its Activity. *J. Pharmacol. Sci.* 112, 320-326.
[122] Zieker, D., Konigsrainer, I., Tritschler, I., Loffler, M., Beckert, S., Traub, F., Nieselt, K., Buhler, S., Weller, M., Gaedcke, J., Taichman, R.S., Northoff, H., Brucher, B.L.D.M., Konigsrainer, A., (2010). Phosphoglycerate kinase 1 a promoting enzyme for peritoneal dissemination in gastric cancer. *Int. J. Cancer.* 126, 1513-1520.
[123] Spoden, G.A., Rostek, U., Lechner, S., Mitterberger, M., Mazurek, S., Zwerschke, W., (2009). Pyruvate kinase isoenzyme M2 is a glycolytic sensor differentially regulating cell proliferation, cell size and apoptotic cell death dependent on glucose supply. *Exp. Cell Res.* 315, 2765-2774.
[124] Wang, Y., Hao, J.R., Zhao, W.W., Yang, Z.J., Wu, W.H., Zhang, Y., Xu, W.T., Luo, Y.B., Huang, K.L., (2013). Comparative proteomics and physiological characterization of Arabidopsis thaliana seedlings in responses to Ochratoxin A. *Plant Mol. Biol.* 82, 321-337.
[125] Lu, B., Yadav, S., Shah, P.G., Liu, T., Tian, B., Pukszta, S., Villaluna, N., Kutejova, E., Newlon, C.S., Santos, J.H., Suzuki, C.K., (2007). Roles for the human ATP-dependent lon protease in mitochondrial DNA maintenance. *J. Biol. Chem.* 282, 17363-17374.
[126] Matsushima, Y., Goto, Y., Kaguni, L.S., (2010). Mitochondrial Lon protease regulates mitochondrial DNA copy number and transcription by selective degradation of mitochondrial transcription factor A (TFAM). *Proc. Natl. Acad. Sci. USA.* 107, 18410-18415.
[127] Ishihara, N., Fujita, Y., Oka, T., Mihara, K., (2006). Regulation of mitochondrial morphology through proteolytic cleavage of OPA1. *The EMBO journal.* 25, 2966-2977.
[128] Arnoult, D., Grodet, A., Lee, Y.J., Estaquier, J., Blackstone, C., (2005). Release of OPA1 during apoptosis participates in the rapid and complete release of cytochrome c and subsequent mitochondrial fragmentation. *J. Biol. Chem.* 280, 35742-35750.

[129] Duvezin-Caubet, S., Jagasia, R., Wagener, J., Hofmann, S., Trifunovic, A., Hansson, A., Chomyn, A., Bauer, M.F., Attardi, G., Larsson, N.G., Neupert, W., Reichert, A.S., (2006). Proteolytic processing of OPA1 links mitochondrial dysfunction to alterations in mitochondrial morphology. *J. Biol. Chem.* 281, 37972-37979.

[130] Baricault, L., Segui, B., Guegand, L., Olichon, A., Valette, A., Larminat, F., Lenaers, G., (2007). OPA1 cleavage depends on decreased mitochondrial ATP level and bivalent metals. *Exp. Cell Res.* 313, 3800-3808.

[131] Jia, L., Patwari, Y., Kelsey, S.M., Srinivasula, S.M., Agrawal, S.G., Alnemri, E.S., Newland, A.C., (2003). Role of Smac in human leukaemic cell apoptosis and proliferation. *Oncogene.* 22, 1589-1599.

[132] Kamp, H.G., Eisenbrand, G., Schlatter, J., Wurth, K., Janzowski, C., (2005). Ochratoxin A: induction of (oxidative) DNA damage, cytotoxicity and apoptosis in mammalian cell lines and primary cells. *Toxicology.* 206, 413-425.

[133] Di Giacomo, C., Acquaviva, R., Piva, A., Sorrenti, V., Vanella, L., Piva, G., Casadei, G., La Fauci, L., Ritieni, A., Bognanno, M., Di Renzo, L., Barcellona, M.L., Morlacchini, M., Galvano, F., (2007). Protective effect of cyanidin 3-O-beta-D-glucoside on ochratoxin A-mediated damage in the rat. *Br. J. Nutr.* 98, 937-943.

[134] Meki, A.R.M.A., Hussein, A.A.A., (2001). Melatonin reduces oxidative stress induced by ochratoxin A in rat liver and kidney. *Comp Biochem Phys C.* 130, 305-313.

[135] Kamp, H.G., Eisenbrand, G., Janzowski, C., Kiossev, J., Latendresse, J.R., Schlatter, J., Turesky, R.J., (2005). Ochratoxin A induces oxidative DNA damage in liver and kidney after oral dosing to rats. *Mol. Nutr. Food Res.* 49, 1160-1167.

[136] Cavin, C., Delatour, T., Marin-Kuan, M., Holzhauser, D., Higgins, L., Bezencon, C., Guignard, G., Junod, S., Richoz-Payot, J., Gremaud, E., Hayes, J.D., Nestler, S., Mantle, P., Schilter, B., (2007). Reduction in antioxidant defenses may contribute to ochratoxin a toxicity and carcinogenicity. *Toxicol. Sci.* 96, 30-39.

[137] Kumar, R., Ansari, K.M., Chaudhari, B.P., Dhawan, A., Dwivedi, P.D., Jain, S.K., Das, M., (2012). Topical Application of Ochratoxin A Causes DNA Damage and Tumor Initiation in Mouse Skin. *PLoS One.* 7.

[138] Dvergsten, J., Manivel, J.C., Correa-Rotter, R., Rosenberg, M.E., (1994). Expression of clusterin in human renal diseases. *Kidney Int.* 45, 828-835.

[139] Yamasaki, T., Seki, N., Yamada, Y., Yoshino, H., Hidaka, H., Chiyomaru, T., Nohata, N., Kinoshita, T., Nakagawa, M., Enokida, H., (2012). Tumor suppressive microRNA-138 contributes to cell migration and invasion through its targeting of vimentin in renal cell carcinoma. *Int. J. Oncol.* 41, 805-817.

[140] Xu, J.H., Zhang, C., Tang, B., Hao, Y.X., Chen, J., Liu, T., Cui, H., (2010). [Effect of JAK2/STAT3/vimentin signaling pathway on proliferation and migration of human colon cancer cells]. *Zhonghua Wei Chang Wai Ke Za Zhi.* 13, 282-285.

[141] Meki, A.R., Hussein, A.A., (2001). Melatonin reduces oxidative stress induced by ochratoxin A in rat liver and kidney. *Comp. Biochem. Physiol. C Toxicol. Pharmacol.* 130, 305-313.

[142] Petrik, J., Zanic-Grubisic, T., Barisic, K., Pepeljnjak, S., Radic, B., Ferencic, Z., Cepelak, I., (2003). Apoptosis and oxidative stress induced by ochratoxin A in rat kidney. *Arch. Toxicol.* 77, 685-693.

[143] Verma, R., Chakraborty, D., (2008). Emblica officinalis aqueous extract ameliorates ochratoxin-induced lipid peroxidation in the testis of mice. *Acta Pol. Pharm.* 65, 187-194.

[144] Domijan, A.M., Peraica, M., Vrdoljak, A.L., Radic, B., Zlender, V., Fuchs, R., (2007). The involvement of oxidative stress in ochratoxin A and fumonisin B1 toxicity in rats. *Mol. Nutr. Food Res.* 51, 1147-1151.

[145] Soyoz, M., Ozcelik, N., Kilinc, I., Altuntas, I., (2004). The effects of ochratoxin A on lipid peroxidation and antioxidant enzymes: a protective role of melatonin. *Cell Biol. Toxicol.* 20, 213-219.

[146] Hoehler, D., Marquardt, R.R., Frohlich, A.A., (1997). Lipid peroxidation as one mode of action in ochratoxin A toxicity in rats and chicks. *Canadian Journal Of Animal Science*. 77, 287-292.

[147] Gautier, J.C., Holzhaeuser, D., Markovic, J., Gremaud, E., Schilter, B., Turesky, R.J., (2001). Oxidative damage and stress response from ochratoxin A exposure in rats. *Free Radic. Biol. Med.* 30, 1089-1098.

[148] Valko, M., Morris, H., Cronin, M., (2005). Metals, toxicity and oxidative stress. *Curr. Med. Chem.* 12, 1161-1208.

[149] Schaaf, G., Nijmeijer, S., Maas, R., Roestenberg, P., De Groene, E., Fink-Gremmels, J., (2002). The role of oxidative stress in the ochratoxin A-mediated toxicity in proximal tubular cells. *Biochimica et Biophysica Acta (BBA)-Molecular Basis of Disease*. 1588, 149-158.

[150] Petrik, J., Žanić-Grubišić, T., Barišić, K., Pepeljnjak, S., Radić, B., Ferenčić, Ž., Čepelak, I., (2003). Apoptosis and oxidative stress induced by ochratoxin A in rat kidney. *Arch. Toxicol.* 77, 685-693.

[151] Finkel, T., (2003). Oxidant signals and oxidative stress. *Curr. Opin. Cell Biol.* 15, 247-254.

[152] Yang, Q., He, X., Li, X., Xu, W., Luo, Y., Yang, X., Wang, Y., Li, Y., Huang, K., (2014). DNA damage and S phase arrest induced by Ochratoxin A in human embryonic kidney cells (HEK 293). *Mutation Research/Fundamental and Molecular Mechanisms of Mutagenesis*. 765, 22-31.

[153] Liu, M., Grigoryev, D.N., Crow, M.T., Haas, M., Yamamoto, M., Reddy, S.P., Rabb, H., (2009). Transcription factor Nrf2 is protective during ischemic and nephrotoxic acute kidney injury in mice. *Kidney Int.* 76, 277-285.

[154] Kovac, S., Angelova, P.R., Holmström, K.M., Zhang, Y., Dinkova-Kostova, A.T., Abramov, A.Y., (2015). Nrf2 regulates ROS production by mitochondria and NADPH oxidase. *Biochimica et Biophysica Acta (BBA)-General Subjects*. 1850, 794-801.

[155] Limonciel, A., Jennings, P., (2014). A review of the evidence that ochratoxin A is an Nrf2 inhibitor: Implications for nephrotoxicity and renal carcinogenicity. *Toxins (Basel)*. 6, 371-379.

[156] Chandel, N.S., Trzyna, W.C., McClintock, D.S., Schumacker, P.T., (2000). Role of oxidants in NF-κB activation and TNF-α gene transcription induced by hypoxia and endotoxin. *The Journal of Immunology*. 165, 1013-1021.

[157] Gilmore, T.D., (2006). Introduction to NF-κB: players, pathways, perspectives. *Oncogene*. 25, 6680-6684.

[158] Hong, J.T., Lee, M.K., Park, K.S., Jung, K.M., Lee, R.D., Jung, H.K., Park, K.L., Yang, K.H., Chung, S.Y., (2002). Inhibitory effect of peroxisome proliferator-activated

receptor gamma agonist on ochratoxin A-induced cytotoxicity and activation of transcription factors in cultured rat embryonic midbrain cells. *Journal of Toxicology and Environmental Health Part A*. 65, 407-418.

[159] Benesic, A., Mildenberger, S., Gekle, M., (2000). Nephritogenic ochratoxin A interferes with hormonal signalling in immortalized human kidney epithelial cells. *Pflügers Archiv*. 439, 278-287.

[160] Guanggang, X., Diqiu, L., Jianzhong, Y., Jingmin, G., Huifeng, Z., Mingan, S., Liming, T., (2013). Carbamate insecticide methomyl confers cytotoxicity through DNA damage induction. *Food Chem. Toxicol.* 53, 352-358.

[161] Fleck, O., Nielsen, O., (2004). DNA repair. *J. Cell Sci.* 117, 515-517.

[162] Bonner, W.M., Redon, C.E., Dickey, J.S., Nakamura, A.J., Sedelnikova, O.A., Solier, S., Pommier, Y., (2008). γH2AX and cancer. *Nature Reviews Cancer*. 8, 957-967.

[163] Yang, Q., Shi, L., Huang, K., Xu, W., (2014). Protective effect of N-acetylcysteine against DNA damage and S-phase arrest induced by ochratoxin A in human embryonic kidney cells (HEK-293). *Food Chem. Toxicol.* 70, 40-47.

[164] Arbillaga, L., Azqueta, A., van Delft, J.H., de Cerain, A.L., (2007). *In vitro* gene expression data supporting a DNA non-reactive genotoxic mechanism for ochratoxin A. *Toxicol. Appl. Pharmacol.* 220, 216-224.

[165] Munaron, L., Antoniotti, S., Lovisolo, D., (2004). Intracellular calcium signals and control of cell proliferation: how many mechanisms? *J. Cell. Mol. Med.* 8, 161-168.

[166] Santella, L., Ercolano, E., Nusco, G., (2005). The cell cycle: a new entry in the field of Ca2+ signaling. *Cellular and Molecular Life Sciences CMLS*. 62, 2405-2413.

[167] Farber, J.L., (1981). The role of calcium in cell death. *Life Sci.* 29, 1289-1295.

[168] Tsien, R., Pozzan, T., Rink, T., (1982). Calcium homeostasis in intact lymphocytes: cytoplasmic free calcium monitored with a new, intracellularly trapped fluorescent indicator. *The Journal of cell biology*. 94, 325-334.

[169] Carafoli, E., (1987). Intracellular calcium homeostasis. *Annu. Rev. Biochem.* 56, 395-433.

[170] Farfariello, V., Iamshanova, O., Germain, E., Fliniaux, I., Prevarskaya, N., (2015). Calcium homeostasis in cancer: a focus on senescence. *Biochimica et Biophysica Acta (BBA)-Molecular Cell Research*.

[171] Khan, S., Martin, M., Bartsch, H., Rahimtula, A.D., (1989). Perturbation of liver microsomal calcium homeostasis by ochratoxin A. *Biochem. Pharmacol.* 38, 67-72.

[172] Hoehler, D., Marquardt, R.R., McIntosh, A.R., Xiao, H., (1996). Free radical generation as induced by ochratoxin A and its analogs in bacteria (Bacillus brevis). *J. Biol. Chem.* 271, 27388-27394.

[173] Dopp, E., Müller, J., Hahnel, C., Schiffmann, D., (1999). Induction of genotoxic effects and modulation of the intracellular calcium level in syrian hamster embryo (SHE) fibroblasts caused by ochratoxin A. *Food Chem. Toxicol.* 37, 713-721.

[174] Mally, A., Decker, M., Bekteshi, M., Dekant, W., (2006). Ochratoxin A alters cell adhesion and gap junction intercellular communication in MDCK cells. *Toxicology*. 223, 15-25.

[175] Chong, X., Rahimtula, A.D., (1992). Alterations in ATP-dependent calcium uptake by rat renal cortex microsomes following ochratoxin A administration *in vivo* or addition *in vitro*. *Biochem. Pharmacol.* 44, 1401-1409.

[176] Gekle, M., Sauvant, C., Schwerdt, G., (2005). Ochratoxin A at nanomolar concentrations: a signal modulator in renal cells. *Mol. Nutr. Food Res.* 49, 118-130.

[177] Rahimtula, A., Chong, X., (1990). Alterations in calcium homeostasis as a possible cause of ochratoxin A nephrotoxicity. *IARC Sci. Publ.*, 207-214.

[178] Arbillaga, L., Vettorazzi, A., Gil, A.G., van Delft, J.H., García-Jalón, J.A., de Cerain, A.L., (2008). Gene expression changes induced by ochratoxin A in renal and hepatic tissues of male F344 rat after oral repeated administration. *Toxicol. Appl. Pharmacol.* 230, 197-207.

[179] Marin-Kuan, M., Ehrlich, V., Delatour, T., Cavin, C., Schilter, B., (2011). Evidence for a role of oxidative stress in the carcinogenicity of ochratoxin A. *J. Toxicol.* 2011.

[180] Eder, S., Benesic, A., Freudinger, R., Engert, J., Schwerdt, G., Drumm, K., Gekle, M., (2000). Nephritogenic ochratoxin A interferes with mitochondrial function and pH homeostasis in immortalized human kidney epithelial cells. *Pflügers Archiv.* 440, 521-529.

[181] Klarić, M.Š., Želježić, D., Rumora, L., Peraica, M., Pepeljnjak, S., Domijan, A.-M., (2012). A potential role of calcium in apoptosis and aberrant chromatin forms in porcine kidney PK15 cells induced by individual and combined ochratoxin A and citrinin. *Arch. Toxicol.* 86, 97-107.

[182] Rached, E., Pfeiffer, E., Dekant, W., Mally, A., (2006). Ochratoxin A: apoptosis and aberrant exit from mitosis due to perturbation of microtubule dynamics? *Toxicol. Sci.* 92, 78-86.

[183] Marin-Kuan, M., Nestler, S., Verguet, C., Bezencon, C., Piguet, D., Delatour, T., Mantle, P., Cavin, C., Schilter, B., (2007). MAPK-ERK activation in kidney of male rats chronically fed ochratoxin A at a dose causing a significant incidence of renal carcinoma. *Toxicol. Appl. Pharmacol.* 224, 174-181.

[184] Nelson, W.J., Nusse, R., (2004). Convergence of Wnt, ß-catenin, and cadherin pathways. *Science*. 303, 1483-1487.

[185] Lin, X., Farooqi, A.A., Ismail, M., (2015). Recent progress in fungus-derived bioactive agents for targeting of signaling machinery in cancer cells. *Drug Des. Devel. Ther.* 9, 1797.

[186] Lancaster, M.A., Schroth, J., Gleeson, J.G., (2011). Subcellular spatial regulation of canonical Wnt signalling at the primary cilium. *Nat. Cell Biol.* 13, 700-707.

[187] Reya, T., Clevers, H., (2005). Wnt signalling in stem cells and cancer. *Nature*. 434, 843-850.

[188] Li, X., Gao, J., Huang, K., Qi, X., Dai, Q., Mei, X., Xu, W., (2015). Dynamic changes of global DNA methylation and hypermethylation of cell adhesion-related genes in rat kidneys in response to ochratoxin A. *World Mycotoxin Journal*, 1-12.

[189] Ortiz, A., (2000). Apoptotic regulatory proteins in renal injury. *Kidney Int.* 58, 467-485.

[190] Schwerdt, G., Freudinger, R., Mildenberger, S., Silbernagl, S., Gekle, M., (1999). The nephrotoxin ochratoxin A induces apoptosis in cultured human proximal tubule cells. *Cell Biol. Toxicol.* 15, 405-415.

[191] Seegers, J., Bohmer, L., Kruger, M., Lottering, M., Dekock, M., (1994). A comparative study of ochratoxin A-induced apoptosis in hamster kidney and HeLa cells. *Toxicol. Appl. Pharmacol.* 129, 1-11.

[192] Chang, L., Karin, M., (2001). Mammalian MAP kinase signalling cascades. *Nature*. 410, 37-40.

[193] Hayakawa, T., Matsuzawa, A., Noguchi, T., Takeda, K., Ichijo, H., (2006). The ASK1–MAP kinase pathways in immune and stress responses. *Microb. Infect.* 8, 1098-1107.

[194] Ichijo, H., Nishida, E., Irie, K., ten Dijke, P., Saitoh, M., Moriguchi, T., Takagi, M., Matsumoto, K., Miyazono, K., Gotoh, Y., (1997). Induction of apoptosis by ASK1, a mammalian MAPKKK that activates SAPK/JNK and p38 signaling pathways. *Science.* 275, 90-94.

[195] Matsuzawa, A., Nishitoh, H., Tobiume, K., Takeda, K., Ichijo, H., (2002). Physiological roles of ASK1-mediated signal transduction in oxidative stress-and endoplasmic reticulum stress-induced apoptosis: advanced findings from ASK1 knockout mice. *Antioxid. Redox Signal.* 4, 415-425.

[196] Chambers, J.W., LoGrasso, P.V., (2011). Mitochondrial c-Jun N-terminal Kinase (JNK) Signaling Initiates Physiological Changes Resulting in Amplification of Reactive Oxygen Species Generation. *J. Biol. Chem.* 286.

[197] Kim, K.Y., Yu, S.N., Lee, S.Y., Chun, S.S., Choi, Y.L., Park, Y.M., Song, C.S., Chatterjee, B., Ahn, S.C., (2011). Salinomycin-induced apoptosis of human prostate cancer cells due to accumulated reactive oxygen species and mitochondrial membrane depolarization. *Biochem. Biophys. Res. Commun.* 413, 80-86.

[198] Kumar, M., Dwivedi, P., Sharma, A.K., Sankar, M., Patil, R.D., Singh, N.D., (2014). Apoptosis and lipid peroxidation in ochratoxin A- and citrinin-induced nephrotoxicity in rabbits. *Toxicol. Ind. Health.* 30, 90-98.

[199] Dorrenhaus, A., Follmann, W., (1997). Effects of ochratoxin A on DNA repair in cultures of rat hepatocytes and porcine urinary bladder epithelial cells. *Arch. Toxicol.* 71, 709-713.

[200] Proudfoot, N.J., Furger, A., Dye, M.J., (2002). Integrating mRNA processing with transcription. *Cell.* 108, 501-512.

[201] Smith, C.W., Valcarcel, J., (2000). Alternative pre-mRNA splicing: the logic of combinatorial control. *Trends Biochem. Sci.* 25, 381-388.

[202] Kuo, P.L., Chen, C.Y., Hsu, Y.L., (2007). Isoobtusilactone a induces cell cycle arrest and apoptosis through reactive oxygen species/apoptosis signal-regulating kinase 1 signaling pathway in human breast cancer cells. *Cancer Res.* 67, 7406-7420.

[203] Tormos, A.M., Talens-Visconti, R., Nebreda, A.R., Sastre, J., (2013). p38 MAPK: A dual role in hepatocyte proliferation through reactive oxygen species. *Free Radical Res.* 47, 905-916.

[204] Roos, W.P., Kaina, B., (2006). DNA damage-induced cell death by apoptosis. *Trends Mol Med.* 12, 440-450.

[205] Norbury, C.J., Zhivotovsky, B., (2004). DNA damage-induced apoptosis. *Oncogene.* 23, 2797-2808.

[206] Poele, R.H.T., Okorokov, A.L., Joel, S.P., (1999). RNA synthesis block by 5,6-dichloro-1-beta-D-ribofuranosylbenzimidazole (DRB) triggers p53-dependent apoptosis in human colon carcinoma cells. *Oncogene.* 18, 5765-5772.

[207] Arbillaga, L., Azqueta, A., Ezpeleta, O., de Cerain, A.L., (2006). Oxidative DNA damage induced by ochratoxin A in the HK-2 human kidney cell line: evidence of the relationship with cytotoxicity. *Mutagenesis.* 22, 35-42.

[208] Okuyama, H., Endo, H., Akashika, T., Kato, K., Inoue, M., (2010). Downregulation of c-MYC protein levels contributes to cancer cell survival under dual deficiency of oxygen and glucose. *Cancer Res.* 70, 10213-10223.

[209] Dietrich, D., Heussner, A.H., O'Brien, E., (2005). Ochratoxin A: comparative pharmacokinetics and toxicological implications (experimental and domestic animals and humans). *Food Addit. Contam.* 22, 45-52.

[210] O'Brien, E., Dietrich, D.R., (2005). Ochratoxin A: the continuing enigma. *Crit. Rev. Toxicol.* 35, 33-60.

[211] Vogel, C., Kienitz, A., Hofmann, I., Müller, R., Bastians, H., (2004). Crosstalk of the mitotic spindle assembly checkpoint with p53 to prevent polyploidy. *Oncogene.* 23, 6845-6853.

[212] Besson, A., Dowdy, S.F., Roberts, J.M., (2008). CDK inhibitors: cell cycle regulators and beyond. *Dev. Cell.* 14, 159-169.

[213] Chang, B.-D., Broude, E.V., Fang, J., Kalinichenko, T.V., Abdryashitov, R., Poole, J.C., Roninson, I.B., (2000). p21Waf1/Cip1/Sdi1-induced growth arrest is associated with depletion of mitosis-control proteins and leads to abnormal mitosis and endoreduplication in recovering cells. *Oncogene.* 19, 2165-2170.

[214] Roninson, I.B., (2002). Oncogenic functions of tumour suppressor p21 Waf1/Cip1/Sdi1: association with cell senescence and tumour-promoting activities of stromal fibroblasts. *Cancer Lett.* 179, 1-14.

[215] Adler, M., Müller, K., Rached, E., Dekant, W., Mally, A., (2009). Modulation of key regulators of mitosis linked to chromosomal instability is an early event in ochratoxin A carcinogenicity. *Carcinogenesis.* 30, 711-719.

[216] Komatsu, M., Ichimura, Y., (2010). Selective autophagy regulates various cellular functions. *Genes Cells.* 15, 923-933.

[217] Novak, I., (2012). Mitophagy: A Complex Mechanism of Mitochondrial Removal. *Antioxid Redox Sign.* 17, 794-802.

[218] Kim, I., Rodriguez-Enriquez, S., Lemasters, J.J., (2007). Selective degradation of mitochondria by mitophagy. *Arch. Biochem. Biophys.* 462, 245-253.

[219] Maiuri, M.C., Zalckvar, E., Kimchi, A., Kroemer, G., (2007). Self-eating and self-killing: crosstalk between autophagy and apoptosis. *Nat Rev Mol Cell Bio.* 8, 741-752.

[220] Antonsson, B., (2001). Bax and other pro-apoptotic Bcl-2 family "killer-proteins" and their victim the mitochondrion. *Cell Tissue Res.* 306, 347-361.

[221] Bossy-Wetzel, E., Green, D.R., (1999). Caspases induce cytochrome c release from mitochondria by activating cytosolic factors. *J. Biol. Chem.* 274, 17484-17490.

[222] Li, H., Zhu, H., Xu, C.J., Yuan, J., (1998). Cleavage of BID by caspase 8 mediates the mitochondrial damage in the Fas pathway of apoptosis. *Cell.* 94, 491-501.

[223] Li, H.L., Yuan, J.Y., (1998). Cleavage of BID by caspase-8 mediates the mitochondrial damage in the FAS pathway of apoptosis. *Mol. Biol. Cell.* 9, 256a-256a.

[224] Lakhani, S.A., Masud, A., Kuida, K., Porter, G.A., Booth, C.J., Mehal, W.Z., Inayat, I., Flavell, R.A., (2006). Caspases 3 and 7: Key mediators of mitochondrial events of apoptosis. *Science.* 311, 847-851.

[225] Deveraux, Q.L., Takahashi, R., Salvesen, G.S., Reed, J.C., (1997). X-linked IAP is a direct inhibitor of cell-death proteases. *Nature.* 388, 300-304.

[226] Deveraux, Q.L., Roy, N., Stennicke, H.R., Van Arsdale, T., Zhou, Q., Srinivasula, S.M., Alnemri, E.S., Salvesen, G.S., Reed, J.C., (1998). IAPs block apoptotic events induced by caspase-8 and cytochrome c by direct inhibition of distinct caspases. *EMBO J.* 17, 2215-2223.

[227] Wurstle, M.L., Laussmann, M.A., Rehm, M., (2012). The central role of initiator caspase-9 in apoptosis signal transduction and the regulation of its activation and activity on the apoptosome. *Exp. Cell Res.* 318, 1213-1220.

[228] Meins, T., Vonrhein, C., Zeth, K., (2008). Crystallization and preliminary X-ray crystallographic studies of human voltage-dependent anion channel isoform I (HVDAC1). *Acta Crystallogr F.* 64, 651-655.

[229] He, C.C., Klionsky, D.J., (2009). Regulation Mechanisms and Signaling Pathways of Autophagy. *Annu. Rev. Genet.* 43, 67-93.

[230] Izzotti, A., Calin, G.A., Arrigo, P., Steele, V.E., Croce, C.M., De Flora, S., (2009). Downregulation of microRNA expression in the lungs of rats exposed to cigarette smoke. *The FASEB Journal.* 23, 806-812.

[231] Wulfken, L.M., Moritz, R., Ohlmann, C., Holdenrieder, S., Jung, V., Becker, F., Herrmann, E., Walgenbach-Brünagel, G., von Ruecker, A., Müller, S.C., (2011). MicroRNAs in renal cell carcinoma: diagnostic implications of serum miR-1233 levels.

[232] Cui, L., Zhou, H., Zhao, H., Zhou, Y., Xu, R., Xu, X., Zheng, L., Xue, Z., Xia, W., Zhang, B., (2012). MicroRNA-99a induces G1-phase cell cycle arrest and suppresses tumorigenicity in renal cell carcinoma. *BMC Cancer.* 12, 546.

[233] Dai, Q., Zhao, J., Qi, X.Z., Xu, W.T., He, X.Y., Guo, M.Z., Dweep, H., Cheng, W.H., Luo, Y.B., Xia, K., Gretz, N., Huang, K.L., (2014). MicroRNA profiling of rats with ochratoxin A nephrotoxicity. *BMC Genomics.* 15.

[234] Marin-Kuan, M., Nestler, S., Verguet, C., Bezencon, C., Piguet, D., Mansourian, R., Holzwarth, J., Grigorov, M., Delatour, T., Mantle, P., Cavin, C., Schilter, B., (2006). A toxicogenomics approach to identify new plausible epigenetic mechanisms of ochratoxin a carcinogenicity in rat. *Toxicol. Sci.* 89, 120-134.

[235] Kumar, R., Alam, S., Chaudhari, B.P., Dwivedi, P.D., Jain, S.K., Ansari, K.M., Das, M., (2013). Ochratoxin A-induced cell proliferation and tumor promotion in mouse skin by activating the expression of cyclin-D1 and cyclooxygenase-2 through nuclear factor-kappa B and activator protein-1. *Carcinogenesis.* 34, 647-657.

[236] Marin-Kuan, M., Nestler, S., Verguet, C., Bezencon, C., Piguet, D., Mansourian, R., Holzwarth, J., Grigorov, M., Delatour, T., Mantle, P., Cavin, C., Schilter, B., (2006). A toxicogenomics approach to identify new plausible epigenetic mechanisms of ochratoxin a carcinogenicity in rat. *Toxicol. Sci.* 89, 120-134.

[237] Hammond, S.M., (2005). Dicing and slicing: the core machinery of the RNA interference pathway. *FEBS Lett.* 579, 5822-5829.

[238] Hutvagner, G., McLachlan, J., Pasquinelli, A.E., Balint, E., Tuschl, T., Zamore, P.D., (2001). A cellular function for the RNA-interference enzyme Dicer in the maturation of the let-7 small temporal RNA. *Science.* 293, 834-838.

[239] Siomi, H., Siomi, M.C., (2010). Posttranscriptional regulation of microRNA biogenesis in animals. *Mol. Cell.* 38, 323-332.

[240] Han, J., Pedersen, J.S., Kwon, S.C., Belair, C.D., Kim, Y.-K., Yeom, K.-H., Yang, W.-Y., Haussler, D., Blelloch, R., Kim, V.N., (2009). Posttranscriptional crossregulation between Drosha and DGCR8. *Cell.* 136, 75-84.

[241] Dai, Q., Zhao, J., Qi, X., Xu, W., He, X., Guo, M., Dweep, H., Cheng, W.-H., Luo, Y., Xia, K., (2014). MicroRNA profiling of rats with ochratoxin A nephrotoxicity. *BMC Genomics.* 15, 333.

[242] Lambrou, A., Baccarelli, A., Wright, R.O., Weisskopf, M., Bollati, V., Amarasiriwardena, C., Vokonas, P., Schwartz, J., (2012). Arsenic Exposure and DNA Methylation Among Elderly Men. *Epidemiology*. 23, 668-676.

[243] Brocato, J., Costa, M., (2013). Basic mechanics of DNA methylation and the unique landscape of the DNA methylome in metal-induced carcinogenesis. *Crit. Rev. Toxicol.* 43, 493-514.

[244] Niedzwiecki, M.M., Hall, M.N., Liu, X., Oka, J., Harper, K.N., Slavkovich, V., Ilievski, V., Levy, D., van Geen, A., Mey, J.L., Alam, S., Siddique, A.B., Parvez, F., Graziano, J.H., Gamble, M.V., (2013). A Dose-Response Study of Arsenic Exposure and Global Methylation of Peripheral Blood Mononuclear Cell DNA in Bangladeshi Adults. *Environ. Health Perspect.* 121, 1306-1312.

[245] Zhang, Y.J., Wu, H.C., Yazici, H., Yu, M.W., Lee, P.H., Santella, R.M., (2012). Global hypomethylation in hepatocellular carcinoma and its relationship to aflatoxin B(1) exposure. *World journal of hepatology*. 4, 169-175.

[246] Wu, H.C., Wang, Q., Yang, H.I., Tsai, W.Y., Chen, C.J., Santella, R.M., (2013). Global DNA methylation in a population with aflatoxin B1 exposure. *Epigenetics*. 8, 962-969.

[247] Zheng, J., Zhang, Y., Xu, W., Luo, Y., Hao, J., Shen, X.L., Yang, X., Li, X., Huang, K., (2013). Zinc protects HepG2 cells against the oxidative damage and DNA damage induced by ochratoxin A. *Toxicol. Appl. Pharmacol.* 268, 123-131.

[248] Selvakumar, P., Owens, T.A., David, J.M., Petrelli, N.J., Christensen, B.C., Lakshmikuttyamma, A., Rajasekaran, A.K., (2014). Epigenetic silencing of Na,K-ATPase beta 1 subunit gene ATP1B1 by methylation in clear cell renal cell carcinoma. *Epigenetics*. 9, 579-586.

[249] Wing, M.R., Devaney, J.M., Joffe, M.M., Xie, D., Feldman, H.I., Dominic, E.A., Guzman, N.J., Ramezani, A., Susztak, K., Herman, J.G., Cope, L., Harmon, B., Kwabi-Addo, B., Gordish-Dressman, H., Go, A.S., He, J., Lash, J.P., Kusek, J.W., Raj, D.S., Chronic Renal Insufficiency Cohort, S., (2014). DNA methylation profile associated with rapid decline in kidney function: findings from the CRIC study. *Nephrol. Dial. Transplant.* 29, 864-872.

[250] Wilson, A.S., Power, B.E., Molloy, P.L., (2007). DNA hypomethylation and human diseases. *Biochim. Biophys. Acta*. 1775, 138-162.

[251] Starlard-Davenport, A., Tryndyak, V.P., James, S.R., Karpf, A.R., Latendresse, J.R., Beland, F.A., Pogribny, I.P., (2010). Mechanisms of epigenetic silencing of the Rassf1a gene during estrogen-induced breast carcinogenesis in ACI rats. *Carcinogenesis*. 31, 376-381.

[252] Takiguchi, M., Achanzar, W.E., Qu, W., Li, G., Waalkes, M.P., (2003). Effects of cadmium on DNA-(Cytosine-5) methyltransferase activity and DNA methylation status during cadmium-induced cellular transformation. *Exp. Cell Res.* 286, 355-365.

[253] Bogdanovic, O., Veenstra, G.J., (2009). DNA methylation and methyl-CpG binding proteins: developmental requirements and function. *Chromosoma*. 118, 549-565.

[254] Majumder, S., Ghoshal, K., Datta, J., Smith, D.S., Bai, S., Jacob, S.T., (2006). Role of DNA methyltransferases in regulation of human ribosomal RNA gene transcription. *J. Biol. Chem.* 281, 22062-22072.

[255] Roll, J.D., Rivenbark, A.G., Jones, W.D., Coleman, W.B., (2008). DNMT3b overexpression contributes to a hypermethylator phenotype in human breast cancer cell lines. *Mol. Cancer*. 7, 15.

[256] Takeichi, M., (1995). Morphogenetic roles of classic cadherins. *Curr. Opin. Cell Biol.* 7, 619-627.

[257] Okegawa, T., Li, Y.M., Pong, R.C., Hsieh, J.T., (2002). Cell adhesion proteins as tumor suppressors. *J. Urol.* 167, 1836-1843.

[258] Paschos, K.A., Canovas, D., Bird, N.C., (2009). The role of cell adhesion molecules in the progression of colorectal cancer and the development of liver metastasis. *Cell. Signal.* 21, 665-674.

[259] Wagner, M.C., Molitoris, B.A., (1999). Renal epithelial polarity in health and disease. *Pediatr. Nephrol.* 13, 163-170.

[260] van Roy, F., Berx, G., (2008). The cell-cell adhesion molecule E-cadherin. *Cell. Mol. Life Sci.* 65, 3756-3788.

[261] Prozialeck, W.C., Lamar, P.C., Appelt, D.M., (2004). Differential expression of E-cadherin, N-cadherin and beta-catenin in proximal and distal segments of the rat nephron. *BMC Physiol.* 4, 10.

[262] Nojima, D., Nakajima, K., Li, L.C., Franks, J., Ribeiro-Filho, L., Ishii, N., Dahiya, R., (2001). CpG methylation of promoter region inactivates E-cadherin gene in renal cell carcinoma. *Mol. Carcinog.* 32, 19-27.

[263] Akintola, A.D., Crislip, Z.L., Catania, J.M., Chen, G., Zimmer, W.E., Burghardt, R.C., Parrish, A.R., (2008). Promoter methylation is associated with the age-dependent loss of N-cadherin in the rat kidney. *American Journal Of Physiology-Renal Physiology.* 294, F170-F176.

[264] Gagliano, N., Donne, I.D., Torri, C., Migliori, M., Grizzi, F., Milzani, A., Filippi, C., Annoni, G., Colombo, P., Costa, F., Ceva-Grimaldi, G., Bertelli, A.A.E., Giovannini, L., Gioia, M., (2006). Early cytotoxic effects of ochratoxin A in rat liver: A morphological, biochemical and molecular study. *Toxicology.* 225, 214-224.

[265] Gagliano, N., Torri, C., Donetti, E., Grizzi, F., Costa, F., Bertelli, A.A., Migliori, M., Filippl, C., Bedoni, M., Panichi, V., Giovannini, L., Gioia, M., (2005). Ochratoxin A-induced renal cortex fibrosis and epithelial-to-mesenchymal transition: Molecular mechanisms of ochratoxin A-injury and potential effects of red wine. *Mol. Med.* 11, 30-38.

[266] Yoshiura, K., Kanai, Y., Ochiai, A., Shimoyama, Y., Sugimura, T., Hirohashi, S., (1995). Silencing Of the E-Cadherin Invasion-Suppressor Gene by Cpg Methylation In Human Carcinomas. *Proc. Natl. Acad. Sci. U. S. A.* 92, 7416-7419.

[267] Kanai, Y., Ushijima, S., Hui, A.M., Ochiai, A., Tsuda, H., Sakamoto, M., Hirohashi, S., (1997). The E-cadherin gene is silenced by CpG methylation in human hepatocellular carcinomas. *Int. J. Cancer.* 71, 355-359.

[268] Van den Bossche, J., Malissen, B., Mantovani, A., De Baetselier, P., Van Ginderachter, J.A., (2012). Regulation and function of the E-cadherin/catenin complex in cells of the monocyte-macrophage lineage and DCs. *Blood.* 119, 1623-1633.

[269] Kuhl, M., Sheldahl, L.C., Park, M., Miller, J.R., Moon, R.T., (2000). The Wnt/Ca2+ pathway: a new vertebrate Wnt signaling pathway takes shape. *Trends Genet.* 16, 279-283.

[270] Stemmer, K., Ellinger-Ziegelbauer, H., Ahr, H.J., Dietrich, D.R., (2007). Carcinogen-specific gene expression profiles in short-term treated Eker and wild-type rats indicative of pathways involved in renal tumorigenesis. *Cancer Res.* 67, 4052-4068.

[271] Gottardi, C.J., Wong, E., Gumbiner, B.M., (2001). E-cadherin suppresses cellular transformation by inhibiting beta-catenin signaling in an adhesion-independent manner. *J. Cell Biol.* 153, 1049-1060.

[272] Lau, M.T., Klausen, C., Leung, P.C., (2011). E-cadherin inhibits tumor cell growth by suppressing PI3K/Akt signaling via beta-catenin-Egr1-mediated PTEN expression. *Oncogene*. 30, 2753-2766.

[273] Chung, H.Y., Sung, B., Jung, K.J., Zou, Y., Yu, B.P., (2006). The molecular inflammatory process in aging. *Antioxid. Redox Signal.* 8, 572-581.

[274] DeCarolis, N.A., Wharton, K.A., Eisch, A.J., (2008). Which way does the Wnt blow? Exploring the duality of canonical Wnt signaling on cellular aging. *Bioessays*. 30, 102-106.

[275] Inuzuka, Y., Okuda, J., Kawashima, T., Kato, T., Niizuma, S., Tamaki, Y., Iwanaga, Y., Yoshida, Y., Kosugi, R., Watanabe-Maeda, K., Machida, Y., Tsuji, S., Aburatani, H., Izumi, T., Kita, T., Shioi, T., (2009). Suppression of Phosphoinositide 3-Kinase Prevents Cardiac Aging in Mice. *Circulation*. 120, 1695-U1119.

[276] Marchand, A., Atassi, F., Gaaya, A., Leprince, P., Le Feuvre, C., Soubrier, F., Lompre, A.M., Nadaud, S., (2011). The Wnt/beta-catenin pathway is activated during advanced arterial aging in humans. *Aging Cell*. 10, 220-232.

[277] Pfohlleszkowicz, A., Chakor, K., Creppy, E.E., Dirheimer, G. (1991) *DNA Adduct Formation in Mice Treated with Ochratoxin-A*. Int Agency Research Cancer, Lyons.

[278] Pfohl-Leszkowicz, A., Grosse, Y., Castegnaro, M., Nicolov, I., Chernozemsky, I., Bartsch, H., Betbeder, A., Creppy, E., Dirheimer, G., (1992). Ochratoxin A-related DNA adducts in urinary tract tumours of Bulgarian subjects. *IARC Sci. Publ.*, 141-148.

[279] Obrecht-Pflumio, S., Grosse, Y., Pfohl-Leszkowicz, A., Dirheimer, G., (1996). Protection by indomethacin and aspirin against genotoxicity of ochratoxin A, particularly in the urinary bladder and kidney. *Arch. Toxicol.* 70, 244-248.

[280] Miljkovic, A., Pfohl-Leszkowicz, A., Dobrota, M., Mantle, P.G., (2003). Comparative responses to mode of oral administration and dose of ochratoxin A or nephrotoxic extract of Penicillium polonicum in rats. *Exp. Toxicol. Pathol.* 54, 305-312.

[281] Obrecht-Pflumio, S., Dirheimer, G., (2000). *In vitro* DNA and dGMP adducts formation caused by ochratoxin A. *Chem. Biol. Interact.* 127, 29-44.

[282] Manderville, R.A., (2005). A case for the genotoxicity of ochratoxin A by bioactivation and covalent DNA adduction. *Chem. Res. Toxicol.* 18, 1091-1097.

[283] Sharma, P., Manderville, R.A., Wetmore, S.D., (2014). Structural and energetic characterization of the major DNA adduct formed from the food mutagen ochratoxin A in the NarI hotspot sequence: influence of adduct ionization on the conformational preferences and implications for the NER propensity. *Nucleic Acids Res.* 42, 11831-11845.

[284] Gautier, J.-C., Richoz, J., Welti, D.H., Markovic, J., Gremaud, E., Guengerich, F.P., Turesky, R.J., (2001). Metabolism of ochratoxin A: absence of formation of genotoxic derivatives by human and rat enzymes. *Chem. Res. Toxicol.* 14, 34-45.

[285] Mally, A., Zepnik, H., Wanek, P., Eder, E., Dingley, K., Ihmels, H., Völkel, W., Dekant, W., (2004). Ochratoxin A: lack of formation of covalent DNA adducts. *Chem. Res. Toxicol.* 17, 234-242.

[286] Wollgast, J., Anklam, E., (2000). Review on polyphenols in Theobroma cacao: changes in composition during the manufacture of chocolate and methodology for identification and quantification. *Food Res. Int.* 33, 423-447.

[287] Manach, C., Scalbert, A., Morand, C., Rémésy, C., Jiménez, L., (2004). Polyphenols: food sources and bioavailability. *The American journal of clinical nutrition.* 79, 727-747.

[288] Cariddi, L., Sabini, M., Escobar, F., Montironi, I., Mañas, F., Iglesias, D., Comini, L., Sabini, L., Dalcero, A., (2015). Polyphenols as possible bioprotectors against cytotoxicity and DNA damage induced by ochratoxin A. *Environ. Toxicol. Pharmacol.* 39, 1008-1018.

[289] Ramyaa, P., Padma, V.V., (2013). Ochratoxin-induced toxicity, oxidative stress and apoptosis ameliorated by quercetin–Modulation by Nrf2. *Food Chem. Toxicol.* 62, 205-216.

[290] Mazière, C., Conte, M.-A., Degonville, J., Ali, D., Mazière, J.-C., (1999). Cellular enrichment with polyunsaturated fatty acids induces an oxidative stress and activates the transcription factors AP1 and NFκB. *Biochem. Biophys. Res. Commun.* 265, 116-122.

[291] Faucet-Marquis, V., Pont, F., Stormer, F.C., Rizk, T., Castegnaro, M., Pfohl-Leszkowicz, A., (2006). Evidence of a new dechlorinated ochratoxin A derivative formed in opossum kidney cell cultures after pretreatment by modulators of glutathione pathways: Correlation with DNA-adduct formation. *Mol. Nutr. Food Res.* 50, 530-542.

[292] Liu, J., Wang, Y., Cui, J., Xing, L., Shen, H., Wu, S., Lian, H., Wang, J., Yan, X., Zhang, X., (2012). Ochratoxin A induces oxidative DNA damage and G1 phase arrest in human peripheral blood mononuclear cells *in vitro*. *Toxicol. Lett.* 211, 164-171.

Chapter 5

OCHRATOXIN A -INDUCED HEPATOTOXICITY

ABSTRACT

OTA is a common mycotoxic contaminant of crops. The toxin has many potential harmful effects on the health of humans and animals. The liver is one of the major target organs of OTA biotransformation. The liver can regulate glycogen storage; bile production; fat, protein, and vitamin metabolism; hormone production; and detoxification. OTA-induced hepatotoxicity has therefore captured the attention of an increasing number of researchers. Chronic OTA toxicity is characterized by a low-dose exposure over a long period that results in cancers and hepatitis. Evidence derived from animal experiments suggests that OTA induces hepatic carcinogenesis. After OTA absorption by human and animals, the toxin can accumulate in the body. Several animals, such as rats, mice, pigs, poultry, broiler chickens and ducks, have been used to assess the toxicity of OTA. The toxin induces abnormal physiological changes in all the animals. OTA hepatotoxicity is studied through traditional and molecular methods. Traditional indexes include blood biochemistry, oxidative stress, cell apoptosis, and Ca2+. Bilirubin, triglycerides, lactate dehydrogenase (LDH), aspartate aminotransferase (AST), alanine aminotransferase (ALT), gamma-glutamyl transferase (GGT), and alkaline phosphatase (ALP) are usually used to reflect the primary metabolic enzymes. Molecular indexes consist of three parts: DNA, RNA, and proteins. These parameters have recently been denoted as new markers for the evaluation of toxic effects because of their sensitivity and clearer mechanisms. Regarding DNA, methylation plays an important role in the regulation of gene expression, transposable elements, genomic imprinting, embryonic development and genomic integrity maintenance. Until now, there have been no reports regarding DNA methylation induced by OTA in the liver. In addition, OTA-induced genotoxicity and its mechanism of action as a carcinogen remain hotly debated. DNA adducts were studied in OTA-induced genotoxicity. DNA damage may be the reason for OTA carcinogenesis. Regarding RNA, miRNAs have become the new way to understand OTA-induced hepatotoxicity. miRNA expression profiling is useful for identifying miRNAs that regulate a large number of biological processes. Additionally, profiling can also be used for system-wide studies of gene regulation. DGE analysis of mRNA was conducted after OTA treatment on HepG-2 cells. The rat liver proteome and membrane proteome of HepG2 cells were evaluated using 2D electrophoresis and iTRAQ techniques. Moreover, the best way to prevent people and animals from developing OTA-induced hepatotoxicity is an important topic for consideration. Recently, many studies have shown that OTA-induced hepatotoxicity is principally due to oxidative stress and DNA damage. Therefore, antioxidant materials such as vitamin E, polyphenols, quercetin, and Zn were evaluated to determine whether they could remit OTA-induced hepatotoxicity in HepG2 cells. OTA-induced hepatotoxicity remain a highly debated topic and should be further explored.

Keywords: OTA, hepatotoxicity, traditional hepatotoxicity, molecular hepatotoxicity

1. OTA and Liver Diseases

1.1. Hepatitis

Mycotoxin can induce either acute or chronic intoxication in people. Usually, acute toxicity rapidly presents with an apparently toxic response, whereas chronic toxicity is caused by low dose toxic substances over a relatively long time, which then leads to chronic diseases, including cancers [1]. OTA principally results in hepatitis or cancers of the liver.

The occurrence of hepatitis, specifically liver inflammation, suggests that there are inflammatory cells in the liver. Hepatitis is sometimes asymptomatic; however, it always results in poor appetite, jaundice and malaise. When hepatitis continues for less than six months, the condition is considered to be acute. If the duration is longer, hepatitis is deemed to be chronic. Acute hepatitis can progress to chronic hepatitis or can result in acute liver failure [2]. Symptoms with chronic hepatitis are not easily detected; the condition progresses to fibrosis and cirrhosis over time. Cirrhosis of the liver increases the risk of developing hepatocellular carcinoma. Some individuals who have hepatitis experience light expansion of the liver and enlarged lymph nodes, along with expansion of the spleen. A small subset of individuals with acute hepatitis are prone to acute liver damage, leading to a failure to remove poisonous substances from the bloodstream. Toxins can trigger liver injury through a variety of mechanisms, including direct cell damage, disturbing cell metabolism, and causing structural changes. The liver responds to damage in different ways, and many response patterns have been identified. Since the histology of chronic hepatitis is better known, histology of acute hepatitis of the liver is rarely performed. The consequence of hepatitis principally depends on the disease or con tion that is causing the symptoms. The liver can regenerate injured cells. Chronic damage would result in fibrosis and nodules that could limit liver function. This phenomenon is called cirrhosis and leads to permanent damage [3].

A survey conducted by Pepeljnjak et al. showed that feeding swine moldy cereals led to pathological changes in the liver (interstitial hepatitis with fatty degeneration of the liver) [4]. Infected OTA-fed mice experienced severe generalized edema, hemorrhages, congestion, hepatic hemosiderosis, and hepatitis [5]. The effects of dietary contamination with various levels of ochratoxin A (OTA) and the potential preventive action of mycotoxin-deactivation product were investigated in two hundred one-day-old male broiler chicken. The birds were fed 5 different diets for 5 weeks. The relative weights of the liver and the kidney, as well as the activity of alanine aminotransferase (ALT), in the groups fed diets containing OTA alone were significantly higher than those in the control group. Increased ALT activity in blood plasma could be a sign of hepatitis, liver cancer, gall bladder inflammation and bile duct and toxic liver damage (intoxication) [6]. Silibinin has a protective effect against chronic liver diseases such as hepatitis and cirrhosis. In a study of the inhibitory effect of silibinin on ochratoxin A (OTA), silibinin was found to protect the liver from the effect of the toxins.

Silibinin can also prevent OTA-mediated TNF-a release [7]. Forty young adult (12 weeks old) female SPF Wag/mbl rats were treated for 26–28 days by gavage with daily doses of 0, 0.07, 0.34 or 1.68 mg/kg OTA/kg body weight. The weight of the liver was affected in some treated groups in young and old rats. The hepatocellular vacuolation in young rats was not as serious that in the older ones. Old rats developed hepatic erythropoiesis, biliary-/oval cell-proliferation, inflammation, and multinucleated giant cell formation [8].

1.2. Hepatocellular Carcinoma

Human mycotoxicoses have been reported since the Middle Ages. The relationship between mycotoxin and disease has always been a topic of discussion. Because both experimental and epidemiological evidence suggest that some mycotoxins are related to several types of human cancer, research has principally focused on carcinogenic and mutagenic fungal toxins. In addition, aflatoxins, ochratoxins and fumonisins have garnered attention globally. Hepatocellular carcinoma (HCC) is becoming a serious threat to human health and even life in Western countries [9,10], Most HCC cases (>80%) occur in Eastern Asia or sub-Saharan Africa. China harbors over half the world's cases of HCC [11]. HCC comprises more than 70% of primary liver cancers, which has been the third highest cause of cancer-related deaths in the world [12]. Dietary exposure to OTA is also an underlying factor in the high occurrence of HCC in sub-Saharan Africa and Asia. OTA can disturb protein synthesis. The liver can metabolize OTA before it is excreted into both bile and the renal proximal tubules after absorption [13]. In animal experiments, OTA causes tumors in the kidney and liver [14, 15]. Several studies in animals have suggested that OTA could induce hepatic carcinogenesis [16, 17].

Studies of the tumorigenicity of OTA in rats and mice have been published. In one study, male ddY mice were fed diets containing 40 ppm OTA for 45 weeks [18]. One group received a single dose of 20 mg aflatoxinB]/kg body weight dissolved in DMSO at the start of OTA feeding. Another group received a single dose of DMSO, and a third group was fed a diet containing OTA without pretreatment. In all groups, hepatic tumors were increased and most of the mice (37/39) developed renal tumors, which indicated that OTA was a renal and hepatic carcinogen for male ddY mice. In 1983, the International Agency for Research on Cancer concluded that there was no hard evidence for the carcinogenicity of OTA in experimental animals because epidemiologic data were lacking, and no assessment of the carcinogenicity to humans could therefore be conducted [19]. In addition, OTA was administered to male and female (C57BL/6J X C3H) F1 (B6C3F1) mice for 24 months [20]. The hepatocellular carcinomas were large, irregular, poorly circumscribed, pale-to-mottled hemorrhagic masses, often resulting in marked enlargement of the liver. Histologically, neoplastic hepatocytes form sheets and trabeculae with minimal or no retention of lobular architecture. The toxin may have been converted to a carcinogenic metabolite in the liver. The incidence of both types of hepatocellular neoplasms was increased in OTA-treated animals compared with controls in this study. However, the incidence in males was within the historic control range for this strain of mouse. A statistically significant increase in hepatocellular carcinomas was evident only in female mice fed 40 ppm OTA. OTA may have acted as a promoter for liver cell tumors, because this strain of mouse has a fairly high spontaneous incidence of liver-cell neoplasms [21]. Additionally, chronic experiments in Fisher 344/N rats have suggested that OTA could induce liver and renal tumors. Although the experimental and epidemiological evidence suggest that OTA mediates carcinogenicity, its concrete mechanisms are unclear. It has been reported that intact hepatocytes could transform OTA into mutagenic metabolites [22]. Many studies focus on the relationship between the carcinogenic effects and DNA damage. In rodents and human patients, [32P] post-labeling techniques were used to observe DNA-adducts in the kidney, liver, and spleen [23, 24]. It was reported that OTA mutagenicity requires a cytochrome P450-dependent activation step in cell lines. The cell line expresses selected human cytochrome P450 forms which use the bacterial lacZ gene as a reporter for mutations [25]. The results showed large deletions in these mutations. In addition, large amounts of DNA single strand breaks induced by OTA were observed in primary cultures of rat hepatocytes through the method of the DNA alkaline comet assay [26, 27]. Primary hepatocytes are able to metabolize OTA to various metabolites. Previously, the research on OTA metabolism showed that the metabolites could increase the mutation frequency by the Salmonella mutation assay [22]. Since intact cells may metabolize OTA in different ways [28], it is assumed that further processing of metabolites would respond to the mutagenesis. Furthermore, binding to glutathione is necessary for triggering mutagenicity apart from cytochrome P450 activation of OTA. In addition, when glutathione-depleted CYP450-expressing cell lines were used to observe the mutagenicity, no mutagenicity was seen. Currently, whether OTA induces the DNA-damaging effects through quinone redox-cycling or through a direct mutagen needs to be determined. When young weanling mice (Mus musculus) were administered ochratoxin A, they developed several hematological changes, in addition to hepatoma and renal carcinoma [29].

The toxicity of more than two mycotoxins has also been evaluated. Ten-week-old male Fisher 344 (F344) rats were orally administered a single dose of a mixture of 0.25 mg/kg bw of aflatoxin B1 (AFB1) and 0.5 mg/kg bw of OTA in $NaHCO_3 \cdot H_2O$ (0.1 M pH 7.4) (0.5% DMSO) [30]. During the necropsies, the livers of the treated animals had lost some color and appeared light red compared to the livers of the control animals after 48, 72 and 96 h. Moreover, the relative weight of the livers of treated animals was significantly higher than that of the control animals after 48, 72 and 96 h. There was much evidence of alterations in the liver, including a progressive lesional state. The liver had the following characteristics: diffuse hepatocyte necrosis at 8 h accompanied by outstanding focal hepatitis. At 24 h, hepatocyte necrosis was observed in the parenchyma and portal spaces. Discreet degeneration began to be simultaneously observed in the cytoplasm of the hepatocytes; several displayed large nuclei or were binucleated. At 48 h, large numbers of hepatocytes showed intense tumefaction and degeneration in the cytoplasm, whereas others presented with necrosis surrounded by inflammatory infiltrates. In the portal spaces, necrosis increased within the bile canaliculi and fibrosis appeared. Diffuse hepatocyte necrosis and the inflammatory response in the parenchyma were maintained at 72 h. Both fibrosis and proliferation of bile canaliculi in the portal spaces, as well as focal fibrosis of the parenchyma, were remarkable. Overall, these signs can be considered as the beginning stages of cirrhosis. At 96 h, hepatocyte necrosis could be observed. The inflammatory response had decreased or stabilized. The cytoplasm of the hepatocytes showed moderate degeneration and tumefaction. Proliferation of portal fibrosis and bile canaliculi appeared, in addition to regeneration of hepatocytes with many binucleated cells. The obtained OTA profile and other experiments and models of OTA kinetics were compared, and the differences were more closely correlated to the age of the rats than to the addition of AFB1 to the system [31, 32].

Table 1. Toxicokinetic parameters for ochratoxin A and ochratoxin B following oral and intravenous administrations of 50 ng/g body weight. The concentration vs time curves used described two-compartment open models[a]

Species	C_{max} (ng/ml)	$t_{1/2}$ (h)	V_1 (ml/kg)	V_d (ml/kg)	*AUC* (ng·h/ml)	*F* (%)	*Cl* (ml/kg·h)
Ochratoxin A							
Fish i.v.	200	8.3	170	690	870		57
Fish p.o.	14	0.68	6.8	57	14	1.6	58
Quail i.v.	590	12	76	1500	600		86
Quail p.o.	260	6.7	42	550	370	62	57
Mouse i.v.	370	48	120	420	8300		6.1
Mouse p.o.	370	39	110	340	8010	97	6.1
Rat i.v.	2100	170	24	230	54 000		0.92
Rat p.o.	390	120	32	160	24 000	44	0.91
Monkey i.v.	980	840	40	200	490 000		0.17
Monkey p.o.	500	510		130	280 000	57	0.18
Pig i.v.[b]	490	150	100	330	15 000		1.5
Ochratoxin B							
Rat i.v.	760	12	39	250	3400		15
Rat p.o.	120	18	140	360	2000	61	14

[a] C_{max} = maxmimal concentration in serum; $t_{1,2}$ = biological half-life of the toxin; V_1 = volume of the central compartment; V_d =apparent volume of distribution; AUC = the area under the curve of plasma concentration vs time from t = 0 to t = x after toxin administration; F = relative bioavailability (AUC oral AUC i.v.); Cl = plasma | clearence.

[b] Data taken from Hult et al.[34].

2. Experimental Animal

In 1965, Scoot et al. [33] isolated *Aspergillus ochraceus* from sorghum, and this fungus could produce ochratoxin A. OTA was weaker than only AFB1. After the absorption of OTA by humans and animals, the toxin can accumulate in the body and induce more toxicity. Oral and intravenous administration are the common modes of OTA intake. The half-periods of OTA by these two modes in different animals are listed in Table 1 [34].

OTA is a common mycotoxic contaminant of crops. The toxin has many potential harmful effects on the health of humans and animals. Therefore, an increasing amount of research is focusing on the mechanism of OTA intoxication.

The kidney is the first target organ of OTA. Since 1978, Balkan nephropathy has been seen to be related to the intake of OTA [35]. Afterwards, there was much research aimed at studying the relationship between OTA and nephrotoxicity. The liver is one of the major target organs of OTA biotransformation. This organ can regulate glycogen storage, bile production, fat metabolism, protein metabolism, vitamin metabolism, hormone production, and detoxification. Several trials have been conducted in different animals, and the pathological changes induced by OTA have been explored in these experimental animals. The following section discusses the hepatotoxicity induced by OTA.

2.1. Hepatotoxicity of the HepG2 cell by OTA-induced

HepG2 cells were selected as the *in vitro* model in many studies for evaluating OTA-induced liver damage to humans.

Zheng et al. [36] found that OTA can induce apparent cell and DNA damage. OTA caused an increase in the activity of lactic dehydrogenase and the production of ROS. The comet assay revealed that OTA could induce DNA strand breaks. Apparent oxidative DNA damage was also detected by HPLC/MS/MS, and the level of 8-OHdG increased. Ehrlich et al. [37] also determined by the micronucleus test and comet assay that OTA could induce apparent DNA damage and chromosome breakage, as well.

2.2. Hepatotoxicity of mammals by OTA-induced

Most research regarding hepatotoxicity induced by OTA in mammals is conducted in rats and mice.

2.2.1. Rat

To assess acute hepatotoxicity, Theron et al. [38] treated thirty weanling male Wistar rats (average weight, 60 g) with OTA by gastric gavage. Six time points (1, 2, 4, 8, 16 and 24 h) were selected for OTA treatment. Acute hepatotoxicity occurred and was evaluated. At the earliest stage, hypertrophy of the smooth ER was observed, and the hypertrophy corresponded to the hyaline degeneration. At a later stage, hypertrophy of the smooth ER was more marked, and the smooth agranular profiles predominated in the cytoplasm of the cell. The hypertrophy of the smooth ER was due to the loss of ribosomes from the granular ER. Smooth membranes were simultaneously arranged in “fingerprint” patterns. At the final stage, degenerative phenomena were clear. Liver cell necrosis was also observed in rats after the administration of OTA for 8-24 h. The necrotic hepatocytes were scattered in groups throughout the liver lobule.

In 2003, Aydın et al. [39] also conducted an assay aimed at evaluating OTA. Twenty-four male Wistar rats were divided into three groups: the control group, a second group at a dose of 289 μg/kg OTA, and a third group with melatonin at a dose of 10 mg/kg body weight, co-administered with OTA. Rats received the tested compounds orally in drinking water for 30 days. Histopathological damage was detected in the OTA-treated group compared to the control group. Proliferation of bile ducts in some portal areas and granular cell degeneration of the liver cells were found. Moreover, this damage also involved the sinusoids and central venous vasodilation, enlargement of periportal areas with mononuclear cell inflammatory infiltration, mild fibroblast tissue hyperplasia and common hepatocyte necrosis. Also in the OTA-treated group, the central veins and sinusoids of the livers were hyperplastic and filled with blood, and this event was induced by hemorrhages of small areas, vacuolated hepatocytes and individual cell necrosis between the hepatic parenchyma. The results were also stated by Abdel Wahhab et al. [40].

In chronic hepatotoxicity, sixteen male Wistar rats were divided into two groups: the control group and the OTA group (289 μg/kg body weight every other day for 90 days by gastric gavage). A mild steatosis was detected, but no fibrosis was observed [41].

No hepatic morphological changes were observed in acute or chronic hepatotoxicity. Palabiyik et al. [42] proved that no pathophysiology was evident in acute hepatotoxicity. Rats were administered 500 μg/kg body weight OTA for 14 days. In addition, male F344 rats were given OTA at a dose of 300 μg/kg body weight. Oxidative DNA damage was found in the liver; however, it did not induce hepatic morphological changes [43]. In a long-term chronic liver toxicity study involving OTA, Rached et al. [44] found that 70 μg/kg and 210 μg/kg body weight OTA for two years could induce hepatocyte proliferation but not induce liver damage.

The liver regulates glycogen storage. Some studies have reported that OTA can affect glycogen metabolism. Purchase and Theron [45] found that the accumulation of glycogen in the cytoplasm of the liver cells is accompanied in some cells by mitochondrial degeneration after a single dose of OTA at 10 mg/kg body weight. The reason for the increase in liver glycogen may be that OTA interferes with the phosphorylase enzyme system to produce this increase in liver glycogen. The inhibition of the phosphorylase enzyme system induced by OTA might be due to interaction of OTA with coenzymes and cofactors, competition inhibition, noncompetitive inhibition, coupling inhibition and binding to the hormones adrenaline and glucagon. In addition, Suzuki and Satoh [46, 47] found that ochratoxin A could induce a decrease in glycogen in rats.

2.2.2. Pig

In 1973, sixteen female, specific-pathogen-free pigs were [48] studied in three trials designed to assess different doses and modes of OTA delivery. Fatty change was demonstrated in all three trials. Other changes in the pigs' livers were not severe. In 2012, OTA was administered to 23 wild boars. The results revealed that gross lesions were present in two samples treated at the highest OTA levels: 1.8 μg/kg and 2 μg/kg. Alternating dark-red areas were reported, and hepatic lobular thinning at the margins also occurred. Moreover, fatty infiltration was observed in a small number of hepatocytes. Vacuolated hepatocytes and necrotic cells were found scattered throughout the liver parenchyma. There was considerable proliferation of the fibroblastic connective tissue around the bile ducts and veins [49].

2.3. Hepatotoxicity of Poultry by OTA-induced

Chickens and ducks were used to study the hepatotoxicity of OTA in poultry.

2.3.1. Broiler Chicken

In 1974, Huff et al. [50] found that the relative weight of the broiler chicken liver was increased after 2.0 mg/kg, 4.0 mg/kg, and 8.0 mg/kg OTA, although a significant increase occurred only with a dose of 8.0 mg/kg OTA ($P < 0.05$). In 1979, liver glycogen levels were significantly elevated ($P < 0.05$) in the high-dose groups (4.0 mg/kg and 8.0 mg/kg) but not in the low-dose group. Moreover, cytoplasmic deposits of glycogen in cells at the periphery of the liver lobes were detected. Warren et al. [51,52] suggested that the accumulation of liver glycogen was due to inhibition of the glycogen phosphorylase system in OTA-treated broiler chicken. Yankui Guo et al. [53] found that OTA could induce apparent hepatotoxicity in the broiler chicken. OTA can result in not only liver degeneration but also the accumulation of hepatic glycogen. Finally, swelling of the liver was observed. Additionally, significant liver endothelial cell proliferation and a smaller space in the liver sinusoid could be induced by OTA.

The synergism between aflatoxin and ochratoxin A in broiler chickens was explored by Huff et al. [50]. Broiler chickens were treated with 0, 2.5 μg/g AFB1, 2.0 μg/g OTA, and 2.5 μg/g AFB1 + 2.0 μg/g OTA. They found that individual toxins significantly altered the size of the liver ($P < 0.05$), but no significant difference was observed in the synergistic group. Aflatoxin, which can induce the liver lipid level, increased significantly ($P < 0.05$), and ochratoxin A decreased the liver lipid level. The result in both mycotoxins groups revealed that ochratoxin A inhibited lipid accumulation normally induced by aflatoxin.

2.3.2. Duck

In 1966, thirty day-old Pekin ducklings (average weight, 50 g) were treated with OTA by gastric gavage. Six time points (1, 2, 4, 8, 16 and 24 h) were selected for OTA treatment. Acute hepatotoxicity occurred. The pathologic changes observed in the liver were as follows: an increase in the fatty vacuolation of the hepatic parenchymal cell and swollen mitochondria, in addition to morphological changes in the endoplasmic reticulum (ER) of the duckling livers, which would be more severe at later stages. Then, cell mitochondria, dilated ER vesicles and lipid droplets were free in the sinusoids. At the final stage, the hepatocyte cytoplasm was comprised of fine, granular ground-glass cytoplasm, with occasional dilated ER vesicles, inconspicuous Golgi lamellae and ribosomal granules. In addition, there appeared to be a significant decrease in the number of ribosomal granules seen [38].

3. The Comprehensive Mechanism of OTA Hepatotoxicity

3.1. Traditional Hepatotoxicity

3.1.1. Blood Biochemistry

Various blood markers can be used to assess liver function. These markers are reflected in the primary metabolic enzymes. Furthermore, these enzymes may reflect more functions. Because these markers are located in other organs besides the liver, different metabolic enzymes can also reveal the function of other systems of the body. The liver has over 500 critical biological functions. If the liver is not correctly operating, then many functions are impeded. Various indexes, such as lactate dehydrogenase (LDH), bilirubin, triglycerides, aspartate aminotransferase (AST), gamma-glutamyl transferase (GGT), alanine aminotransferase (ALT), and alkaline phosphatase (ALP), can jointly provide insight into liver function.

Bilirubin

Bilirubin is a by-product of the breakdown of hemoglobin. Bone marrow cells and the spleen transport bilirubin to the liver. It is then conjugated and excreted through bile. If the total bilirubin is increased excessively, then the liver does not properly function or the biliary tract may be obstructed.

Triglycerides

Triglycerides could be the fuel for the muscles of the body. When the levels are too high or too low, human health is potentially placed at risk, resulting in a series of metabolic problems. A fasting blood test is used to detect the levels of triglyceride synthesized by the liver; therefore, triglycerides levels can reflect fatty liver and/or liver congestion if they are higher than normal.

Lactate Dehydrogenase (LDH)

LDH is a series of metabolic enzymes that mirror carbohydrate metabolic activity. Moreover, LDH is distributed in all tissues of the body. LDH levels that are too high can suggest congestion in the liver. If liver congestion does occur, then ALP, ALT, AST, GGT and bilirubin can also increase.

Alanine Aminotransferase (ALT)

ALT is also metabolic enzyme that is distributed in the liver, kidney, heart and muscle cells. When there is damage to these specific cells, high levels of ALT will be released into the blood. Elevated ALT can suggest different stages of liver breakdown and dysfunction.

Aspartate Aminotransferase (AST)

AST is a less sensitive metabolic enzyme for the assessment of liver issues. Elevated AST can be considered as a cardiovascular risk factors. Nonetheless, increased AST possibly indicates damage to the liver when considered with other blood chemistry factors such as ALT, ALP, GGT and LDH.

Gamma-Glutamyl Transferase (GGTP)

GGT is metabolic enzyme that is mainly located in liver cells. It is a sensitive factor for evaluating liver damage. GGT will increase with regular alcohol consumption. GGT levels will also increase when taking pharmaceutical drugs because the medicine tends to exert a toxic effect on the liver. In humans, GGT levels greater than 30 are closely associated with hepatic-related issues.

Alkaline Phosphatase (ALP)

Higher ALP levels than those seen in the normal range represent different degrees of liver dysfunction. ALP is a metabolic enzyme that is distributed in liver, bone and the intestines. Hence, abnormal values are possibly related to various types of dysfunction. If these three parameters are all high, then some issues involving GGT or ALT likely exist in the liver.

A study on the individual and combined toxic effects of OTA, aflatoxin B1 (AFB1) and T-2 toxin (T-2) was conducted, and serum biochemistry, organ morphology, hematology of broiler chickens, and the efficacy of esterified-glucomannan (E-GM) were assessed. Two dietary doses of AFB1 (0 and 0·3 mg/kg), OTA (0 and 2 mg/kg), T-2 (0 and 3 mg/kg) and E-GM (0 and 1 g/kg) were tested on a total 960 broiler chickens from 1 to 35 d of age. All the mycotoxins could suppress body weight and food intake. OTA was the most toxic during early life. AF increased the liver weight by 17.8%. OTA increased the gizzard weight by 14·6% and decreased the bone ash content by 8.1%. AF and OTA decreased the serum cholesterol content and increased the activity of serum GGT. AF decreased the serum protein content.

Several variables were not consistent: the activities of serum alanine amino transferase, blood urea nitrogen (BUN) content, and aspartate aminotransferase. AFB1 and T-2 suppressed blood hemoglobin; however, OTA prolonged the blood coagulation time. In addition, the additive effects on food intake, body weight and serum GGT activity were significantly different between any two toxins at 21 days [54].

OTA was fed to 240 day-old WL cockerels. Two doses of OTA (1000 and 2000 μg/kg of feed) were given to cockerels for up to 7 weeks. At week 7 of the experiment, the erythrocyte count of the OTA group was significantly lower than that of the control group. At week 3, the total protein value of the OTA groups was significantly lower than that of the control group. Additionally, in the OTA groups, creatinine was significantly higher than that in the control group. At week 7, the creatinine concentration in the OTA groups was significantly higher than that in the control group. At week 3, the ALT concentration in the OTA groups was significantly higher than in the control group. At week 7, the ALT concentration in the OTA groups was still significantly higher than that in the control group. At week 3 of the experiment, the urea concentration in the OTA groups was significantly higher than that in the control group. At week 7, the urea concentration in the OTA groups was significantly higher than that in the control group [55].

Male Wistar rats were gavaged with OTA (289 μg/kg body weight) daily for 90 days. AST, ALT, LDH and albumin (ALB) in plasma were assayed. Serum ALT, LDH and AST were increased after OTA treatment. ALB was decreased in OTA-treated rats ($p < 0.05$ versus CT) [17]. OTA (0, 0.07, 0.34 or 1.68 mg/kg) was administered to forty 12-week old female Wag/mbl rats for 26–28 days. The urinary volume, osmolarity, specific gravity, pH, creatinine and urea were significantly affected by OTA in young rats. Serum ALP was elevated in all the treated groups. In the older rats, urinary volume, osmolarity, specific gravity, pH, creatinine and urea were changed by OTA treatment. The clearance of urea and creatinine was significantly decreased. Serum ALT, AST, GGT, ALP and creatinine increased, as well, while urea decreased. Red blood cells (RBC), hemoglobin (Hb), mean corpuscular volume (MCV) and blood platelet (PLT) count decreased in the young 1.68 mg/kg group. Mean corpuscular hemoglobin concentration (MCHC) was minimally but significantly increased ($P<0.05$). Bone marrow cellularity and WBC were not significantly changed in the young rats. The hematology and bone marrow analyses were not significantly changed in the older rats [56]. OTA at 0, 21, 70, or 210 μg/kg bw doses was administered to male F344/N rats (6–7 weeks old) for 14, 28, or 90 days, respectively. OTA did not change the serum toxicity marker enzymes that reflect liver damage. However, OTA led to a slight decrease in alkaline phosphatase activity for the 28-day treatment. Serum triglycerides or potassium were not thought of as related compounds due to the lack of dose response [57].

OTA at doses of 0, 70 or 210 μg/kg bw was administered to male F344 rats for 2, 4, 13 and 26 weeks, respectively. LDH of the 2- and 13-week groups, and AST of the 13-week group decreased in the high-dose group compared with the control rats. In contrast, ALB activity of the c group increased slightly in the high-dose group compared with the control group. At 13 weeks, the liver damage was more serious in the high-dose group than in the medium dose group [58]. Sixty male Sprague–Dawley rats weighing 150 ± 6 g were fed an OTA-contaminated diet (3 mg/kg bw) for 15 days. Serum TP and albumin were significantly decreased in animals treated with OTA; ALP increased significantly in animals treated with

OTA. Gamma-glutamyl transpeptidase (GGT) was significantly higher in animals given OTA [59].

3.2. Oxidativc Stress

Oxidative stress has been recognized as a potential factor related to aging and some diseases. Free radicals are major forms of reactive oxygen species (ROS) *in vivo*. There are several types of free radicals, like hydroxyl radicals, superoxide anion radicals, lipid-free radicals, nitrogen dioxide, nitric oxide free radicals, hydrogen peroxide, singlet oxygen and ozone. Radicals are necessary because reactive ROS participate in signal transduction to maintain life and combat bacteria, parasites and toxins. The adverse factors of the surroundings compel the body to produce more ROS, leading to DNA mutations, which are related to disease and aging, and the loss of serum, cell membrane, cell mutation and anti-protease activity. Many studies have shown that cancer and aging, inflammation, illnesses of the heart, skin, liver, lung and many more diseases are all tightly associated with the reduction of free radical scavenging. Recently, an increasing number of studies have illustrated that ROS and reactive nitrogen species (RNS) result in liver damage, reduction of bile and production of harmful substances. Drugs and toxins in our body are metabolized in the liver; liver cells simultaneously keep the production of oxygen species in balance. If the redox reaction is not balanced, oxidative damage can result in many diseases such as inflammatory necrotic hepatitis, subclinical hepatitis without jaundice, liver cirrhosis and cancer.

3.2.1. General Parameters of Oxidative Stress

Many studies are investigating the relationship between O2 free radicals and pathogenesis of toxic substances. *In vivo*, the levels of these reactive oxygen species are controlled by a series of antioxidant enzymes; for example, catalase (CAT), superoxide dismutase (SOD), glutathione reductase (GR), glutathione peroxidase (GSPx), and glutathione-S-transferase (GST). In addition, there are non-enzymatic scavengers like reduced glutathione (GSH). When the system of antioxidant defense weakens or reactive oxygen species production is increased, an oxidative stress may occur. Apart from that directly inflicted on DNA by oxidative damage other indirect possibilities to cause genome destruction also work. Additionally, lipid peroxides enter the nucleus where they react with Fe^{2+} to generate the alkoxyl radical which attacks DNA. Additionally, oxidative damage to membranes can result from an increase in intracellular calcium levels. Calcium enters the nucleus where it can activate nucleases that cause DNA strand breaks [60].

The major α-tocopherol (TOCO) oxidation product, α-tocopherylquinone (α-TQ) is thought to be a sensitive biomarker of oxidative stress in membranes.α-TQ formation occurs before lipid peroxidation. Some proteins such as heat shock proteins are also early markers of oxidative stress; they induce cellular defense mechanisms. When obvious oxidative damage and toxicity are still not observed, oxidative and/or chemical stressors induce these proteins. Heat shock protein 32, namely heme oxygenase-1 (HO-1), is of particular relevance regarding oxidative stress [61]. Male rats that were treated with OTA (up to 2 mg/ 24 h exposure) did not have a higher level of biomarkers of oxidative damage; for example, malondialdehyde (MDA), which is the lipid peroxidation marker, or 8-oxo-7,8-dihydro-2' deoxyguanosine in kidney DNA, which is the DNA damage marker. However, OTA treatment (1 mg/kg) reduced the α-tocopherol plasma level by 22% and induced the expression of the oxidative stress responsive protein heme oxygenase-1, specifically in the kidney.

3.2.2. Oxidative Stress Induces DNA Damage

Another generally used biomarker of oxidative stress and DNA damage is 8-oxo-7,8-dihydro-2'-deoxyguanosine (8-oxodG). It is always increased in rats upon exposure to several oxidants and renal toxins. As shown in Table 2 [62], diverse biomarkers demonstrate that OTA induces DNA damage. Male F344 rats were treated with OTA for 4 weeks (0, 0.03, 0.1, 0.3 mg/kg bw per day). Ultimately, the comet assay was used to detect oxidative DNA damage in both the kidney and the liver with/without use of the repair enzyme, formamidopyrimidine-DNA-glycosylase (FPG). OTA did not work on basic DNA damage (determined without FPG); this repair enzyme specifically identifies oxidized purine and pyrimidine bases, excises the lesions and nicks the DNA at the respective sites, resulting in additional DNA damage which is thought to be oxidation-specific. However, OTA-mediated oxidative damage was detected with FPG treatment in kidney and liver DNA of all dose groups [16].

3.2.3. Oxidative Stress Induces Protein Damage

Oxidative pathways can lead to carbonylation of proteins. Carbonylation is a key protein modification related to changes in protein misfolding, protein function, proteolysis, and protein fate. If the carbonyl protein of tissues were increased, a number of pathological disorders will begin to appear [62].

3.2.4. Oxidative Stress Induces Lipid Damage

Lipid peroxidation is the oxidative degradation of lipids and has recently been widely investigated. Free radicals are possibly the main driver of the process. The by-products, the tautomer malondialdehyde (MDA), 4-hydroxy-2-nonenal (HNE), acrolein and crotonaldehyde, have been broadly studied. The ability of these reactive electrophiles to modify DNA bases, yielding promutagenic lesions, is thought to be the basis of mutagenesis and carcinogenicity. MDA HNE and HNE MDA have increasingly been implicated in carcinogenesis [62].

3.2.5. Prevention of OTA-Induced Oxidative Stress

Much attention has been paid to natural antioxidants as candidates for the prevention of oxidative damage caused by free radicals.

Melatonin exists in many living things, for example, animals, plants, bacteria and fungi. It is a hormone in animals that plays an important role in the daily onset of darkness. Melatonin was investigated in rats in the setting of the protective effect on OTA-induced oxidative stress. Four groups of rats were used in the experiment: controls, melatonin-treated rats, OTA-treated rats and melatonin + OTA-treated rats. A lipid peroxidation product (LPO), malondialdehyde (MDA), was detected in the serum, liver and kidney after 4 weeks of treatment. Additionally, the levels of glutathione (GSH) and the activity of glutathione reductase (GR), glutathione peroxidase (GSPx), superoxide dismutase (SOD), catalase (CAT) and glutathione-S-transferase (GST) in the liver and kidney were determined. In OTA-treated rats, LPOs were significantly increased in the liver, kidney and serum. Meanwhile, the GSH value and the enzyme activity of SOD, CAT, GSPx and GR in both the kidney and liver were significantly decreased. In the melatonin + OTA group, rat LPO in liver, kidney and serum were not statistically different from those of the controls. The levels of GSPx, GR and GST activity in both the liver and kidney tissues were significantly increased compared with those of the controls. GSPx, GR and GST activity was also increased in melatonin-treated rats compared with controls. In a word, melatonin protected rats from OTA toxicity through stimulation of GST activity and inhibition of oxidative damage [60].

α-tocopherol is another antioxidant. However, some studies have shown that OTA may exert its toxic effects without being directly modulated by α-TOC or polyphenols. α-tocopherol (α-TOC) and different polyphenols – catechin (CAT), daidzein (DAI), epicatechin (EC), epigallocatechin gallate (EGCG), genistein (GEN), and quercetin (QUE) – were examined for their effects on OTA-induced cytotoxicity in HepG2 liver cells. Incubation of HepG2 cells with α-TOC as well as with different polyphenols could not prevent OTA toxicity [107].

To explore whether an aqueous extract of *Emblica officinalis* has a protective effect on ochratoxin-induced lipid peroxidation in the liver and kidney of mice, adult male albino mice were gavaged with 50 μg or 100 μg of ochratoxin for 45 days. The results showed that the administration of *Emblica officinalis* aqueous extract (2 mg/animal/day) and ochratoxin for 45 days caused significant relief in ochratoxin-induced lipid peroxidation in the kidney and the liver [108].

CoQ10 has protective effects against OTA toxicity in liver tissue, whereas it has no effect in kidney tissue. Rats were treated for 4 weeks with OTA (289 μg/kg/day) or OTA+CoQ10 (289μg/kg/day OTA + 1 mg/100 g/day body weight (bw) CoQ10). In the OTA-treated group, the levels of MDA and Hyp in both the liver and kidney were significantly increased compared with the control, whereas GPx activity decreased. In the OTA+CoQ10–treated group, the levels of MDA and Hyp decreased compared with the levels of OTA-treated rats, whereas GPx activity increased in the liver. In the OTA+CoQ10-treated group, the levels of MDA, Hyp, and GPx were not significantly changed in the kidney when compared to those in the OTA-treated group.

Lactobacillus rhamnosus GG (LGG) can be protective against OTA-induced oxidative stress in mice. The mice were treated with LGG (1×10^{10} CFU), OTA (1.8 mg/kg bw) or given LGG two hours before OTA gavage. The levels of malondialdehyde (MDA), glutathione (GSH) and superoxide dismutase (SOD) activity were detected in the liver and kidney. The bone marrow micronucleus test, mitotic and meiotic activity and chromosomal aberrations in spermatocytes were performed to assess the genotoxicity. OTA significantly increased the tissue levels of MDA; however, the levels of GSH and SOD activity were significantly lower than those in the control group in both the liver and the kidney. The frequency of MNPCEs in the bone marrow and structural and numerical aberrations in spermatocytes was significantly higher in the OTA group. Furthermore, the mitotic and meiotic activity of somatic and germ cells significantly decreased. In addition, OTA reduced sperm motility and cauda epididymal sperm count, and increased sperm abnormalities. In mice that received LGG before OTA gavage, GSH and SOD activity were increased significantly in the liver and kidney. The frequency of MNPCEs in bone marrow and chromosomal aberrations in spermatocytes significantly decreased, and mitotic and meiotic activity significantly recovered in the LGG administration group. Furthermore, all sperm parameters were recovered when gavaging LGG before OTA treatment [109].

Table 2. OTA oxidative stress-related studies

Model	Gender	Via	Time treatment	Dose	Aim	Results/Conclusion	Ref.
BALB/c Macrophage 1774a cell line			24,48,72 h	30 nM-100 μM	OTA immunotoxicity and modulation inflammatory process	Induction of iNOs, COX-2 and NF-κb expression by OTA. OTA is an immuotoxic compound	[63]
Porcine kidney tubuli cells LL-PK1			6-24 h	1-100 μmol/L	Characterization effect of OTA ON Nrf2 response	Nrf2 potential signal transduction pathway by which OTA impairs its own detoxification	[64]
Porcine kidney tubuli cells LL-PK1			24 h	1-100 μmol/L	Impact of OTA on Nrf2, AP-1activity, antioxidant enzymes and GST	Enhanced production of ROS, GST impairment. Nrf2 and AP-1 disruption by OTA. Impairment of the detoxification machinery	[65]
Rat Sprague-Dawley	male	diet	15 days	3.0 mg/kg b.w.	Oxidative stress protection study	OTA-induced oxidative stress chemoprotection by *Inula crithmoides*	[66]
Rat F344	male	gavage	7 and 21 days	0.5 mg/kg b.w.	Mechanism of action study-microarrays	Oxidative stress, calcium homeostasis, cytoskeleton structure	[67]
Human hepatocytes HepG2; Monkey kidney Vero cells				0-100 μM	Decrease GSH	No induction of heat shock protein (HSP)	[68]
Rat Wistar	male	diet	15 days	5 ng/g b.w. 50 ng/kg b.w.	Oxidative damage study (proteins and lipids)	Malondialdehyde (MDA) and protein carbonylation (PC) increase in kidney > liver	[69]
Chinese Hamster lung V79 cells; Lymphoma mouse LY5178 cells				0-438 μM	OTA mutagenicity	OTA is mutagenic at cytotoxic doses in mammalian cells via oxidative DNA damage induction.	[70]
Rats Sprague- Dawley	male	diet	4 weeks	200 ppb	Oxidative stress protection study	OTA-induced oxidative stress and DNA damage chemoprotection cyaniding 3-O-β-D-glucoside OTA-induced ROS.	[71]
Pig kidney cell line LLC-PK1			24 h	0, 10, 15, 20 μM	Oxidative stress protection study	Scavenging by cathechins (epigallocathechin gallate (EGCG) and Epicatechin gallat (ECG))	[72]

Model	Gender	Via	Time treatment	Dose	Aim	Results/Conclusion	Ref.
Human epithelial colorectal adenocarcioma Caco-2 cells				100 μM	Effect of OTA on barrier function impairment	Loss of microdomains associated with tight junctions maybe due to oxidative events	[73]
Neural stem/progenitor cells (NSCs)				0.01-100 μg/mL	Vulnerability of brains mouse cells to OTA	Robust increased in total and mitochondrial SOD activity. OTA impaired hippocampus neurogenesis	[74]
Rats	Male/liver and kidney	Diet (Drinking water)	4 weeks	289 μg/kg	Oxidative stress protection study	Melatonin protection against OTA-induced oxidative damage in liver and kidney. CoQ protective in liver.	[75]
Human renal cell line HK-2			6-24 h	50 μM	Mechanism of action study-microarrays	Significant increase in ROS level and oxidative DNA damage.	[67]
Human renal proximal tubular epithelial cell line HK-2				50–800 μM	single-strand DNA breaks and oxidative damage induction by OTA	Oxidative stress precedes cytotoxicity and genotoxicity	[76]
Eker and wild type rats	male	gavage	1-14 days	210 μg/kg b.w.	Early carcinogen-specific gene expression study	Oxidative DNA damage response genes, general stress response, and cell proliferation	[77]
Wistar rats		gavage	90 days	289 μg/kg b.w.	Early effects of chronic OTA administration in liver	Reduction in the ability to counteract oxidative stress in liver	[41]
Swiss ICR	male	i. p	6, 24, 72 h	0–6 mg/kg b.w.	Oxidative stress and OTA neurotoxicity	Acute depletion of striate DA on a background of globally increased oxidative stress and transient inhibition of oxidative DNA repair	[78]
Swiss mice	male	I.p; infusion	2 weeks	Acute 3.5 mg/kg; chronic 4, 8, 16 mg/kg	Effect of chronic low dose OTA exposure on regional brain oxidative stress and stratial DA metabolism	Low doses exposure caused global oxidative stress	[79]
F344 rats	male	diet	7 and 21 d; 4, 7 and 12 months	300 mg/kg b.w.	OTA mechanism of action-microarrays study in liver and kidney	Oxidative stress and metabolic response modulated involving mainly Nrf2 and HNF4α pathway disruption	[80]

Table 2. (Continued)

Model	Gender	Via	Time treatment	Dose	Aim	Results/Conclusion	Ref.
Swiss mice	male	oral	24 h	10 mg/kg	Immune cells response after acute OTA exposure	OTA-induced oxidative stress response responsible of its own toxicity.	[81]
Wistar rats	female	Intraperitoneally	7, 14 and 21 days	0.5 mg/kg b.w./day	Genotoxic potential of OTA measuring DNA strand breaks (comet assay) in the kidney	OTA-induced DNA strand breaks were detected, OTA concentration in the kidney and duration of the treatment correlated with severity of the DNA damage	[82]
Wistar rats	male	oral	15 days	5 ng; 0.05 mg; 0.5 mg/bw	Effect of OTA on DNA damage	Oxidative stress responsible for OTA-DNA damage as shown by Fpg-modified comet assay	[83]
Pig kidney microsomes, human bronchial epithelial cells, human kidney cells			Cells: 2, 7, 24 h	0.5, 1.0, 2.5 μM	Genotoxicity of the hydroquinone (OTHQ) metabolite of OTA	OTQ-mediated adduct spots form in a dose and time dependent manner	[84]
Wistar rats	female	oral	7, 14 and 21 days	0.5 mg/kg b.w.	Effect of OTA on protein oxidation in kidney and liver	Increased protein carbonyls in the kidney and liver	[85]
F344 rats	male	gavage		0.03, 0.10, 0.30 mg/kg b.w.	Evaluate relevance of OTA-induced oxidative damage on nephrotoxicity and carcinogenicity	Tumours in rat kidney may be attributable to oxidative DNA damage in combination with cell-specific cytotoxic and proliferation-stimulating effects as cell-signaling response	[43]
V79 (Chinese hamster lung fibroblasts) cells, CV-1 (African green monkey, kidney) cells, primary rat kidney cells			1-24 h	2.5, 100 μM OTA	Relevance of OTA-induced oxidative damage in nephrotoxicity and carcinogenicity	Cytotoxicity and oxidative DNA damage already at low doses could be a relevant factor for the nephrocarcinogenicity	[86]

Model	Gender	Via	Time treatment	Dose	Aim	Results/Conclusion	Ref.
Rat lymphoid cells			1 h	0.5, 2, 20 μM	OTA immune function modification	Protein synthesis inhibition, oxidative metabolism of OTA, prostaglandin synthesis implicated in NK cells toxicity	[87]
Human hepatoma—derived cell lines (HepG2), human colonic adenocarcinoma cell line (Caco-2)			24, 48, 72 h	0–100 μM	Oxidative stress protection study	OTA-induced oxidative stress damage. Protective effect byCyanidin-3-O-β glucopyranoside(C-3-G)	[88]
F344 Fischer rats	male	gavage	2 weeks	0–2000 μg/kg b.w.	Genotoxicity of OTA	DNA strand breaks in target and nontarget tissues probably involving oxidative stress mechanism	[89]
Human hepatoma—derived cell line (HepG2)			48–72 h	35–10 mM	Oxidative damage protection study	No cytotoxicity protection observed with Vitamine E, polyphenols	[90]
Sprague-Dawley	male	diet	15 days	3 mg/kg	Oxidative stress protection *in vivo* study	Preventive effect against OTA-induced oxidative stress and lipid peroxidation by melatonin	[91]
Human fibroblast cells			48–72 h	0–50 μM	Oxidative stress protection study	Reduction of free radical Species production and DNA damage prevention by cyanin 3-O-β-D-glucoside (C3G)	[92]
Fetal rat Telencephalon aggregating cells			24–48 h, 9 days	0–20 nM	Adverse effect of OTA in brain	Brain inflammatory response induction of HO-1, iNOs, PPARγ, cytoskeletal damage	[93]
Human hepatoma-derived cell line (HepG2)			24 h	0 – 40 μg/mL	Genotoxicity of OTA	Dose-dependent induction of DNA single strand breaks (comet assay) and micronuclei(MN)	[94]
Primary proximal tubules renal (PT)cells, proximal tubular cell line(LLC-PK1)			0-24 h	0–100 μM	OTA mediated oxidative stress response in proximal tubular cells, oxidative stress protection	Oxidative stress contributes to tubular toxicity. Antioxidants (α-tocopherol, N-acetyl-Lcysteine (NAC) treatment prevents OTA toxicity	[95]

Table 2. (Continued)

Model	Gender	Via	Time treatment	Dose	Aim	Results/Conclusion	Ref.
Wistar rats	male	gavage	10, 30, 60 days	120 μg/kg b.w.	Kidney low dose OTA response: sequence of events leading to cell death	Low dose induces oxidative stress, apoptosis in proximal, and distal tubule kidney cells	[96]
Human hepatoma—derived cell line (HepG2)			1 h, 24 h	0–50 μg/mL	Genotoxicity of OTA	No inductions of mutations in the Ames assay, a dose-dependent induction of micronuclei in the MN assay, and DNA migration (comet assay) were detected	[37]
Proximal tubular Cells (PTC), Wistar rats	male	gavage	24–72 h (*in vivo* and *in vitro*)	5.0 μM, 12.5 μM *in vitro*; 1 and 10 mg/kg b.w.	*In vivo* and *in vitro* gene expression comparative study	*In vitro* and *in vivo* gene expression data were comparable. Response to oxidative stress-related genes hypoxia-inducible factor 1 and catalase was observed	[97]
Dark Agouti (DA),Lewis rats	male	Intragastricintubation	0.4 mg/kg b.w.	Life-time	Life-time study to evaluate if MESNA leads to a more effective reduction of OTA-induced tumour development orurinary tract damage	Lack of effect of mesna on OTA-induced urinary tract damage or renal tumor development	[98]
Dark Agouti (DA), Lewis rats	male	Intragastricintubation	0.4 mg/ kg b.w.	2 years	Life-time study to evaluate the Potential protective effect of 2 mercaptoethane sulfonate (MESNA) and N-acetyl cysteine(NAC)	MESNA decreased karyomegalies in kidney, but had no beneficial effect on renal tumour incidence	[99]
Fischer rats	male	gavage	4, 8, 24, 48 h	0–2.0 mg/kg b.w.	Chemical and Biological markers induced by OTA exposure associated with oxidative stress	Oxidative stress may contribute to mechanism of OTA renal toxicity and carcinogenicity in rats over long term exposure	[100]

Model	Gender	Via	Time treatment	Dose	Aim	Results/Conclusion	Ref.
Bronchial epithelial cells incubated with microsomes of seminal vesicles of pig			4 h	10 μM	Roles of cyclooxygenase and lipoxygenase in ochratoxin A genotoxicity inhuman epithelial lung cells	OTA is biotransformed into genotoxic metabolite via a lipoxygenase, whereas prostaglandin—Hsynthetase (PGHS)decreases OTA genotoxicity	[101]
Sprague-Dawley liver microsomes, liver mitochondria and hepatocytes cells	female			2.5 mM	Free radical generation by OTA in hepatocytes, mitochondria, and microsomes using electron paramagnetic resonance (EPR)	Oxidative damage maybe one of the manifestations of cellular damage in the toxicity of OTA	[102]
Bacillus brevis			10 min	1mg/mL	Study free radical generation in bacteria as model system	OTA incuces free radical production, enhancing permeability of the cellular membrane toCa^{2+}	[103]
Swiss mice	male	Gastric intubation	48 h	2 mg/kg b.w.	Effects of vitamins on genotoxicity of OTA	Vitamins E, A, and C also reduced OTA-DNA adduct formation in mice kidney	[104]
Wistar rat	male	Gastric intubation	Every 48 h/3 weeks	289 μg/kg b.w.	Protective effect of superoxide dismutase (SOD)and catalase	SOD + catalase prevents the nephrotoxicity induced by OTA in rats	[105]
Wistar rat liver microsome, kidney microsomes	male			6 mg/kg b.w.	Lipid peroxidation induction by OTA	Lipid peroxidation may play a role in the observed toxicity of OTA	[106]

Table 3. Experimental conditions of the main gene expression (GE)a studies performed with ochratoxin A

Reference	Model system	Aim	Timepoints	Dosage	Microarray
In vitro					
Gennari et al. (2004)[112]	LLC-PK1	Mechanistic	24 h	Phenotypic assays: 0–1–5–10–50 μM (DMSO)	cDNA macroarrays (Atlas array membrane<comma> Clontech)
	(porcine renal proximal tubular epithelial cell line)	Renal damage at epithelial barrier			GEa: 50 μM (n = 3) (DMSO) b
Arbillaga et al. (2007)[67]	HK-2	Mechanistic	6 h and 24 h	MTT<comma>LDH: 1<comma> 10<comma> 50<comma> 100<comma> 200 μM	Human Genome U133A 2.0 GeneChip
	(human renal proximal tubular epithelial cell line)			Comet<comma> ROS (H2DCF) and	(Affymetrix)
				GEa: 50 μM (n = 3) (NaHCO3) b	
Hundhausen et al. (2008)[113]	HepG2	Mechanistic	24 h	Intracellular Ca2+: 2.5 μM.	Human Genome U133A 2.0 GeneChip (Affymetrix)
	(human liver cell line)			GEa<comma> PCR and WB: 0.25 μM<comma> 2.5 μM (n = 2)	
			(vehicle not indicated)c		
Radford et al. (2012)[114]	RPTEC/TERT1	Evaluation (OTA<comma> KBrO3<comma> Nifedipine)	Resazurin assay: 72 h<comma> increasing doses (n = 3)Loss of primary cilium and GE a: 6 h<comma> 24 h<comma> 72 h 300 nM (n = 3) (0.1% DMSO) (n = 3) Effects on tubulin microtubule network and cell cycle: 72 h 300 nM (n = 3) (0.1% DMSO) c		Human Genome U133 Plus 2.0 GeneChip
	(human renal proximal tubular epithelial cell line)	Mechanistic Primary cilium level	(Affymetrix)		
In vivo					
Marin-Kuan et al. (2006) [80]	Male F344 rats (liver<comma> kidney)	Mechanistic	2 years study	Daily diet	Rat Genome U34A GeneChip (Affymetrix)
				300 μg/kg b.w .until 333 g b.w was reached	
			GEa timepoints:	Afterwards: 100 μg/rat	
			Early: 7 and 21 days	(phenotypic<comma> n = 4) (GE a<comma> n = 3)	
			Late: 4<comma> 7 and 12 months		

Reference	Model system	Aim	Timepoints	Dosage	Microarray
Stemmer et al. (2007) [115]	Male Eker rats (Tsc2+/−<comma> Long Evans)		1<comma> 3<comma> 7<comma> 14 days	Daily gavage	Rat Genome RAE230A GeneChip (Affymetrix)
	Male Wt Long Evans	Evaluation		210 μg/kg b.w (n = 3)	
	(kidney)	(OTA vs AA)		(NaHCO3 0.1 M)	
Arbillaga et al. (2008) [116]	Male F344 rats (liver<comma> kidney)	Mechanistic	7 and 21 days	Daily gavage	Rat Genome 230 2.0 GeneChip (Affymetrix)
				500 μg/kg b.w (n = 5)	
				(NaHCO3 0.1 M)	
Stemmer et al. (2009) [117]	Male Eker ratsd	Evaluation	3 and 6 month	Gavage 5 days per week (phenotypic<comma> n = 10) (GE a<comma> n = 3)	Rat Genome RAE230A_2.0 GeneChip (Affymetrix)
	(Tsc2+/−<comma> Long Evans) (preneoplasic lesions from male kidneys)	(OTA vs AA)	GEa timepoint: 6 month	210 μg/kg b.w	
		Mechanistic		(NaHCO3 0.1 M)	
Taniai et al.<comma> 2012 [118]	Male F344/NSlc rats (kidney<comma> OSOMe)	Prediction	28 days	Daily gavage	Rat microarray Euk Expr 4x72 K (Roche NimbleGen)
				210 μg/kg b.w	
				(phenotypic<comma> n = 10) (GE a = 4)	
				(NaHCO3 0.1 M)	
Hibi et al. (2013) [119]	Male F344/NSlc-Tg (gpt delta) rats	Mechanistic	28 days	Daily diet	4 × 44 K whole rat genome microarray (Agilent Technologies)
	(kidney CORf and OMg)			5 ppm	
				(RT-qPCR<comma> n = 5) (GE a = 3)	
In vitro–in vivo comparison					
Luhe et al. (2003) [97]	(1) PR-PTC (primary rat proximal tubular cells)	Evaluation (*in vitro* vs *in vivo*)	24 h and 72 h	Low dose: 5 μM (n = 2) (DMSO) c	cDNA microarrays (Phase-1<comma> Santa Fe)
	(male Wistar rats)	Mechanistic	(OTA containing medium exchanged daily)	High dose: 12.5 μM (n = 2) (DMSO) c	

Table 3. (Continued)

Reference	Model system	Aim	Timepoints	Dosage	Microarray
Luhe et al. (2003) [97]	(2) Male Wistar rats (kidney cortex)		1 and 3 days	Dosage: p.o<comma> daily dose	
				Dosage: p.o<comma> daily dose	
				(corn oil)	
				High dose: 10 mg/kg b.w (n = 5/timepoint)	
				(corn oil)	
				Control: corn oil (n = 4/timepoint)	
Jennings et al. (2012) [120]	(1) PH-PTC (primary human proximal tubular cells)	Evaluation (*In vitro* vs *in vivo* comparison)	1 and 3 days		Human data:
	(2) RPTEC/TERT1	Mechanistic		PH-PTC: 5 µM; RPTEC/TERT1: 0.3 µM; HK-2<comma> NRK<comma> PR-PTC: 0.5 µM (n = 3) (except PH-PTC<comma> n = 4)	Human Genome U133 Plus 2.0 GeneChip
	(human renal proximal tubular epithelial cell line)			Vehicle: DMSO (0.1%)c	(Affymetrix)
	(3) HK-2 (human renal proximal tubular epithelial cell line)				Rat primary and *in vivo*:
	(4) PR-PTC (primary rats proximal tubular cells)				Rat Genome RAE230A GeneChip (Affymetrix)
	(5) NRK-52E (rat renal epithelial cell line)				
	(6) Rat renal cortex (*in vivo*) (mined from Luhe et al.<comma> 2003)		1<comma> 3 and 7 days	Dosage: p.o<comma> daily dose	NRK-52E Rat Genome 230 2.0 GeneChip (Affymetrix)
				Dose: 3 mg/kg b.w<comma> (corn oil) (n = 3)	
				Control: corn oil (n = 3)	

[a] GE: gene expression studies (microarrays).b Exposure performed in serum-free medium.c Exposure in serum-free medium not specifically indicated.d Stemmer et al., 2009.e OSOM: outer stripe of the outer medulla . This study was carried out in both sexes, but gene expression analysis was only performed in males.f COR: cortex.g OM: outer medulla.

3.3. Cell Apoptosis

Apoptosis, namely, programmed cell death, is a special form of cell death that is different from cell necrosis in many biochemical and morphological aspects. Like mitosis or differentiation, apoptosis is a normal cell phenomenon. During this process, the expression of some genes will change to induce or inhibit apoptosis. However, apoptosis can be caused by many external factors as well. Apoptosis has been associated with many diseases. Because of the variety of apoptosis-controlling genes, apoptosis conditions can vary.

Apoptosis occurs in the liver cells: first in the normal state during liver development and subsequently in the adult liver for liver organogenesis and the renewal of hepatocytes, respectively. However, apoptosis is also present in various viral, immunological, malignant or drug-induced human liver diseases. Additionally, in the animal, hepatocyte apoptosis can be caused either *in vitro* or *in vivo* or by many toxic agents [110].

Transcriptomic studies involving OTA (4 *in vitro*, 6 *in vivo*, 2 *in vitro/in vivo*, Table 3) showed that apoptosis genes were deregulated *in vitro* but were less affected *in vivo* [111].

OTA (0.5 mg/kg body weight/day) was orally administered to male Sprague-Dawley rats for 14 days. Apoptotic cell death of hepatocytes was evaluated by the TdT-mediated dUTP nick-end labeling (TUNEL) assay. TUNEL analysis revealed a significant ~2.7-fold increase in the number of TUNEL-positive liver cells of rats exposed to OTA compared to the control group [42].

Reportedly, apoptosis occurred after application of 12.5 μmol/L OTA. In total, 130 to 260 μmol/L antioxidative flavonolignan silibinin prevented chromatin condensation, apoptotic DNA fragmentation and caspase-3 activation, processes that were caused by OTA. OTA stimulated lipid peroxidation on cultured immortalized rat liver HPCT cells that was revealed by malondialdehyde (MDA) production. These reactions were also inhibited by silibinin pretreatment. We conclude that the anti-apoptotic activity of silibinin against OTA is induced *in vitro* by the antioxidative effects of the flavonolignan. Moreover, cytotoxicity of the pro-apoptotic toxins was detected by the MTT-test. The concentrations of OTA used induced a great decrease in cell viability within 36 h, which was principally prevented by silibinin. Silibinin can effectively protect cells from apoptosis and cytotoxicity caused by OTA [121]. In addition, another study also reported that silibinin at 130 μmol/L showed significant hepatoprotective and antiapoptotic effects against OTA-mediated cell damage in cultured rat hepatocytes. Hepatocytes prepared under ETDA perfusion were treated with 12.5 μmol/L OTA. After 24–36 h, a DNA-ladder was developed. A low concentration (26 μmol/L) of silibinin administered prior to OTA slightly prevented OTA-mediated DNA-laddering, whereas 130 μmol/L silibinin completely suppressed OTA-mediated apoptosis. In the same case, caspase-3 activity in hepatocytes was increased in a time-dependent manner under OTA exposure within 12–24 h but was prevented by 130 μmol/L silibinin [122].

3.3.1. Ca^{2+}

Hepatic microsomal calcium sequestration induced by ochratoxin A was studied both *in vivo* and *in vitro*. The rate of ATP-dependent calcium uptake was suppressed by 42–45% in intoxicated rats administrated with ochratoxin A. With NADPH, the addition of ochratoxin A (2.5 to 100 μM) caused a concentration-dependent inhibition of calcium uptake (28–94%) by untreated rat liver microsomes. The rate of NADPH-dependent lipid peroxidation was also enhanced to a great extent by OTA. Different agents that inhibited OTA enhanced lipid peroxidation and could block the destruction of calcium uptake activity. Lipid peroxidation increased by OTA occurred with calcium leakage from calcium-loaded microsomes. OTA disrupts microsomal calcium homeostasis by impairing the endoplasmic reticulum membrane, most likely via enhanced lipid peroxidation [123].

4. The Molecular Mechanism of OTA Hepatotoxicity

The parameters of traditional hepatotoxicity were always applied to studies on hepatotoxicity induced by mycotoxin. Furthermore, the parameters of molecular hepatotoxicity have been increasingly studied over time. Examples include DNA methylation, DNA adducts, mRNA modification and non-coding RNA (miRNA, lncRNA). Changes in these parameters can affect the normal expression of genes, further affecting their function. Therefore, these parameters have become new markers for the evaluation of toxic effects. The different parameters of molecular hepatotoxicity will be introduced in the following sections.

4.1. DNA

4.1.1. DNA Methylation

DNA methylation is one of the most common modes of modification of the DNA of the eukaryotic genome DNA. The mechanism of DNA methylation was through the methyl group providing S-adenosylmethionine (SAM) combined with DNA cytosine 5 carbon atoms to form 5-mC, along with the catalysis of DNA methyltransferase (DNMT). DNA methylation often occurs in three different nucleotide sequences: the symmetrical CpG sequence, CHG sequence and the unsymmetrical CHH sequence (H represents the basis of A, T or C). DNA methylation plays an important role in regulation of gene expression, transposable element silence, genomic imprinting, embryonic development and the maintenance of genomic integrity. DNA methylation is prominent in researching for a mycotoxin mechanism.

HepG2 cells were treated with OTA (2.5-150 μmol/L) for 24 h. DNA methylation was detected by HPLC/MS/MS. The level of DNA methylation decreased after OTA treatment. The most apparent decrease occurred with 5 μmol/L OTA, but there were no significant differences with 25-150 μmol/L OTA [124]. In contrast, Kouadio et al. [125] found that DON, ZEA, FB1 could increase the level of DNA methylation. Until now, there had been no reports regarding DNA methylation in the liver induced by OTA, and this phenomenon will be the new mechanism for further study on OTA-induced hepatotoxicity *in vivo*.

4.1.2. DNA Adduct

OTA-induced genotoxicity and its mechanism of action as a carcinogen are still being debated. Since 1990 [126], DNA adducts have been studied in OTA-induced genotoxicity.

A DNA adduct is a piece of DNA covalently bonded to a (cancer-causing) chemical in molecular genetics, which is always considered a critical event in carcinogenesis [127]. Either directly or following metabolic activation to DNA-reactive intermediates, the formation of DNA adducts was the cause of many mutagenic and carcinogenic events [128, 129]. The formation of DNA adducts could be the start of a cancerous cell or carcinogenesis. Because

DNA replication was conducted after chemical modification of DNA, the repairable lesion was converted into a heritable mutation. Mutations of the critical sequences in DNA may inhibit the activity of tumor suppressor genes and activate oncogenes [130].

Swenberg et al.[131] determined that the events involving absorption of the agent, distribution in different tissues, metabolic activation, detoxification and DNA repair, could affect the formation of DNA adducts, as well as potential sites of DNA adduction [131] as shown in Figure 1. Moreover, the reactive species and nucleophilicity of the DNA sites and steric factors directly affect the specificity of the reactions. Guanine and adenine were the most nucleophilic and exocyclic. Their N3 and N7 sites were easier to combine with the chemical substances forming the DNA adducts [132-134]. However, not all DNA adducts induced the mutation. Mutations induced by DNA adducts, which were involved in hydrogen base-paring, occurred due to erroneous incorporation during DNA replication., Some DNA adducts would simultaneously not induce the mutation because base-paring had not happened in alkylation of positions, such as N7 of guanine [130]. Furthermore, conformational changes of DNA induced by DNA adducts derived from bulky aromatic carcinogens also induced mutations. In the DNA adduct 2-acetylaminofluorene, C8 of guanine was the site, and it was not involved in the base-pairing. This adduct can induce a change in DNA structure, GC to TA base-pair substitutions and frame shift mutations [135]. Thus, the roles of DNA adducts in the genotoxicity of mycotoxin (AFB1 and OTA) are the focus of many studies.

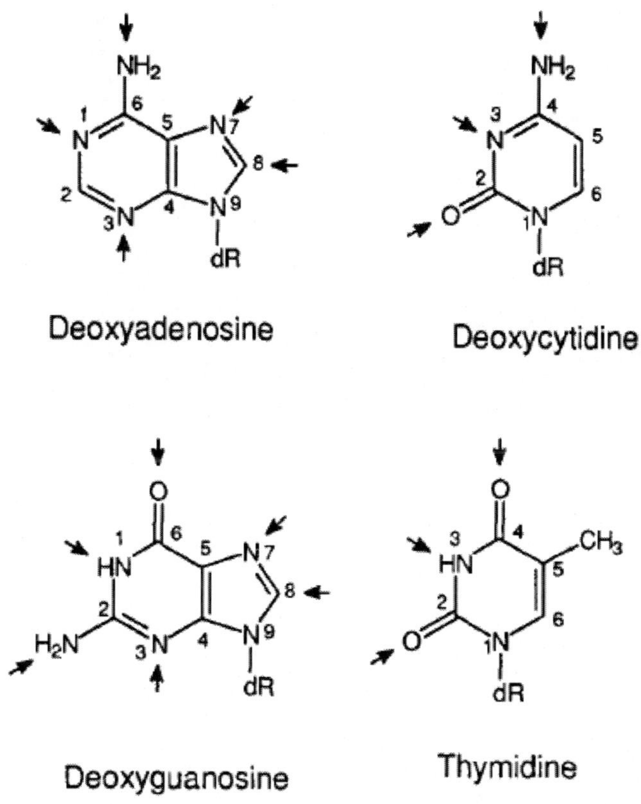

Figure 1. Potential sites of DNA adduct [130].

OTA can reportedly induce hepatocellular carcinomas in mice and rats [136-139]. The DNA adduct is always considered to be the beginning of a cancerous cell or carcinogenesis. Based on these factors, OTA-DNA adducts were studied in mice, rats, pigs, and humans [126, 140-143]. Although OTA-DNA adducts were found in some studies, whether genotoxicity is induced by OTA remains controversial.

Some studies supported the notion that OTA-DNA adducts were the cause of OTA genotoxicity [144, 145]. The process has been previously summarized [146]. Three pathways for the bioactivation of OTA were proposed. The first proposed pathway is that for providing the quinone species, ochratoxin quinone (OTQ), the parachlorophenolic moiety of OTA, would undergo oxidative dechlorination, as shown in Figure 2. OTQ can combine with DNA to form the OTA-DNA adduct. Also, OTQ can become OTHQ in the present of reducing agents. In OTA oxidation, OTQ was the intermediacy [147-149]. Further, ochratoxin A-glutathione conjugate (OTA-GSH) was also found, which can be produced in the reaction in which rat liver microsomes are incubated with 100 μM OTA in the presence of 5 mM GSH for 1 h [150]. The second pathway is that conducted in the presence of an H-donor in which photo-irradiation of OTA generates the nonchlorinated OTB derivative (included the interaction of the phenoxyl radical with DNA to form the OTA-DNA adduct). The last pathway involves the photoreactivity of OTA. Photoreactivity measurements revealed that C-Cl bond hemolysis could be induced by the interaction of OTA with a solvated electron. This action can produce an OTA carbon-centered radical and Cl- [151, 152]; an H-donor can combine with the carbon-centered radical to form the nonchlorinated OTB derivative [152]. Interaction of the carbon-centered radical with DNA can form the OTA-DNA adduct.

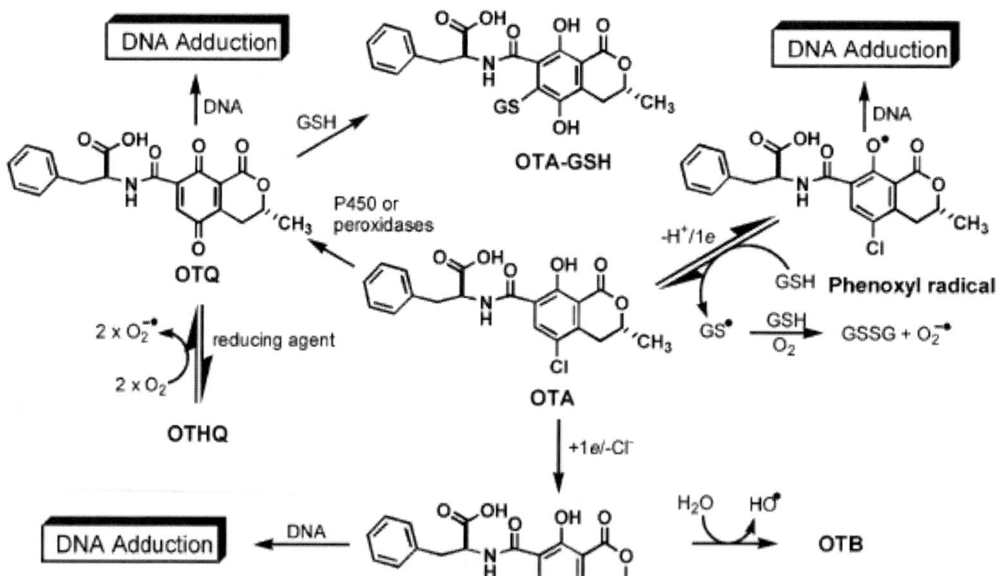

Figure 2. Proposed Pathways for the bioactivation of OTA.[146].

Pfohl-Leszkowicz et al. [126] treated mice with OTA at 0.6, 1.2, and 2.5 mg/kg body weight. After treating for 24, 48, and 72 h, the kidney, the liver and the spleen were excised.

OTA-DNA adducts were detected in all three tissues. The results that suggested there existed various amounts of adducts in different tissues. After 72 h in the high-dose group, in the kidney, 40 adducts/109 nucleotides were found; in the liver, 7 adducts/109 nucleotides were found. In 1993, another study was explored by Pfohl-Leszkowicz et al. [143]. Swiss male mice weighing 25 + 2 g and aged 7 weeks were treated with OTA at 2.5 mg/kg body weight. They were sacrificed after 8, 24, 48, 72, 120, 192 or 384 h. DNA adducts levels in kidney, liver and spleen were detected by the nuclease P1 enrichment version of the ^{32}P-postlabeling method. The total amount of DNA adducts in the three tissues was related to time under experimentation. At 24 h, 40.6, 19.4 and 2.1 adducts per 10^9 nucleotides were detected in the kidney, liver and spleen, respectively; at 48 h, there were 103, 42 and 2.2 adducts per 10^9 nucleotides in the kidney, liver and spleen, respectively. This was the maximum. After being in the kidneys for 8 days and 16 days after OTA administration, 23.3 and 7.7 adducts per 10^9 nucleotides were still detected; in liver, they were repaired within 8 days, whereas in the spleen, within 72 h, all the adducts were repaired. During the study, 15 DNA adducts were detected in the liver. Eleven adducts are also observed in the kidney. Four of them were specific to liver DNA. At 24 h, 4 DNA adducts reached their highest level. At 48 h, 9 DNA adducts reached the maximum. All the others peaked at 72 h. Furthermore, the major adduct represents 26% of the total at 24 h. At 48 h, two adducts were the major ones. They represented 15% and 14% of the total, respectively. Gradually, DNA adducts disappeared after 72 h. By 192 h, all DNA adducts disappeared. This result also proved that the kidney was the first target organ of OTA, followed by the liver and finally the spleen. There are two explanations for the quantitative and qualitative differences of DNA adducts in the three organs. One explanation was that more efficient repair occurred in the liver and spleen after the DNA adducts appeared. Another was that adducts in the three organs came from different metabolism pathways of OTA. In addition, the mechanisms of toxicity were also different.

Petkova-Bocharova et al. [142] also explored the OTA-DNA adduct in the female mice. Fifteen female mice weighing 35–40 g, 65 days old, were divided into three groups. Groups I and II were given a single dose of 2 and 0.5 mg/kg b.w. OTA by gastric gavage on days 17 or 18 of gestation. Group III was the control group. At 48 h after OTA, two mothers were killed, and at 30 days after delivery, three mothers and both male and female offspring from each group were sacrificed. The kidney and liver of the mothers and fetuses were collected. The 32P-post-labeling method was also applied for detecting the DNA adduct. After the treatment of 2 and 0.5 mg/kg b.w. OTA, several characteristic DNA adducts were detected in the kidney and liver. Sixteen fold fewer adducts were formed in the livers and four times less in the kidneys from mothers treated with the lower dose and killed 48h after treatment. After 30 days after delivery, in the kidneys of mothers, there are 4 adducts in the high dose, and 2 adducts in the low dose; In the liver, 7 adducts were detected in the low dose, and 10 adducts were detected in the high dose. Also, 3 times lower was in the total adduct in the low dose than high dose. In the fetuses from the low-dose mother, there are 4 adducts in the kidney and 2 adducts in the liver. And 2 times lower was in the total adduct in the low dose than high dose. In offspring, fewer adducts were formed in liver and kidney in the low dose than the high dose. Furthermore, in the female offspring, only 8 adducts were detected in the liver, whereas in the male offspring, 2 adducts and 4 adducts were observed in livers and kidneys, respectively. Based on these results, DNA adduct can be formed after OTA treatment. Moreover, DNA adduct in kidneys from male offspring was more than from female offspring and their mothers. In contrast, more DNA adduct were found in livers from female offspring. These results indicated that the pattern of DNA adducts was related to sex and age in the kidney and liver of the mouse progeny. Furthermore, the formation of DNA-adduct in fetuses is due to transplacental penetration of the parent compound or reactive metabolites.

However, another study considered whether OTA-DNA adducts couldn't be formed. Mally et al. [153] explored the role of DNA binding of OTA and whether derivatives of OTA can be formed *in vivo* and *in vitro*. *In vitro*, horseradish peroxidase can active OTA and OTB, its derivative, to yield OTHQ. However, using LC-MS/MS, no postulated OTA−dG adduct was detected. Also consistent with this finding was that OTA-related DNA adducts were not observed by ^{32}P-postlabeling. *In vivo*, male F344 rats were administered OTA (2 mg/kg b.w.) for 2 weeks, and OTHQ was detected in the urine. OTA−dG was not detected by LC-MS/MS in the liver and kidney DNA. Moreover, male rats were fed a single dose of ^{14}C-OTA or ^{14}C-OTB and analyzed using accelerator mass spectrometry. The ^{14}C content in the liver and kidney was not significantly different from controls.

In 2005, male F344 rats were repeatedly administered with OTA (0, 250, 500, 1000, and 2000 μg/kg b.w.) or the non-chlorinated analogue OTB (2000 μg/kg of b.w.) for 2 weeks (5 days/week). The comet assay was used to detect the break down DNA breakage. However, there is a lack of evidence for the formation of lipophilic DNA adducts [89].

Until now, there has not been an exact conclusion regarding the existence of OTA-DNA adducts in the liver. Further studies on the DNA adduct mechanism of OTA are needed.

4.1.3. DNA Damage

In 1985, male Balb/c mice were injected i.p. with 2.5 mg/kg OTA. They were killed at 6, 12, 24, 48 or 72 h after injection. DNA single-strand breaks appeared in the liver after 24 h-48 h, and the breaks were repaired afterwards [154]. In 1986, male Wistar rats were fed with OTA (288.8 μg/kg, corresponding to an intake of 4 ppm in feed each 48 h) for 12 weeks. DNA single-strand breaks were detected [155]. The formation of DNA single-strand breaks may correspond either to the excision repair of the DNA-adduct by an endonuclease [156] or to other kinds of lesions, such as the induction of alkali-labile sites. The DNA single-strand or double-strand breaks have a relationship with the mutation [157]. Therefore, DNA damage may be the reason for OTA carcinogenesis.

4.2. OTA and RNA

4.2.1. OTA Influenced the Expression of mRNA in Livers

Mycotoxins, such as OTA and AFB1, can induce changes in genes, and the transcriptome has been used in studying the mode of action of AFB1 [158]. However, the transcriptome after OTA treatment has not been fully utilized. The organ-specificity induced by OTA can be simultaneously reflected in the expression of genes; thus, study of the transcriptome after OTA treatment is necessary.

In 2012, the gene expression profiles after non-genotoxic carcinogens, genotoxic carcinogens and non-carcinogens treatment included rat liver and the primary rat hepatocytes. Gene expression was compared in an *in vivo* and *in vitro* model. The results revealed that the major discriminating cellular pathways found *in vivo* in livers were not apparent *in vitro*

[159]. Thus, the entire analysis of mRNA can more useful for the exploration of the mode of action of these toxicogenomics.

At present, researchers have studied the comprehensive and unbiased *in vivo* and *in vitro* transcriptional profile induced by OTA. In 2008, Hundhausen et al. [113] applied a microarray to detect HepG2 liver cell transcriptome. HepG2 cells were treated with OTA at a dose of 0, 0.25, 2.5 μmol/L for 24 h. All the mRNA levels were analyzed. Moreover, the genes were enriched. The relevant pathways were predicted. The results revealed that OTA can decrease the mRNA level of genes that encode for key proteins of liver cell metabolism. Previous mRNA sequencing was conducted on the livers of rat administered OTA. By the Illumina/Solexa digital gene expression (DGE) system, a comprehensive and unbiased transcriptional profile during the early phase toxicity of OTA in HepG2 cells was performed. Using Illumina RNA-seq technology, Zhang et al. compared the accumulation of transcription products in HepG2 cells treated with 12.5 μM OTA for 4 hours [160]. Using the $P < 0.05$ and a fold-change ≥ 2, 2330 genes were identified as differentially expressed between the control group and the OTA treatment group. Among these genes, 1790 genes were up-regulated and 530 genes were down-regulated. Most up-regulated genes exhibited a 2-5-fold level change. Regarding the function of the genes, three categories, including biological pathways (BP), molecular function (MF) and cellular component (CC) function, were devised. On the DAVID website, the up-regulated genes could be identified for more clusters than down-regulated genes after OTA exposure. This finding indicated that the up-regulated genes may influence more cellular functions than the genes down-regulated by OTA. However, OTA-induced genes activate transcription factors and repressors. In molecular biology and genetics, transcription factors (sequence-specific DNA-binding factors) include a protein that can bind to specific DNA sequences. In this way, transcription factors can control the transcription of genetic information from DNA to mRNA [161, 162]. To perform this function, transcription factors form a complex with the DNA either alone or with other proteins. This complex promotes or blocks the recruitment of RNA polymerase, which performs the transcription of genetic information from DNA to RNA [163]. OTA induced changes in dozens of transcription factors in HepG2 cells, and these transcription factors can be classified into four groups: zinc finger proteins, Really Interesting New Gene (ring) finger proteins, leucine zipper transcription factors, and other transcription factors [160]. The roles of zinc finger structures include DNA recognition, RNA packaging, protein folding and assembly, transcriptional activation, lipid binding, and apoptosis regulation.

Zhang et al. [160] also found that all of the ring finger proteins were decreased after OTA treatment. A ring finger domain is a protein structural domain that contains a Cys3HisCys4 amino acid motif that binds two zinc cations [164, 165]. In the down-regulated group of genes, the zinc finger proteins were involved in some processes, including transcription repression, ion binding, nucleic acid binding, and transcription factor activity. However, in the up-regulated group, zinc fingers were only involved in ion binding. Among the zinc fingers, the most regulated zinc exporter is SLC30A1, and its expression was decreased by 5.4-fold after OTA treatment. The protein is located in the plasma membrane. This result is consistent with the result of Overbeck et al. which found that cellular zinc depletion can induce the down-regulation of SLC30A1 [166]. Moreover, the major endogenous zinc uptake transporters, SLC39A1, SLC39A10 and SLC39A6, were up-regulated after OTA treatment [167].

Another noteworthy issue is heterogeneous nuclear ribonucleoproteins (hnRNPs). The hnRNPs play an important role in RNA metabolism [168]. HnRNPs are composed of the ribonucleoprotein particle, which consists of mRNA precursors and the specific protein. HnRNPs exist in the cell nucleus. These proteins are associated with pre-mRNAs in the nucleus and appear to influence pre-mRNA processing and other aspects of mRNA metabolism and transport. HnRNPs play a role by binding to nascent RNA polymerase II transcripts. And the functions of hnRNPs include 1) to prevent the formation of pre-mRNA into secondary structures that may inhibit its interactions with other proteins, 2) to associate with the splicing apparatus, and 3) to transport mRNA out of the nucleus.

Zhang et al. [160] also determined that after OTA exposure, 11 genes, belonging to the group of hnRNPs, are up-regulated. These genes include *HNRNPH1*, *HNRNPA1*, *HNRNPK*, *HNRNPA2B1*, *HNRNPH2*, *RALY*, *HNRNPL*, *HNRNPC*, *HNRNPH3*, *HNRNPA3* and *HNRNPM*. Among these genes, *HNRNPL* interacted with *APEX* (nuclease multifunctional DNA repair enzyme 1), and after exposure to OTA, *APEX1* expression increased 3.3-fold. This result suggested that more DNA damage occurred due to OTA. The *APEX1* gene encodes the enzyme DNA-(apurinic or apyrimidinic site) lyase. DNA molecules are treated by spontaneous hydrolysis, by DNA glycosylases or by DNA damaging agents that remove specific abnormal bases and can induce the formation of apurinic-apyrimidinic (AP) sites (also called "abasic sites"). Another hnRNP is *HNRNPH3*, which participates in the splicing process and in the arrest of early heat shock-induced splicing. Additionally, *AHSA1*, *HSPA5*, *HSP90B1*, *HSPB8*, *HSPB7*, *HSPA9*, *HSPD1* and *HSP90AB1* are all up-regulated by 2- to 6.5-fold. Moreover, *HSBP1L1*, a heat shock factor, was also up-regulated by 7-fold. Heat shock proteins (HSPs) comprise a family proteins that perform a chaperone function by stabilizing new proteins to ensure correct folding or by helping to refold proteins that were damaged by the cell stress [169]. Thus, the expression of HSPs is increased when cells are exposed to elevated temperatures or other stressors [169]. In the heat shock response, these heat shock proteins are dramatically up-regulated. In addition, the up-regulation can be induced primarily by heat shock factor (HSF) [170]. In 2004, Barisic et al. used Western blot to analyze the level of Hsp70 expression in rat kidneys, LLC-PK1 cells and MDCK cells exposed to OTA [171]. The result was that OTA can affect the relative distribution of the Hsp70 isoforms, but the total amount of Hsp70 was not changed in rat kidneys [171]. Additionally, after OTA treatment, polypyrimidine tract binding protein 1 (PTBP1), an RNA-binding protein also known as PTB or hnRNP I, was up-regulated. The main role of PTBP1 is a multifunctional regulator. Moreover, it participates in the control of alternative processing and translation of various RNAs, such as alternative 3' end processing, mRNA stability and RNA localization [172].

The transcription process needs energy, and mitochondria provide it. The mitochondria play the key role in the cell's movement, division, and production of secretory products. As early as 1991, Aleo et al. determined that mitochondrial dysfunction is an early event during the development of OTA toxicity [173]. Further, in 2015, Zhang et al. [160] found that nearly one hundred mitochondrial genes have differential expression levels after OTA was used to treat HepG2 cells. These genes can be divided into two groups: 1) those involved in ATPase H^+ and ATPase Na^+/K^+ transportation and 2) those involved with mitochondria and ribosomes. ATPase-dependent organelle acidification is necessary in such intracellular processes as protein sorting, zymogen activation, receptor-mediated endocytosis, and synaptic vesicle proton gradient generation.

Based on these conclusions, the transcriptomic studies can provide some evidence regarding the mechanism of OTA, but it still unclear. The many hypotheses of the different mechanisms of OTA should be verified. Moreover, dose- and time- points should also be considered. The *in vivo* and *in vitro* study of mRNA can provide direct evidence for gene changes induced by OTA, which will be useful for exploring the mode of action of OTA and other mycotoxins.

4.2.2. miRNA

MicroRNAs (miRNAs) are a type of endogenous, conserved, non-coding and single-stranded small RNA. miRNAs are 20 to 25 nucleotides derived from 70- to 100- base pair hairpin-shaped precursors. miRNAs play an important role in stress resistance and metabolism in animals and plants. The principal role of miRNAs is to negatively regulate gene expression by either translational inhibition or exonucleolytic messenger RNA (mRNA) decay. miRNAs are also crucial regulators of gene expression in a wide range of processes via the post-transcriptional regulation of mRNA translation and stability, including the induction or maintenance of the cell fate in normal, stem and cancerous cells. Based on these functions, miRNAs have been extensively studied as biomarkers for the diagnosis and therapy in a diverse range of diseases, as well as for their impact on gene regulation. The miRNA expression profiling and the special miRNA in tissues have been evaluated and can identify the over trends of mRNA changes in regulating a wide range of biological processes. Therefore, miRNAs have been considered as the new avenue to understanding the mechanisms of mycotoxins, such as AFB1 and OTA.

Qi et al. [58] studied the biological pathways in early OTA hepatotoxicity by using miRNA, mRNA and proteomic profiling. Male F344 rats were treated with OTA at 0, 70 or 210 μg/kg b.w. for 2, 4, 13 and 26 weeks (marked as a, b, c and d). A total of 304, 301 and 304 known mature miRNAs were identified in 0, 70 or 210 μg/kg b.w. groups, respectively. Of these miRNAs, 272 mature miRNAs were expressed in all the three groups. Among these, 47 up- and 26 down-regulated miRNAs were found only in the 210 μg/kg b.w. group ($p < 0.05$) compared to the control livers. In addition, 59 up- and 47 down-regulated miRNAs were found only in the 70 μg/kg b.w. group. This result proved that different doses of OTA could induce different responses. Additionally, there were 29 up- and 14 down-regulated miRNAs in both the 210 μg/kg b.w. and 70 μg/kg b.w. groups, respectively. In addition, compared with the three groups that received OTA treatment, 8 up-regulated miRNAs and 4 down-regulated miRNAs were commonly detected respectively. The up-regulated miRNAs are *rno-let-7a-5p*, *let-7c-5p*, *let-7e-5p*, *let-7f-5p*, *miR-92a-3p*, *miR-126a-3p*, *miR-92b-3p*, and *miR-3596c*; the down-regulated miRNAs are *rno-miR-19a-3p*, *miR-19b-3p*, *miR-29b-3p*, and *miR-532-5p*. Compared to the control group using KEGG pathway enrichment, there are 136 and 145 pathways in the high- and medial-dose groups, respectively. Among these, a total of 133 were included as overlapping pathways, and "pathways in cancer," "MAPK signaling pathway," and "metabolic pathways" were significantly enriched in the OTA-treated groups.

Qi et al. [58] also explored the changes in mRNA profiling, the target genes of miRNAs, in the livers in the study. Compared with the control group, there were 28 up-regulated and 17 down-regulated genes in the 210 μg/kg bw group; and there was 6 up-regulated and 3 down-regulated genes in the 70 μg/kg b.w. group. Only one gene was differentially expressed in the high-dose compared to the 70 μg/kg b.w. group. Signaling pathways also appeared to be enriched by KEGG. The result was that "amino acid metabolism," "xenobiotics biodegradation and metabolism," "energy metabolism," and "environmental information processing" were enriched.

Moreover, proteomics were detected in the 210 μg/kg bw group. A total 61 proteins were significantly different between the 210 μg/kg bw and control groups. By the pathway enrichment, there are five pathways. These pathways include cysteine and methionine metabolism, PPAR signaling, primary bile acid biosynthesis, arginine and proline metabolism, and metabolism of xenobiotics by cytochrome P450. The changes in the proteins in the five pathways were detected.

In vitro, Zhao et al. (Human and Experimental Toxicology, Manuscript ID HET-15-0443.R4, Accept) applied HEK293 and HepG2 cells as the model to study OTA-induced cell toxicity via miRNA profiling. In this study, HEK293 and HepG2 cells were treated with 25 μM OTA for 24 h. High-throughput miRNA profiling and KEGG analysis were conducted. By high-throughput miRNA profiling, in the HEK293 cell line, the rRNA levels slightly decreased after OTA treatment, while the miRNA and tRNA levels slightly increased. However, HepG2 exhibited different changes. The rRNA and miRNA levels decreased after exposure to OTA; alterations in HepG2 were much more obvious than those in HEK293. In addition, the tRNA level rose significantly in both the GT and KT groups. By the RT-PCR assay, the mRNA of *Drosha*, *Dicer1* and *DGCR8*, which were involved in miRNA processing, were down-regulated in both the HEK293 and HepG2 cell lines. Therefore, the miRNA biogenesis processes have been hampered by OTA. *Dicer1*, *Drosha* and *DGCR8* were all significantly down-regulated after the OTA-treated group in HEK293, while only *DGCR8* was significantly down-regulated in the OTA group in HepG2. Based on these findings, the key regulators of the miRNA maturation process in HEK293 were more sensitive to OTA than those of HepG2.

In the HEK293 cells, 17241302 and 18234755 reads were acquired in the control and OTA groups. In the HepG2 cells, 16937975 and 14682556 reads were obtained in the control and OTA groups, respectively. There are different criteria among these reads, such as the ability to form a stable hairpin structure, a sequence starting with a 5' uridine, lower minimal free energies (MFEs) for the hairpin structures of its precursors, and the detection of miRNAs [174]. The hairpin structures of novel miRNAs precursors was made by the RNAfold web server (http://rna.tbi.univie.ac.at/cgi-bin/RNAfold.cgi). Twelve novel miRNAs were identified: m0115-3p, m0006-5p, m0168-5p, m0104-5p, m0019-5p, m0044-5p, m0125-3p, m0147-3p, m0156-5p, m0186-3p, m0120-5p, and m0095-5p. Among these miRNAs, m104-5p, m0044-5p, m0125-3p, m0120-5p, and m0095-5p existed only in KT; m0147-3p and m0156-5p were observed in GC; and m0115-3p was only detected after OTA treatment. It appears that most of the novel miRNAs existed in HEK 293 or the OTA treatment group.

Moreover, in this study, 76 up-regulated and 28 down-regulated miRNAs in HEK293 were found, and 25 up-regulated and 40 down-regulated miRNAs in HepG2 were detected. The target genes were also predicted with MiRanda, miRDB, miRWalk, PITA, TargetScan and RNAhybrid. In HEK293, there were 310408 putative targets enriched in up-regulated miRNAs and 85513 targets in the down-regulated group, while in HepG2, 84110 putative targets were up-regulated, and 173692 putative targets were down-regulated. Using KEGG pathway analysis, there were 96 and 81 pathways enriched in the up-regulated and down-regulated group, respectively, in HEK293, while in HepG2, 89 pathways were enriched the up-regulated group and 90 pathways were enriched the down-regulated group. Eight of the top ten most enriched pathways were enriched in each group. In the two cell lines, "Pathways in cancer," "Endocytosis," "Axon guidance," "Glioma," "Prostate cancer," "mitogen-activated protein kinases (MAPK) signaling pathway", "Chronic myeloid leukemia," and the "Neurotrophin signaling pathway" were all found. Specifically, 40 up-regulated miRNAs in HEK293 and 25 down-regulated in HepG2 were related to the "MAPK signaling pathway". MAPK pathways are known to play important roles in cell survival, anti-apoptotic activity and renal cancer development.

Moreover, *hsa-miR-342-5p, hsa-miR-195-5p, hsa-miR-515-5p, hsa-miR-520g-3p* and *hsa-miR-524-5p* were found to regulate the greatest number of pathways; the numbers of pathways related to these five miRNAs are 52, 47, 46, 41 and 38, respectively. The regulation of these pathways may be strongly associated with these miRNAs, and they may play a significant role in OTA toxicity.

Furthermore, in the two cell lines, the common and different expression of miRNA was also explored. Nine miRNAs showed common expression trends after OTA treatment in the two cell lines (up-regulated: *hsa-miR-512-3p*, *hsa-miR-184*, *hsa-miR-1*, *hsa-miR-122-5p*; down-regulated: *hsa-miR-125b-5p*, *hsa-miR-92a-1-5p*, *hsa-miR-3529-3p*, hsa-miR-365a-5p, *hsa-miR-151a-5p*,). Therefore, there are thirty miRNAs demonstrating common expression trends. Based on the 9 commonly expressed miRNAs and 30 differentially expressed miRNAs, 32 pathways were identified. "Commonly expressed miRNAs" were related to signal transduction, and "differentially-expressed miRNAs" were more relevant to human cancer.

Compared with the *in vivo* results [175], the decrease of *Dicer1*, *Drosha*, *DGCR8* expression was in agreement with those found *in vitro*. In the KEGG analysis, the results in the *in vivo* model were similar to those found *in vitro*. Most pathways were associated with human cancer and signal transduction. However, signal transduction and genetic information processing are more common *in vivo* compared to *in vitro*. In the kidney, pathways related to cellular processes were more common. While in the liver, pathways related to organismal systems were more common.

However, the differentially expressed miRNAs are not alike *in vivo* and *in vitro*. Only *miR-92a-3p* and *miR-378b* exist in both hepatocytes and the liver or in kidney cells and the kidney. Viewed from the expression of modulated miRNAs, there was a weak relationship between the *in vivo* and *in vitro* models.

Moreover, miRNAs profiling can provide the chosen objects. In the liver, *miR-122* was the most specific miRNA. *miR-122* accounts for approximately 70% of all miRNAs in the liver. *miR-122* appears to play a key role in normal and diseased livers, including regulating hepatocyte development, differentiation, lipid metabolism, and stress response. In the mammalian liver, the most well-known function of *miR-122* is regulating lipid and cholesterol metabolism [176]. Nassirpour et al. [177] have reported that *miR-122* is involved in cancer cell metabolism, apoptosis, and transfer and can also be considered a tumor suppressor in the liver. There is sufficient evidence supporting that the aberrant expression of *miR-122* was found in many hepatic diseases [178-180].

Zhu et al. [181] focuses on studying the role of *miR-122* in ochratoxin A-induced hepatocyte apoptosis *in vivo* and *in vitro*. *In vivo*, rats were treated with 0, 70 or 210 μg/kg b.w. OTA for 4 and 13 weeks. The TUNEL results indicated that the hepatocyte apoptosis had a different relationship with the OTA-treatment doses at different time points. *In vitro*, HepG2 cells were treated with different concentrations of OTA for 24 h. Cell viability and cell apoptosis were detected by the CCK8 assay and flow cytometry, respectively. The percentage of viable cells was decreased, and cell apoptosis was increased with OTA treatment in a dose-dependent manner.

OTA first affected the expression of *miR-122. miR-122* was up-regulated after OTA treatment at 13 weeks *in vivo* and *in vitro*. CCNG1, a member of the cyclin family and a protein family that regulates the cell cycle, is the target gene of *miR-122* [182]. It was lower in the high-dose group at 13 weeks in rats and in HepG2 cells after OTA-treatment. *p53*, the downstream gene of *CCNG1*, was lower in the high-dose group *in vivo* and higher *in vitro* after OTA treatment. *CCNG1* and *p53* have a feedback relationship [183]. Thus, the *CCNG1/p53* pathway was activated by the change of *miR-122* induced by OTA. Furthermore, *Bcl-w* is one of anti-apoptotic genes, and the other is a target gene of *miR-122. Caspase-3* is a key protein that is involved in cell apoptosis and the downstream genes of *Bcl-w*. Zhu et al. [181] found that *Bcl-w* was significantly increased in the high-dose OTA group in rats and in the HepG2 cells. Also, *caspase-3* increased at 4 weeks and decreased at 13 weeks in the high-dose rats. *Caspase-3* significantly decreased in the treatment group compared with the control group in HepG2 cells. Therefore, the study determined that hepatocyte apoptosis induced by OTA may occur via regulation of the mitochondrial pathway because Burchard et al. [180] have stated that *miR-122* was a regulator of the mitochondrial metabolic gene network in hepatocellular carcinoma. *Bcl-w* and *caspase-3* are also involved in the mitochondrial pathway. There are studies reporting that OTA can induce the activities *caspase-3* and result in the apoptosis of neuronal cells accompanied by the disappearance of the mitochondrial membrane potential (MMP). After the disappearance of the MMP, cytochrome c is released into the cytosol and binds to Apaf-1, resulting in the recruitment of pro-caspases, which further activates the effector caspases [184]. However, in both HepG2 and HEK293 cells, *caspase-3* was down-regulated after OTA treatment. Caspase-3 was decreased in HEK293 cells after the OTA treatment. Shen et al. [185] supported that via inhibition of the expression of apoptotic genes; cells respond to OTA by induction of cell survival programs. The inhibitor-of-apoptosis (IAP) family of genes was evolutionarily conserved in regulating programmed cell death. The inhibition of caspase-3 may be affected by the inhibitor-of-apoptosis (IAP) family of genes, which regulate apoptosis, block the activity of procaspase-9 induced by cytochrome c, and further prevent the activity of *caspase-3*; however, IAPs cannot prevent the activity of *caspase-8* [186, 187]. In summary, OTA induced hepatocyte apoptosis via the *CCNG1/p53* pathway and *Bcl-w/caspase-3* pathway *in vivo* and *in vitro*, which was regulated by the change of *miR-122*. To further verify the conclusion, after *miR-122* was inhibited, the HepG2 cells were treated with OTA, and the changes in *CCNG1/P53* and *Bcl-w/caspase-3* were all relieved. Based on this study, *miR-122* played a key role in OTA-induced cell apoptosis.

Based on these factors, the roles of some miRNAs should be investigated in the mechanism of OTA-induced toxicity. miRNAs have been the new avenue in studies about the molecular mechanism of OTA. More comprehensive studies should be conducted.

4.3. Protein

OTA can enter the cell by transfer through membrane proteins. Further, it induces toxicity. Thus, cell membrane proteins are the first cellular barrier against xenobiotics. Further, the plasma membrane is a biological membrane, and it can separate the interior of cells from the outside environment [188]. The cell membrane also has important roles, such as selective permeation for ions and organic molecules and the control of substances moving into and out of cells [189]. The changes in membrane fluidity, integrity and composition due to cytotoxicity induced by chemicals must be considered [190]. The cell membrane has also been of concern in the study of mycotoxins. The toxic effect of OTA on the plasma membrane *in vivo* and *in vitro* has been shown [190]. Both the structure and function of the cell membrane are changed by OTA. Gekle et al. has reported that OTA could block fractional anion conductance of the plasma membrane both *in vivo* and *in vitro* [191]. Regarding the structure of the cell membrane, plasma membrane lipid rafts primarily consist of cholesterol, (glyco) sphingolipids and specific membrane proteins [192]. Lipid rafts play a vital role in mediating cytotoxicity. In drug-induced liver injury, the function of gap junctions is to alter the sphingolipid acyl chain length composition [190]. Moreover, lipid rafts also involved an ethanol-induced oxidative stress of the liver [190]. Thus, for exploring the mode of action of OTA, the role of cell membrane proteins also needs to be explored.

In 2015, Zhang's study [193] applied HepG2 cells as the *in vitro* model. The cells were incubated with OTA for 12, 24 or 36 h. Plasma membrane was isolated, and iTRAQ proteomics were used for analyzing the protein changes. The viability of HepG2 cells decreased in a time-dependent manner after OTA treatment at 10 or 100 μM. Previous studies demonstrated that MβCD is able to remove cholesterol from the plasma membrane. In this study, Zhang found that 5 mM MβCD treated HepG2 cells for 30 min did not cause any significant loss of cell viability. Compared to the cells without MβCD pretreatment, OTA significantly decreased the cell survival rate and significantly increased LDH leakage after MβCD pretreatment. This evidence proves that the loss of membrane integrity in HepG2 cells enhanced OTA-induced toxicity. MβCD also significantly affected the intracellular SOD and ROS levels. The membrane integrity protected the cells from OTA-induced oxidative damage. Thus, membrane integrity is important for the cell to resist OTA-induced toxicity. Based on *the proteome* results by iTRAQ in Zhang et al. [193], most proteins were down-regulated after OTA treatment for 12 and 24 h. However, cytochrome c oxidase subunit 7C and transhydrogenase were up-regulated. After OTA treatment for 36 h, 19 proteins were up-regulated and 13 proteins were down-regulated.

Through the protein interaction networks, three proteins were found at all the time points. These proteins are Q16850, P04004 and Q13423. Q16850 is cytochrome P450 (CYP), which belongs to the superfamily of hemoproteins and exists in all kinds of life, such as, animals, plants, fungi, protists, bacteria, archaea, and even in viruses. The role of CYPs is to catalyze the oxidation of organic substances. These proteins can metabolize thousands of endogenous and exogenous chemicals. In humans, cytochrome P450 enzymes are important in hormone synthesis and breakdown, vitamin D metabolism, and cholesterol synthesis. Cytochrome P450 enzymes also function to metabolize potentially toxic compounds, principally in the liver. P04004 is the vitronectin. Vitronectin is an abundant glycoprotein that can be found in serum, the extracellular matrix and bone. The functions of vitronectin are promoting cell adhesion and spreading, inhibiting the membrane-damaging effect of the terminal cytolytic complement pathway, and binding to several serpins. Vitronectin has been speculated to be involved in hemostasis and tumor malignancy. Q13423 is nicotinamide nucleotide transhydrogenase. It participates in transhydrogenation between NADH and NADP. Under most physiological conditions, the enzyme can generate high concentrations of NADPH using energy from the mitochondrial proton gradient. Then, NADPH is used for biosynthesis, and reactions inside the mitochondria are required to remove reactive oxygen species. Based on proteomics, there are 5, 3 and 2 pathways enriched by KEGG analysis compared to the control group in 12 h, 24 h and 36 h time points, respectively.

In 12 h, five pathways include "Parkinson's disease," "Alzheimer's disease," "oxidative phosphorylation," "Huntington's disease," and "Cardiac muscle contraction:" in 24 h, three pathways include "SNARE interactions in vesicular transport", "lysosome", and "cell adhesion molecules;" in 36 h, two pathways are "glycolysis/gluconeogenesis" and "pentose phosphate pathway." At the same time, GO classification also be applied. "Cell adhesion" appeared in the 24 h group. CD47, cxadr, Rhoa, and Alcam, which have a relationship with cell adhesion, were inhibited by OTA treatment. This result is consistent with enrichment in "cell adhesion molecules" by KEGG. It proves that OTA can affect the integrity and function of the cell membrane.

Of note, OTA can induce fast and permanent damage to the cell monolayer [194]. In summary, the impairment of the biochemical functions of cells can be induced by changes in the plasma membrane structure and alterations in the transport of enzymes and ions [195]. This effect on transport proteins may contribute to the development of hepatotoxicity induced by OTA. Therefore, the study of membrane proteins needs further exploration to clarify the mechanism of OTA-induced hepatotoxicity.

5. The Protective Mechanism of Hepatotoxicity

OTA-induced hepatotoxicity is apparent. OTA comprises a large variety of agricultural commodities. How can people and animal be protected from OTA-induced hepatotoxicity? In recent years, some researchers found that some materials (vitamin E and polyphenols, quercetin, Zn) can reverse OTA-induced hepatotoxicity. These studies were conducted in HepG2 cell. Many studies have revealed that OTA-induced hepatotoxicity is mainly due to oxidative stress. Thus, anti-oxidation is considered as a protective role to protect against hepatotoxicity induced by OTA.

Quercetin is a ubiquitous flavonoid (3, 3', 4', 5, 7-pentahydroxyflavone). Its antioxidant effect was through scavenging ROS, including O2.-; RNS, as well as NO and ONOO-. It also exerts other bioactive effects, such as, anti-inflammatory [69, 70] and antimutageneic effects [71]. In 2014, Ramyaa et al. [196] found that quercetin can remit OTA-induced hepatotoxicity. The result indicated that OTA induced calcium release and ROS generation, as well as activated NF-κB nuclear translocation and expression. With the pre-treatment with quercetin, the production of ROS, the release of calcium, and the induction and expression of NF-κB were ameliorated. Additionally, quercetin can induce Nrf-2 nuclear translocation and expression and down-regulate Cox-2. Moreover, quercetin also prevents DNA damage and micronucleus formation. In summary, quercetin plays a role in modulating OTA-induced oxidative stress and redox-signaling in HepG2 cells.

Zinc is an essential trace element, and it is also the second most abundant transition metal ion in living organisms. It binds to 10% of proteins in the mammalian proteome. It has been considered a potential antioxidant *in vitro*. Also, zinc is an essential component of Cu/Zn superoxide (SOD1). Through the antagonism of redox-active transition metals (Fe, Cu), zinc can protect protein sulfhydryl groups against oxidation and decrease the formation of hydroxyl radicals [197]. Moreover, through the activation of the metal response transcription factor-1 (MTF-1), zinc can regulate the expression of many genes, including metallothionein (MT), glutathione peroxidase (GPx) and glutamylcysteine synthetase (GCS), in the antioxidant process [198]. The protective role of zinc in OTA-induced toxicity has been explored. In 2012, Zheng et al. [36] reported that OTA can induce the transcriptional changes in ZnT (SLC30) and Zip (SLC39), to the members of zinc transporters. Liuzzi et al. [199] revealed that the functions of ZnT (SLC30) and Zip (SLC39) include an effort 1) to become involved in the specific transport of zinc across the biological membrane and 2) to control the intra- and intercellular translocation of zinc.

For the first time, Zheng et al. [36] proved that the addition of zinc can reduce the OTA toxicity of contaminated feeds by applying HepG2 cells as an *in vitro* model. HepG2 cells were incubated by 25 μM OTA and 50 μM $ZnSO_4$ for different times (2, 6, 12 and 24 h). The changes in the intracellular zinc concentration were detected by ICP-MS. The intracellular zinc concentration decreased in a time-dependent manner after OTA, whereas they increased after $ZnSO_4$. The addition of zinc significantly decreased the level of ROS and the activity of SOD induced by OTA, but the mRNA expression of Cu/Zn SOD (SOD1) in OTA-treated cells did not change. Thus, zinc plays an important role in the function of SOD but has no obvious effect on the transcription of SOD1. However, there was no effect on the mitochondrial membrane potential ($\Delta\psi m$). Moreover, the comet assay and HPLC/ESI/MS/MS experiments were performed. The result revealed the formation of 8-hydroxy-2'-deoxyguanosine (8-OHdG), the biomarker of DNA oxidative damage, and the decrease in the 5mdC level after OTA treatment. Based on this finding, zinc helped to maintain DNA damage through the reduction of OTA-induced DNA strand breaks and 8-OHdG formation. Additionally, zinc supplementation partially recovered the percentage of 55mdC to antagonize OTA-induced damage to the DNA stability. However, for zinc deficiency, N,N,N,N'-Tetrakis (2-pyridylmethyl) ethylenediamine (TPEN) was applied. TPEN is a membrane-permeable zinc chelator. After TPEN treatment, the cell viability was inhibited, and LDH release increased. There was a connection between zinc homeostasis and OTA toxicity. These actions all proved that zinc can protect the maintenance of DAN integrity because of its antioxidant function. Zheng et al. [36] also determined that zinc activated the MTF-1 and further activated the transcription of MT. Ranaldi et al. [7] also revealed that OTA exposure modulated the gene expression of zinc transporters and metallothioneins.

Moreover, Pérez et al. [200] revealed that zinc not only protects HepG2 cells against arachidonic acid-induced oxidative damage but also inhibits approximately 50% of the DNA fragmentation. In addition, zinc finger proteins or zincis are involved in DNA-methylation because of the presence of many zinc-associated enzymes, such as DNA methyltransferases and some histone lysine methyltransferases [201, 202]. In general, zine can reduce the OTA-induced cytotoxicity via its antioxidant properties and its involvement in redox signaling.

Until now, oxidative damage and DNA damage have been the main mechanisms of OTA-induced hepatotoxicity. Thus, antioxidants were considered protective agents. However, in 2005 [90] , it was found that incubation with α-tocopherol (α-TOC) and different polyphenols - catechin (CAT), daidzein (DAI), epicatechin (EC), epigallocatechin gallate (EGCG), genistein (GEN), and quercetin (QUE) did not counteract OTA-induced cytotoxicity. Therefore, the controversy regarding OTA-induced hepatotoxicity caused by oxidative damage still exists.

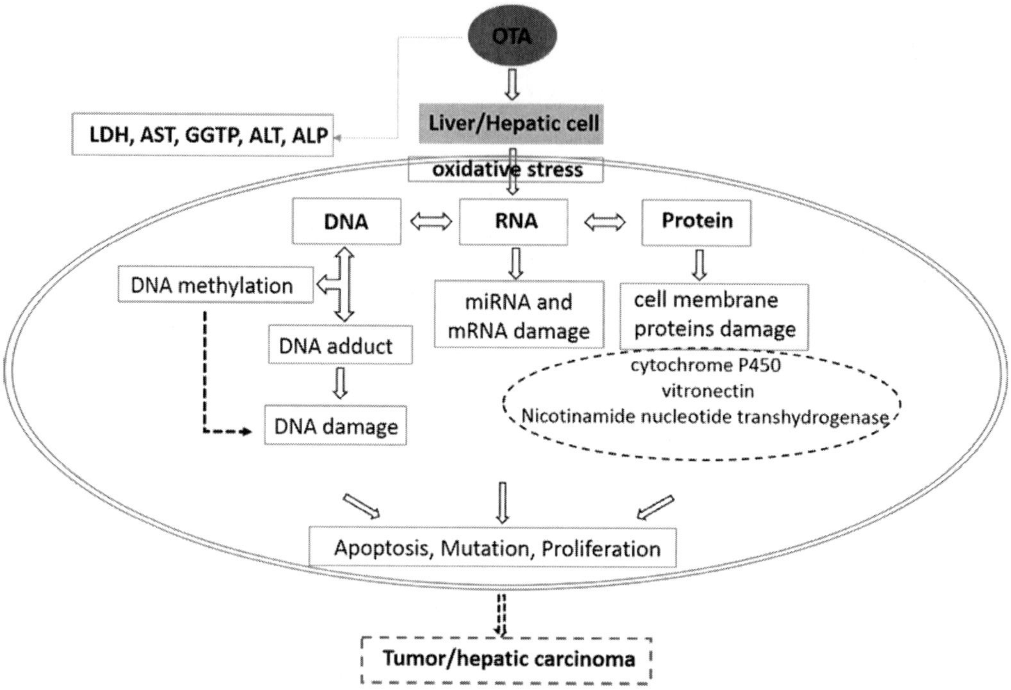

Figure 3. OTA-induced Hepatotoxicity.

There are still too many unknowns regarding OTA-induced hepatotoxicity. Therefore, further studies are needed (Figure 3).

REFERENCES

[1] James, B., Cardwell, K., Edorh, M., Hell, K., Hounsa, A. (2000), *Proceedings of the communications workshop, Lomé, Togo*, pp. 7-9.

[2] Bernal, W., Wendon, J., (2013). Acute liver failure. *N Engl J Med*. 369, 2525-2534.

[3] Schuppan, D., Afdhal, N.H., (2008). Liver cirrhosis. *The Lancet*. 371, 838-851.

[4] Pepeljnjak, S., Cvetnic, Z., (1985). The mycotoxicological chain and contamination of food by ochratoxin A in the nephropathic and non-nephropathic areas in Yugoslavia. *Mycopathologia*. 90, 147-153.

[5] Kibugu, J.K., Ngeranwa, J.J., Makumi, J.N., Gathumbi, J.K., Kagira, J.M., Mwangi, J.N., Muchiri, M.W., Mdachi, R.E., (2009). Aggravation of pathogenesis mediated by ochratoxin A in mice infected with Trypanosoma brucei rhodesiense. *Parasitology*. 136, 273-281.

[6] Joo, Y.D., Kang, C.W., An, B.K., Ahn, J.S., Borutova, R., (2013). Effects of Ochratoxin A And Preventive Action of a Mycotoxin-Deactivation Product In Broiler Chickens. *Veterinarija ir Zootechnika*. 61.

[7] Al - Anati, L., Essid, E., Reinehr, R., Petzinger, E., (2009). Silibinin protects OTA - mediated TNF - α release from perfused rat livers and isolated rat Kupffer cells. *Molecular nutrition & food research*. 53, 460-466.

[8] Dortant, P., Peters-Volleberg, G., Van Loveren, H., Marquardt, R., Speijers, G., (2001). Age-related differences in the toxicity of ochratoxin A in female rats. *Food Chem. Toxicol.* 39, 55-65.

[9] El-Serag, H.B., Lau, M., Eschbach, K., Davila, J., Goodwin, J., (2007). Epidemiology of hepatocellular carcinoma in Hispanics in the United States. *Archives of internal medicine*. 167, 1983-1989.

[10] Wild, C., Turner, P., (2002). The toxicology of aflatoxins as a basis for public health decisions. *Mutagenesis*. 17, 471-481.

[11] El–Serag, H.B., Rudolph, K.L., (2007). Hepatocellular carcinoma: epidemiology and molecular carcinogenesis. *Gastroenterology*. 132, 2557-2576.

[12] Arkin, D., Bray, F., Ferlay, J., (2000). Estimaing the world cancer burden. *GLOBOCAN*. 2001, 153-156.

[13] Reddy, L., Bhoola, K., (2010). Ochratoxins—Food contaminants: Impact on human health. *Toxins*. 2, 771-779.

[14] Pfohl-Leszkowicz, A., Manderville, R.A., (2007). Ochratoxin A: An overview on toxicity and carcinogenicity in animals and humans. *Mol Nutr Food Res*. 51, 61-99.

[15] Boorman, G.A., McDonald, M.R., Imoto, S., Persing, R., (1992). Renal lesions induced by ochratoxin A exposure in the F344 rat. *Toxicol Pathol*. 20, 236-245.

[16] Kamp, H.G., Eisenbrand, G., Janzowski, C., Kiossev, J., Latendresse, J.R., Schlatter, J., Turesky, R.J., (2005). Ochratoxin A induces oxidative DNA damage in liver and kidney after oral dosing to rats. *Mol Nutr Food Res*. 49, 1160-1167.

[17] Gagliano, N., Donne, I.D., Torri, C., Migliori, M., Grizzi, F., Milzani, A., Filippi, C., Annoni, G., Colombo, P., Costa, F., Ceva-Grimaldi, G., Bertelli, A.A., Giovannini, L., Gioia, M., (2006). Early cytotoxic effects of ochratoxin A in rat liver: a morphological, biochemical and molecular study. *Toxicology*. 225, 214-224.

[18] Kanisawa, M., Suzuki, S., (1978). Induction of renal and hepatic tumors in mice by ochratoxin A, a mycotoxin. *Gan*. 69, 599-600.

[19] IARC, (1983). International Agency for Research on Cancer. Some food additives, feed additives and naturally occurring substances. IARC Monogr Eval Carcinog Risk Chern Man. 31, 191-206.

[20] Bendele, A.M., Carlton, W.W., Krogh, P., Lillehoj, E.B., (1985). Ochratoxin A carcinogenesis in the (C57BL/6J X C3H)F1 mouse. *J Natl Cancer Inst*. 75, 733-742.

[21] Ward, J.M., Goodman, D.G., Squire, R.A., Chu, K.C., Linhart, M.S., (1979). Neoplastic and nonneoplastic lesions in aging (C57BL/6N x C3H/HeN)F1 (B6C3F1) mice. *J Natl Cancer Inst*. 63, 849-854.

[22] Hennig, A., Fink-Gremmels, J., Leistner, L., (1991). Mutagenicity and effects of ochratoxin A on the frequency of sister chromatid exchange after metabolic activation. *IARC Sci Publ*, 255-260.

[23] Pfohl-Leszkowicz, A., Grosse, Y., Kane, A., Creppy, E.E., Dirheimer, G., (1993). Differential DNA adduct formation and disappearance in three mouse tissues after treatment with the mycotoxin ochratoxin A. *Mutat Res*. 289, 265-273.

[24] Pfohl-Leszkowicz, A., Grosse, Y., Castegnaro, M., Nicolov, I.G., Chernozemsky, I.N., Bartsch, H., Betbeder, A. M., Creppy, E.E., Dirheimer, G., (1993). Ochratoxin A-related DNA adducts in urinary tract tumours of Bulgarian subjects. *IARC Sci Publ*, 141-148.

[25] de Groene, E.M., Hassing, I.G., Blom, M.J., Seinen, W., Fink-Gremmels, J., Horbach, G.J., (1996). Development of human cytochrome P450-expressing cell lines: application in mutagenicity testing of ochratoxin A. *Cancer Res.* 56, 299-304.

[26] de Groene, E.M., Jahn, A., Horbach, G.J., Fink-Gremmels, J., (1996). Mutagenicity and genotoxicity of the mycotoxin ochratoxin A. *Environ Toxicol Pharmacol.* 1, 21-26.

[27] Fink-Gremmels, J., Jahn, A., Blom, M.J., (1995). Toxicity and metabolism of ochratoxin A. *Natural toxins.* 3, 214-220; discussion 221.

[28] Fink-Gremmels, J., Blom, M., Woutersen van Nijnanten, F., Scudamore, K. (1993), *Proceedings of UK Workshop Occurrence and Significance of Mycotoxins, London*, pp. 124-127.

[29] Jeswal, P., (1998). Antidotal effect of grape juice (Vitis vinifera) on ochratoxin A caused hepatorenal carcinogenesis in mice (Mus musculus). *Cytobios.* 93, 123-128.

[30] Corcuera, L.A., Vettorazzi, A., Arbillaga, L., Gonzalez-Penas, E., Lopez de Cerain, A., (2012). An approach to the toxicity and toxicokinetics of aflatoxin B1 and ochratoxin A after simultaneous oral administration to fasted F344 rats. *Food Chem Toxicol.* 50, 3440-3446.

[31] Vettorazzi, A., de Troconiz, I.F., Gonzalez-Penas, E., Arbillaga, L., Corcuera, L.A., Gil, A.G., de Cerain, A.L., (2011). Kidney and liver distribution of ochratoxin A in male and female F344 rats. *Food Chem Toxicol.* 49, 1935-1942.

[32] Vettorazzi, A., Troconiz, I.F., Gonzalez-Penas, E., Corcuera, L.A., Arbillaga, L., Gil, A.G., Nagy, J.M., Mantle, P.G., de Cerain, A.L., (2010). Effects of fasting and gender on ochratoxin A toxicokinetics in F344 rats. *Food Chem Toxicol.* 48, 3159-3166.

[33] Van der Merwe, K., Steyn, P., Fourie, L., Scott, D.B., Theron, J., (1965). Ochratoxin A, a toxic metabolite produced by Aspergillus ochraceus Wilh.

[34] Hagelberg, S., Hult, K., Fuchs, R., (1989). Toxicokinetics of ochratoxin A in several species and its plasma - binding properties. *J. Appl. Toxicol.* 9, 91-96.

[35] Krogh, P., (1972). Mycotoxic porcine nephropathy: a possible model for Balkan endemic nephropathy. *Endemic nephropathy*, 266-277.

[36] Zheng, J., Zhang, Y., Xu, W., Luo, Y., Hao, J., Shen, X.L., Yang, X., Li, X., Huang, K., (2013). Zinc protects HepG2 cells against the oxidative damage and DNA damage induced by ochratoxin A. *Toxicol. Appl. Pharmacol.* 268, 123-131.

[37] Ehrlich, V., Darroudi, F., Uhl, M., Steinkellner, H., Gann, M., Majer, B., Eisenbauer, M., Knasmüller, S., (2002). Genotoxic effects of ochratoxin A in human-derived hepatoma (HepG2) cells. *Food Chem. Toxicol.* 40, 1085-1090.

[38] Theron, J., Van der Merwe, K., Liebenberg, N., Joubert, H., Nel, W., (1966). Acute liver injury in ducklings and rats as a result of ochratoxin poisoning. *The Journal of pathology and bacteriology.* 91, 521-529.

[39] Aydın, G., Ōzçelik, N., Cicek, E., Soyöz, M., (2003). Histopathologic changes in liver and renal tissues induced by Ochratoxin A and melatonin in rats. *Hum. Exp. Toxicol.* 22, 383-391.

[40] Abdel-Wahhab, M., Nada, S., Arbid, M., (1999). Ochratoxicosis: prevention of developmental toxicity by L-methionine in rats. *J. Appl. Toxicol.* 19, 7-12.

[41] Gagliano, N., Dalle Donne, I., Torri, C., Migliori, M., Grizzi, F., Milzani, A., Filippi, C., Annoni, G., Colombo, P., Costa, F., (2006). Early cytotoxic effects of ochratoxin A

in rat liver: a morphological, biochemical and molecular study. *Toxicology*. 225, 214-224.

[42] Palabiyik, S., Erkekoglu, P., Zeybek, N., Kızılgun, M., Sahin, G., Giray, B.K., (2012). Ochratoxin A causes oxidative stress and cell death in rat liver. *World Mycotoxin Journal*. 5, 377-384.

[43] Kamp, H.G., Eisenbrand, G., Janzowski, C., Kiossev, J., Latendresse, J.R., Schlatter, J., Turesky, R.J., (2005). Ochratoxin A induces oxidative DNA damage in liver and kidney after oral dosing to rats. *Mol. Nutr. Food Res.* 49, 1160-1167.

[44] Rached, E., Hard, G.C., Blumbach, K., Weber, K., Draheim, R., Lutz, W.K., Özden, S., Steger, U., Dekant, W., Mally, A., (2007). Ochratoxin A: 13-week oral toxicity and cell proliferation in male F344/N rats. *Toxicol. Sci.* 97, 288-298.

[45] Purchase, I., Theron, J., (1968). The acute toxicity of ochratoxin A to rats. *Food Cosmet. Toxicol.* 6, 479-483.

[46] Suzuki, S., SATOH, T., (1973). Effects of ochratoxin A on tissue glycogen levels in rats. *The Japanese Journal of Pharmacology*. 23, 415-419.

[47] Suzuki, S., Satoh, T., Yamazaki, M., (1975). Effect of ochratoxin A on carbohydrate metabolism in rat liver. *Toxicol. Appl. Pharmacol.* 32, 116-122.

[48] Szczech, G., Carlton, W., Tuite, J., Caldwell, R., (1973). Ochratoxin A toxicosis in swine. *Veterinary Pathology Online*. 10, 347-364.

[49] Bozzo, G., Ceci, E., Bonerba, E., Di Pinto, A., Tantillo, G., De Giglio, E., (2012). Occurrence of Ochratoxin A in the Wild Boar (Sus scrofa): Chemical and Histological Analysis. *Toxins (Basel)*. 4, 1440-1450.

[50] Huff, W.E., Wyatt, R., Tucker, T., Hamilton, P., (1974). Ochratoxicosis in the broiler chicken. *Poult. Sci.* 53, 1585-1591.

[51] Warren, M.F., Hamilton, P., (1980). Inhibition of the glycogen phosphorylase system during ochratoxicosis in chickens. *Appl. Environ. Microbiol.* 40, 523-525.

[52] WARREN, M.F., HAMILTON, P.B., (1981). Glycogen storage disease type X caused by ochratoxin A in broiler chickens. *Poult. Sci.* 60, 120-123.

[53] Guo, Y., Niu, Zh., Sun, H., (1992). The observation of microstructure in liver of poison chicken with Ochratoxin A. *J. South China Agr. Univ. (In Chinese)*. 1, 8-10.

[54] Raju, M.V., Devegowda, G., (2000). Influence of esterified-glucomannan on performance and organ morphology, serum biochemistry and haematology in broilers exposed to individual and combined mycotoxicosis (aflatoxin, ochratoxin and T-2 toxin). *Br Poult Sci*. 41, 640-650.

[55] Ahmad, M., Saleemi, M.K., Khan, M.Z., Muhammad, F., Hassan, Z., Khatoon, A., Bhatti, S.A., Abbas, R.Z., Rizvi, F., Ahmed, I., (2012). Effects of ochratoxin A feeding in white leghorn cockerels on hematological and serum biochemical parameters and its amelioration with silymarin and vitamin E. *Pak Vet J*. 32, 520-524.

[56] Dortant, P.M., Peters-Volleberg, G.W., Van Loveren, H., Marquardt, R.R., Speijers, G.J., (2001). Age-related differences in the toxicity of ochratoxin A in female rats. *Food Chem Toxicol*. 39, 55-65.

[57] Rached, E., Hard, G.C., Blumbach, K., Weber, K., Draheim, R., Lutz, W.K., Ozden, S., Steger, U., Dekant, W., Mally, A., (2007). Ochratoxin A: 13-week oral toxicity and cell proliferation in male F344/n rats. *Toxicol Sci*. 97, 288-298.

[58] Qi, X., Yang, X., Chen, S., He, X., Dweep, H., Guo, M., Cheng, W.-H., Xu, W., Luo, Y., Gretz, N., (2014). Ochratoxin A induced early hepatotoxicity: new mechanistic insights from microRNA, mRNA and proteomic profiling studies. *Sci. Rep.*

[59] Abdel-Wahhab, M.A., Abdel-Galil, M.M., El-Lithey, M., (2005). Melatonin counteracts oxidative stress in rats fed an ochratoxin A contaminated diet. *Journal of pineal research*. 38, 130-135.

[60] Meki, A.R., Hussein, A.A., (2001). Melatonin reduces oxidative stress induced by ochratoxin A in rat liver and kidney. *Comp Biochem Physiol C Toxicol Pharmacol*. 130, 305-313.

[61] Gautier, J.C., Holzhaeuser, D., Markovic, J., Gremaud, E., Schilter, B., Turesky, R.J., (2001). Oxidative damage and stress response from ochratoxin a exposure in rats. *Free Radic Biol Med*. 30, 1089-1098.

[62] Marin-Kuan, M., Ehrlich, V., Delatour, T., Cavin, C., Schilter, B., (2011). Evidence for a role of oxidative stress in the carcinogenicity of ochratoxin a. *Journal of toxicology*. 2011, 645361.

[63] Ferrante, M., Raso, G.M., Bilancione, M., Esposito, E., Iacono, A., Meli, R., (2008). Differential modification of inflammatory enzymes in J774A. 1 macrophages by ochratoxin A alone or in combination with lipopolysaccharide. *Toxicol. Lett.* 181, 40-46.

[64] Boesch - Saadatmandi, C., Wagner, A., Graeser, A., Hundhausen, C., Wolffram, S., Rimbach, G., (2009). Ochratoxin A impairs Nrf2 - dependent gene expression in porcine kidney tubulus cells. *J. Anim. Physiol. Anim. Nutr. (Berl.)*. 93, 547-554.

[65] Boesch-Saadatmandi, C., Loboda, A., Jozkowicz, A., Huebbe, P., Blank, R., Wolffram, S., Dulak, J., Rimbach, G., (2008). Effect of ochratoxin A on redox-regulated transcription factors, antioxidant enzymes and glutathione-S-transferase in cultured kidney tubulus cells. *Food Chem. Toxicol.* 46, 2665-2671.

[66] Abdel-Wahhab, M.A., Abdel-Azim, S.H., El-Nekeety, A.A., (2008). Inula crithmoides extract protects against ochratoxin A-induced oxidative stress, clastogenic and mutagenic alterations in male rats. *Toxicon*. 52, 566-573.

[67] Arbillaga, L., Azqueta, A., van Delft, J.H., de Cerain, A.L., (2007). *In vitro* gene expression data supporting a DNA non-reactive genotoxic mechanism for ochratoxin A. *Toxicol. Appl. Pharmacol.* 220, 216-224.

[68] Hassen, W., Ayed-Boussema, I., Bouslimi, A., Bacha, H., (2007). Heat shock proteins (Hsp 70) response is not systematic to cell stress: case of the mycotoxin ochratoxin A. *Toxicology*. 242, 63-70.

[69] Domijan, A.M., Peraica, M., Vrdoljak, A.L., Radić, B., Žlender, V., Fuchs, R., (2007). The involvement of oxidative stress in ochratoxin A and fumonisin B1 toxicity in rats. *Mol. Nutr. Food Res.* 51, 1147-1151.

[70] Palma, N., Cinelli, S., Sapora, O., Wilson, S.H., Dogliotti, E., (2007). Ochratoxin A-induced mutagenesis in mammalian cells is consistent with the production of oxidative stress. *Chem. Res. Toxicol.* 20, 1031-1037.

[71] Giacomo, C.D., Acquaviva, R., Piva, A., Sorrenti, V., Vanella, L., Piva, G., Casadei, G., Fauci, L.L., Ritieni, A., Bognanno, M., (2007). Protective effect of cyanidin 3-O-β-d-glucoside on ochratoxin A-mediated damage in the rat. *Br. J. Nutr.* 98, 937-943.

[72] Costa, S., Utan, A., Cervellati, R., Speroni, E., Guerra, M., (2007). Catechins: natural free-radical scavengers against ochratoxin A-induced cell damage in a pig kidney cell line (LLC-PK1). *Food Chem. Toxicol.* 45, 1910-1917.

[73] Lambert, D., Padfield, P.J., McLaughlin, J., Cannell, S., O'Neill, C.A., (2007). Ochratoxin A displaces claudins from detergent resistant membrane microdomains. *Biochem. Biophys. Res. Commun.* 358, 632-636.

[74] Sava, V., Velasquez, A., Song, S., Sanchez-Ramos, J., (2007). Adult hippocampal neural stem/progenitor cells *in vitro* are vulnerable to the mycotoxin ochratoxin-A. *Toxicol. Sci.* 98, 187-197.

[75] Sutken, E., Aral, E., Ozdemir, F., Uslu, S., Alatas, O., Colak, O., (2007). Protective role of melatonin and coenzyme Q10 in ochratoxin A toxicity in rat liver and kidney. *Int. J. Toxicol.* 26, 81-87.

[76] Arbillaga, L., Azqueta, A., Ezpeleta, O., de Cerain, A.L., (2006). Oxidative DNA damage induced by ochratoxin A in the HK-2 human kidney cell line: evidence of the relationship with cytotoxicity. *Mutagenesis.* 22, 35-42.

[77] Stemmer, K., Ellinger-Ziegelbauer, H., Ahr, H.-J., Dietrich, D.R., (2007). Carcinogen-specific gene expression profiles in short-term treated Eker and wild-type rats indicative of pathways involved in renal tumorigenesis. *Cancer Res.* 67, 4052-4068.

[78] Sava, V., Reunova, O., Velasquez, A., Harbison, R., Sanchez-Ramos, J., (2006). Acute neurotoxic effects of the fungal metabolite ochratoxin-A. *Neurotoxicology.* 27, 82-92.

[79] Sava, V., Reunova, O., Velasquez, A., Sanchez-Ramos, J., (2006). Can low level exposure to ochratoxin-A cause parkinsonism? *J. Neurol. Sci.* 249, 68-75.

[80] Marin-Kuan, M., Nestler, S., Verguet, C., Bezencon, C., Piguet, D., Mansourian, R., Holzwarth, J., Grigorov, M., Delatour, T., Mantle, P., (2006). A toxicogenomics approach to identify new plausible epigenetic mechanisms of ochratoxin a carcinogenicity in rat. *Toxicol. Sci.* 89, 120-134.

[81] Ferrante, M.C., Bilancione, M., Raso, G.M., Esposito, E., Iacono, A., Zaccaroni, A., Meli, R., (2006). Expression of COX-2 and hsp72 in peritoneal macrophages after an acute ochratoxin A treatment in mice. *Life Sci.* 79, 1242-1247.

[82] Zeljezic, D., Domijan, A.-M., Peraica, M., (2006). DNA damage by ochratoxin A in rat kidney assessed by the alkaline comet assay. *Braz. J. Med. Biol. Res.* 39, 1563-1568.

[83] Domijan, A.-M., Želježić, D., Kopjar, N., Peraica, M., (2006). Standard and Fpg-modified comet assay in kidney cells of ochratoxin A-and fumonisin B 1-treated rats. *Toxicology.* 222, 53-59.

[84] Tozlovanu, M., Faucet-Marquis, V., Pfohl-Leszkowicz, A., Manderville, R.A., (2006). Genotoxicity of the hydroquinone metabolite of ochratoxin A: structure-activity relationships for covalent DNA adduction. *Chem. Res. Toxicol.* 19, 1241-1247.

[85] Domijan, A.-M., Rudes, K., Peraica, M., (2005). The effect of ochratoxin A on the concentration of protein carbonyls in rats. *Arh. Hig. Rada Toksikol./Arch. Ind. Hyg. Toxicol.* 56, 311-315.

[86] Kamp, H.G., Eisenbrand, G., Schlatter, J., Würth, K., Janzowski, C., (2005). Ochratoxin A: induction of (oxidative) DNA damage, cytotoxicity and apoptosis in mammalian cell lines and primary cells. *Toxicology.* 206, 413-425.

[87] Alvarez-Erviti, L., Leache, C., Gonzalez-Penas, E., De Cerain, A.L., (2005). Alterations induced *in vitro* by ochratoxin A in rat lymphoid cells. *Hum. Exp. Toxicol.* 24, 459-466.

[88] Guerra, M., Galvano, F., Bonsi, L., Speroni, E., Costa, S., Renzulli, C., Cervellati, R., (2005). Cyanidin-3-O-β-glucopyranoside, a natural free-radical scavenger against aflatoxin B1-and ochratoxin A-induced cell damage in a human hepatoma cell line (Hep G2) and a human colonic adenocarcinoma cell line (CaCo-2). *Br. J. Nutr.* 94, 211-220.

[89] Mally, A., Pepe, G., Ravoori, S., Fiore, M., Gupta, R.C., Dekant, W., Mosesso, P., (2005). Ochratoxin A causes DNA damage and cytogenetic effects but no DNA adducts in rats. *Chem. Res. Toxicol.* 18, 1253-1261.

[90] Hundhausen, C., Bösch-Saadatmandi, C., Augustin, K., Blank, R., Wolffram, S., Rimbach, G., (2005). Effect of vitamin E and polyphenols on ochratoxin A-induced cytotoxicity in liver (HepG2) cells. *J. Plant Physiol.* 162, 818-822.

[91] Abdel - Wahhab, M.A., Abdel - Galil, M.M., El - Lithey, M., (2005). Melatonin counteracts oxidative stress in rats fed an ochratoxin A contaminated diet. *J. Pineal Res.* 38, 130-135.

[92] Russo, A., La Fauci, L., Acquaviva, R., Campisi, A., Raciti, G., Scifo, C., Renis, M., Galvano, G., Vanella, A., Galvano, F., (2005). Ochratoxin A-induced DNA damage in human fibroblast: protective effect of cyanidin 3-O-β-D-glucoside. *The Journal of nutritional biochemistry.* 16, 31-37.

[93] Zurich, M.-G., Lengacher, S., Braissant, O., Monnet-Tschudi, F., Pellerin, L., Honegger, P., (2005). Unusual astrocyte reactivity caused by the food mycotoxin ochratoxin A in aggregating rat brain cell cultures. *Neuroscience.* 134, 771-782.

[94] Knasmuller, S., Cavin, C., Chakraborty, A., Darroudi, F., Majer, B.J., Huber, W.W., Ehrlich, V.A., (2004). Structurally related mycotoxins ochratoxin A, ochratoxin B, and citrinin differ in their genotoxic activities and in their mode of action in human-derived liver (HepG2) cells: implications for risk assessment. *Nutr. Cancer.* 50, 190-197.

[95] Schaaf, G., Nijmeijer, S., Maas, R., Roestenberg, P., De Groene, E., Fink-Gremmels, J., (2002). The role of oxidative stress in the ochratoxin A-mediated toxicity in proximal tubular cells. *Biochimica et Biophysica Acta (BBA)-Molecular Basis of Disease.* 1588, 149-158.

[96] Petrik, J., Žanić-Grubišić, T., Barišić, K., Pepeljnjak, S., Radić, B., Ferenčić, Ž., Čepelak, I., (2003). Apoptosis and oxidative stress induced by ochratoxin A in rat kidney. *Arch. Toxicol.* 77, 685-693.

[97] Lühe, A., Hildebrand, H., Bach, U., Dingermann, T., Ahr, H.-J., (2003). A new approach to studying ochratoxin A (OTA)-induced nephrotoxicity: expression profiling *in vivo* and *in vitro* employing cDNA microarrays. *Toxicol. Sci.* 73, 315-328.

[98] Son, W.-C., Kamino, K., Lee, Y.-S., Kang, K.-S., (2003). Lack of effects of sodium 2-mercaptoethane sulfonate (mesna) on Ochratoxin A induced renal tumorigenicity following life-time oral administration of Ochratoxin A in DA and Lewis rats. *Toxicol. Lett.* 142, 19-27.

[99] Pfohl-Leszkowicz, A., Bartsch, H., Azémar, B., Mohr, U., Estève, J., Castegnaro, M., (2002). MESNA protects rats against nephrotoxicity but not carcinogenicity induced by ochratoxin A, implicating two separate pathways. *Facta Univ. Ser. Med. Biol.* 9, 57-63.

[100] Gautier, J.-C., Holzhaeuser, D., Markovic, J., Gremaud, E., Schilter, B.t., Turesky, R.J., (2001). Oxidative damage and stress response from ochratoxin A exposure in rats. *Free Radic. Biol. Med.* 30, 1089-1098.

[101] Pinelli, E., El Adlouni, C., Pipy, B., Quartulli, F., Pfohl-Leszkowicz, A., (1999). Roles of cyclooxygenase and lipoxygenases in ochratoxin A genotoxicity in human epithelial lung cells. *Environ. Toxicol. Pharmacol.* 7, 95-107.

[102] Hoehler, D., Marquardt, R.R., McIntosh, A.R., Hatch, G.M., (1997). Induction of free radicals in hepatocytes, mitochondria and microsomes of rats by ochratoxin A and its analogs. *Biochimica et Biophysica Acta (BBA)-Molecular Cell Research.* 1357, 225-233.

[103] Hoehler, D., Marquardt, R.R., McIntosh, A.R., Xiao, H., (1996). Free radical generation as induced by ochratoxin A and its analogs in bacteria (Bacillus brevis). *J. Biol. Chem.* 271, 27388-27394.

[104] Grosse, Y., Chekir-Ghedira, L., Huc, A., Obrecht-Pflumio, S., Dirheimer, G., Bacha, H., Pfohl-Leszkowicz, A., (1997). Retinol, ascorbic acid and α-tocopherol prevent DNA adduct formation in mice treated with the mycotoxins ochratoxin A and zearalenone. *Cancer Lett.* 114, 225-229.

[105] Baudrimont, I., Betbeder, A.-M., Gharbi, A., Pfohl-Leszkowicz, A., Dirheimer, G., Creppy, E., (1994). Effect of superoxide dismutase and catalase on the nephrotoxicity induced by subchronical administration of ochratoxin A in rats. *Toxicology.* 89, 101-111.

[106] Rahimtula, A., Béréziat, J.-C., Bussacchini-Griot, V., Bartsch, H., (1988). Lipid peroxidation as a possible cause of ochratoxin A toxicity. *Biochem. Pharmacol.* 37, 4469-4477.

[107] Hundhausen, C., Bosch-Saadatmandi, C., Augustin, K., Blank, R., Wolffram, S., Rimbach, G., (2005). Effect of vitamin E and polyphenols on ochratoxin A-induced cytotoxicity in liver (HepG2) cells. *J Plant Physiol.* 162, 818-822.

[108] Chakraborty, D., Verma, R., (2010). Ameliorative effect of Emblica officinalis aqueous extract on ochratoxin-induced lipid peroxidation in the kidney and liver of mice. *International journal of occupational medicine and environmental health.* 23, 63-73.

[109] Farag, I., Abdel-Aziz, K., Nada, S., Tawfek, N., Farouk, T., Darwish, H., (2010). Modulation of ochratoxin-induced oxidative stress, genotoxicity and spermatotoxic alterations by Lactobacillus rhamnosus GG in male albino mice. *J. Am. Sci.* 6, 575-587.

[110] Feldmann, G., (1997). Liver apoptosis. *J Hepatol.* 26 Suppl 2, 1-11.

[111] Vettorazzi, A., van Delft, J., de Cerain, A.L., (2013). A review on ochratoxin A transcriptomic studies. *Food Chem. Toxicol.* 59, 766-783.

[112] Gennari, A., Pazos, P., Boveri, M., Callaghan, R., Casado, J., Maurici, D., Corsini, E., Prieto, P., (2004). New insights into the mechanisms involved in renal proximal tubular damage induced *in vitro* by ochratoxin A. *J. Biochem. Mol. Toxicol.* 18, 43-49.

[113] Hundhausen, C., Boesch-saadatmandi, C., Matzner, N., Lang, F., Blank, R., Wolffram, s., Blaschek, W., Rimbach, G., (2008). Ochratoxin a lowers mRNA levels of genes encoding for key proteins of liver cell metabolism. *Cancer Genomics-Proteomics.* 5, 319-332.

[114] Radford, R., Slattery, C., Jennings, P., Blacque, O., Pfaller, W., Gmuender, H., Van Delft, J., Ryan, M.P., McMorrow, T., (2012). Carcinogens induce loss of the primary cilium in human renal proximal tubular epithelial cells independently of effects on the cell cycle. *American Journal of Physiology-Renal Physiology.* 302, F905-F916.

[115] Stemmer, K., Ellinger-Ziegelbauer, H., Ahr, H.J., Dietrich, D.R., (2007). Carcinogen-specific gene expression profiles in short-term treated Eker and wild-type rats indicative of pathways involved in renal tumorigenesis. *Cancer Res.* 67, 4052-4068.

[116] Arbillaga, L., Vettorazzi, A., Gil, A.G., van Delft, J.H., García-Jalón, J.A., de Cerain, A.L., (2008). Gene expression changes induced by ochratoxin A in renal and hepatic tissues of male F344 rat after oral repeated administration. *Toxicol. Appl. Pharmacol.* 230, 197-207.

[117] Stemmer, K., Ellinger-Ziegelbauer, H., Ahr, H.-J., Dietrich, D.R., (2009). Molecular characterization of preneoplastic lesions provides insight on the development of renal tumors. *The American journal of pathology.* 175, 1686-1698.

[118] Taniai, E., Yafune, A., Hayashi, H., Itahashi, M., Hara-Kudo, Y., Suzuki, K., Mitsumori, K., Shibutani, M., (2012). Aberrant activation of ubiquitin D at G 2 phase and apoptosis by carcinogens that evoke cell proliferation after 28-day administration in rats. *The Journal of toxicological sciences.* 37, 1093-1111.

[119] Hibi, D., Kijima, A., Kuroda, K., Suzuki, Y., Ishii, Y., Jin, M., Nakajima, M., Sugita-Konishi, Y., Yanai, T., Nohmi, T., (2013). Molecular mechanisms underlying ochratoxin A-induced genotoxicity: global gene expression analysis suggests induction of DNA double-strand breaks and cell cycle progression. *The Journal of toxicological sciences.* 38, 57-69.

[120] Jennings, P., Weiland, C., Limonciel, A., Bloch, K.M., Radford, R., Aschauer, L., McMorrow, T., Wilmes, A., Pfaller, W., Ahr, H.J., (2012). Transcriptomic alterations induced by Ochratoxin A in rat and human renal proximal tubular *in vitro* models and comparison to a rat *in vivo* model. *Arch. Toxicol.* 86, 571-589.

[121] Essid, E., Dernawi, Y., Petzinger, E., (2012). Apoptosis induction by OTA and TNF-alpha in cultured primary rat hepatocytes and prevention by silibinin. *Toxins.* 4, 1139-1156.

[122] Essid, E., Petzinger, E., (2011). Silibinin pretreatment protects against ochratoxin A-mediated apoptosis in primary rat hepatocytes. *Mycotoxin research.* 27, 167-176.

[123] Khan, S., Martin, M., Bartsch, H., Rahimtula, A.D., (1989). Perturbation of liver microsomal calcium homeostasis by ochratoxin A. *Biochem Pharmacol.* 38, 67-72.

[124] Zheng, J., Zhang,Y., Xu, W., Yang, X., Shen, X., Huang, K., (2013). Ochratoxin A-induced Cell Damage and DNA Hypomethylation in HepG2 Cell Line. *Journal of Agricultural Biotechnology (In Chinese).* 21, 588-594.

[125] Kouadio, J.H., Dano, S.D., Moukha, S., Mobio, T.A., Creppy, E.E., (2007). Effects of combinations of Fusarium mycotoxins on the inhibition of macromolecular synthesis, malondialdehyde levels, DNA methylation and fragmentation, and viability in Caco-2 cells. *Toxicon.* 49, 306-317.

[126] Pfohl-Leszkowicz, A., Chakor, K., Creppy, E., Dirheimer, G., (1990). DNA adduct formation in mice treated with ochratoxin A. *IARC Sci. Publ.*, 245-253.

[127] Harris, C.C., (1991). Chemical and physical carcinogenesis: advances and perspectives for the 1990s. *Cancer Res.* 51, 5023s-5044s.

[128] Miller, E.C., Miller, J.A., (1981). Searches for ultimate chemical carcinogens and their reactions with cellular macromolecules. *Cancer.* 47, 2327-2345.

[129] Singer, B., Grunberger, D. (1983), *Molecular biology of mutagens and carcinogens.* Springer, pp. 97-141.

[130] La, D.K., Swenberg, J.A., (1996). DNA adducts: biological markers of exposure and potential applications to risk assessment. *Mutation Research/Reviews in Genetic Toxicology*. 365, 129-146.

[131] Swenberg, J., Fedtke, N., Fennell, T., Walker, V. (1990), *Progress in predictive toxicology*. Elsevier Amsterdam, pp. 161-184.

[132] Lawley, P., (1974). Some chemical aspects of dose-response relationships in alkylation mutagenesis. *Mutation Research/Fundamental and Molecular Mechanisms of Mutagenesis*. 23, 283-295.

[133] Pullman, B., Pullman, A. (1980), *Carcinogenesis: Fundamental Mechanisms and Environmental Effects*. Springer, pp. 55-66.

[134] Saffhill, R., Margison, G.P., O'Connor, P.J., (1985). Mechanisms of carcinogenesis induced by alkylating agents. *Biochimica et Biophysica Acta (BBA)-Reviews on Cancer*. 823, 111-145.

[135] Heflich, R.H., Neft, R.E., (1994). Genetic toxicity of 2-acetylaminofluorene, 2-aminofluorene and some of their metabolites and model metabolites. *Mutation Research/Reviews in Genetic Toxicology*. 318, 73-174.

[136] Kanisawa, M., Suzuki, S., (1978). Induction of renal and hepatic tumors in mice by ochratoxin A, a mycotoxin. *Gan*. 69, 599.

[137] Kanisawa, M., (1984). Synergistic effect of citrinin on hepatorenal carcinogenesis of ochratoxin A in mice. *Developments in food science*.

[138] Bendele, A.M., Carlton, W.W., Krogh, P., Lillehoj, E., (1985). Ochratoxin A carcinogenesis in the (C57BL/6J× C3H) F1 mouse. *J. Natl. Cancer Inst.* 75, 733-742.

[139] Boorman, G., (1989). NTP Technical report on the toxicology and carcinogenesis studies of ochratoxin A (CAS No. 303-47-9) in F344/N Rats (Gavage Studies)(NIH Publication No. 89-2813). *National Toxicology Program, US Department of Health and Human Services: Research Triangle Park, NC.*

[140] Grosse, Y., Monje, M., Macé, K., Pfeifer, A., Pfohl-Leszkowicz, A., (1997). Use of bronchial epithelial cells expressing human cytochrome P 450 for study on metabolism and genotoxicity of ochratoxin A. *In vitro Toxicology. 1997.*

[141] El Adlouni, C., Pinelli, E., Azemar, B., Zaoui, D., Beaune, P., Pfohl-Leszkowicz, A., (2000). Phenobarbital increases DNA adduct and metabolites formed by ochratoxin A: Role of CYP 2C9 and microsomal glutathione-S-transferase. *Environ. Mol. Mutagen.* 35, 123-131.

[142] Petkova-Bocharova, T., Stoichev, I., Chernozemsky, I., Castegnaro, M., Pfohl-Leszkowicz, A., (1998). Formation of DNA adducts in tissues of mouse progeny through transplacental contamination and/or lactation after administration of a single dose of ochratoxin A to the pregnant mother. *Environ. Mol. Mutagen.* 32, 155-162.

[143] Pfohl-Leszkowicz, A., Grosse, Y., Kane, A., Creppy, E.E., Dirheimer, G., (1993). Differential DNA adduct formation and disappearance in three mouse tissues after treatment with the mycotoxin ochratoxin A. *Mutation Research/Fundamental and Molecular Mechanisms of Mutagenesis*. 289, 265-273.

[144] Malaveille, C., Brun, G., Bartsch, H., (1990). Genotoxicity of ochratoxin A and structurally related compounds in Escherichia coli strains: studies on their mode of action. *IARC Sci. Publ.*, 261-266.

[145] Malaveille, C., Brun, G., Bartsch, H., (1994). Structure-activity studies in E. coli strains on ochratoxin A (OTA) and its analogues implicate a genotoxic free radical and a

cytotoxic thiol derivative as reactive metabolites. *Mutation Research/Fundamental and Molecular Mechanisms of Mutagenesis*. 307, 141-147.

[146] Manderville, R.A., (2005). A case for the genotoxicity of ochratoxin A by bioactivation and covalent DNA adduction. *Chem. Res. Toxicol.* 18, 1091-1097.

[147] Gillman, I.G., Clark, T.N., Manderville, R.A., (1999). Oxidation of ochratoxin A by an Fe-porphyrin system: model for enzymatic activation and DNA cleavage. *Chem. Res. Toxicol.* 12, 1066-1076.

[148] Calcutt, M.W., Gillman, I.G., Noftle, R.E., Manderville, R.A., (2001). Electrochemical oxidation of ochratoxin A: correlation with 4-chlorophenol. *Chem. Res. Toxicol.* 14, 1266-1272.

[149] Manderville, R.A., Calcutt, M.W., Dai, J., Park, G., Gillman, I.G., Noftle, R.E., Mohammed, A.K., Dizdaroglu, M., Rodriguez, H., Akman, S.A., (2003). Stoichiometric preference in copper-promoted oxidative DNA damage by ochratoxin A. *J. Inorg. Biochem.* 95, 87-96.

[150] Dai, J., Park, G., Wright, M.W., Adams, M., Akman, S.A., Manderville, R.A., (2002). Detection and characterization of a glutathione conjugate of ochratoxin A. *Chem. Res. Toxicol.* 15, 1581-1588.

[151] Il'ichev, Y.V., Perry, J.L., Manderville, R.A., Chignell, C.F., Simon, J.D., (2001). The pH-dependent primary photoreactions of ochratoxin A. *The Journal of Physical Chemistry B*. 105, 11369-11376.

[152] Brow, M.E., Dai, J., Park, G., Wright, M.W., Gillman, I.G., Manderville, R.A., (2002). Photochemically Catalyzed Reaction of Ochratoxin A with d - and l - cysteine. *Photochem. Photobiol.* 76, 649-656.

[153] Mally, A., Zepnik, H., Wanek, P., Eder, E., Dingley, K., Ihmels, H., Völkel, W., Dekant, W., (2004). Ochratoxin A: lack of formation of covalent DNA adducts. *Chem. Res. Toxicol.* 17, 234-242.

[154] Creppy, E., Kane, A., Dirheimer, G., Lafarge-Frayssinet, C., Mousset, S., Frayssinet, C., (1985). Genotoxicity of ochratoxin A in mice: DNA single-strand break evaluation in spleen, liver and kidney. *Toxicol. Lett.* 28, 29-35.

[155] Kane, A., Creppy, E.E., Roth, A., Röschenthaler, R., Dirheimer, G., (1986). Distribution of the [3H]-label from low doses of radioactive ochratoxin A ingested by rats, and evidence for DNA single-strand breaks caused in liver and kidneys. *Arch. Toxicol.* 58, 219-224.

[156] KOHN, K.W., (1979). DNA as target in cancer chemotherapy. Measurement of macromolecular DNA damage produced in mammalian cells by anticancer agents and carcinogens. *Meth. Cancer Res.* 16, 297-345.

[157] Kuroda, K., Hibi, D., Ishii, Y., Takasu, S., Kijima, A., Matsushita, K., Masumura, K.-i., Watanabe, M., Sugita-Konishi, Y., Sakai, H., (2013). Ochratoxin A induces DNA double-strand breaks and large deletion mutations in the carcinogenic target site of gpt delta rats. *Mutagenesis*, get054.

[158] Jennen, D.G., Magkoufopoulou, C., Ketelslegers, H.B., van Herwijnen, M.H., Kleinjans, J.C., van Delft, J.H., (2010). Comparison of HepG2 and HepaRG by whole genome gene expression analysis for the purpose of chemical hazard identification. *Toxicol. Sci.*, kfq026.

[159] Doktorova, T.Y., Ellinger-Ziegelbauer, H., Vinken, M., Vanhaecke, T., van Delft, J., Kleinjans, J., Ahr, H.-J., Rogiers, V., (2012). Comparison of hepatocarcinogen-induced gene expression profiles in conventional primary rat hepatocytes with *in vivo* rat liver. *Arch. Toxicol.* 86, 1399-1411.

[160] Zhang, Y., Qi, X., Zheng, J., Luo, Y., Huang, K., Xu, W., (2015). High - Throughput Tag - Sequencing Analysis of Early Events Induced by Ochratoxin A in HepG - 2 Cells. *Journal of biochemical and molecular toxicology*.

[161] Latchman, D.S., (1997). Transcription factors: an overview. *The international journal of biochemistry & cell biology*. 29, 1305-1312.

[162] Karin, M., (1990). Too many transcription factors: positive and negative interactions. *The New biologist*. 2, 126-131.

[163] Lee, T.I., Young, R.A., (2000). Transcription of eukaryotic protein-coding genes. *Annual review of genetics*. 34, 77-137.

[164] Freemont, P.S., Hanson, I.M., Trowsdale, J., (1991). A novel cysteine-rich sequence motif. *Cell*. 64, 483-484.

[165] Lovering, R., Hanson, I.M., Borden, K.L., Martin, S., O'Reilly, N.J., Evan, G.I., Rahman, D., Pappin, D.J., Trowsdale, J., Freemont, P.S., (1993). Identification and preliminary characterization of a protein motif related to the zinc finger. *Proc Natl Acad Sci U S A*. 90, 2112-2116.

[166] Overbeck, S., Uciechowski, P., Ackland, M.L., Ford, D., Rink, L., (2008). Intracellular zinc homeostasis in leukocyte subsets is regulated by different expression of zinc exporters ZnT-1 to ZnT-9. *J Leukoc Biol*. 83, 368-380.

[167] Franklin, R., Ma, J., Zou, J., Guan, Z., Kukoyi, B., Feng, P., Costello, L., (2003). Human ZIP1 is a major zinc uptake transporter for the accumulation of zinc in prostate cells. *J Inorg Biochem*. 96, 435-442.

[168] Choi, Y.D., Grabowski, P.J., Sharp, P.A., Dreyfuss, G., (1986). Heterogeneous nuclear ribonucleoproteins: role in RNA splicing. *Science*. 231, 1534-1539.

[169] De Maio, A., (1999). Heat shock proteins: facts, thoughts, and dreams. *Shock*. 11, 1-12.

[170] Wu, C., (1995). Heat shock transcription factors: structure and regulation. *Annual review of cell and developmental biology*. 11, 441-469.

[171] Barisic, K., Petrik, J., Rumora, L., Cepelak, I., Grubisic, T.Z., (2002). Expression of Hsp70 in kidney cells exposed to ochratoxin A. *Arch Toxicol*. 76, 218-226.

[172] Valcárcel, J., Gebauer, F., (1997). Post-transcriptional regulation: the dawn of PTB. *Curr Biol*. 7, R705-R708.

[173] Aleo, M.D., Wyatt, R.D., Schnellmann, R.G., (1991). Mitochondrial dysfunction is an early event in ochratoxin A but not oosporein toxicity to rat renal proximal tubules. *Toxicol Appl Pharm*. 107, 73-80.

[174] Ambros, V., Bartel, B., Bartel, D.P., Burge, C.B., Carrington, J.C., Chen, X., Dreyfuss, G., Eddy, S.R., Griffiths-Jones, S., Marshall, M., (2003). A uniform system for microRNA annotation. *RNA*. 9, 277-279.

[175] Dai, Q., Zhao, J., Qi, X., Xu, W., He, X., Guo, M., Dweep, H., Cheng, W.-H., Luo, Y., Xia, K., (2014). MicroRNA profiling of rats with ochratoxin A nephrotoxicity. *BMC Genomics*. 15, 333.

[176] Lewis, A., Jopling, C., (2010). Regulation and biological function of the liver-specific *miR-122*. *Biochem. Soc. Trans.* 38, 1553-1557.

[177] Nassirpour, R., Mehta, P.P., Yin, M.-J., (2013). *miR-122* regulates tumorigenesis in hepatocellular carcinoma by targeting AKT3.

[178] Kutay, H., Bai, S., Datta, J., Motiwala, T., Pogribny, I., Frankel, W., Jacob, S.T., Ghoshal, K., (2006). Downregulation of miR - 122 in the rodent and human hepatocellular carcinomas. *J. Cell. Biochem.* 99, 671-678.

[179] Cheung, O., Puri, P., Eicken, C., Contos, M.J., Mirshahi, F., Maher, J.W., Kellum, J.M., Min, H., Luketic, V.A., Sanyal, A.J., (2008). Nonalcoholic steatohepatitis is associated with altered hepatic MicroRNA expression. *Hepatology*. 48, 1810-1820.

[180] Burchard, J., Zhang, C., Liu, A.M., Poon, R.T., Lee, N.P., Wong, K.F., Sham, P.C., Lam, B.Y., Ferguson, M.D., Tokiwa, G., (2010). MicroRNA - 122 as a regulator of mitochondrial metabolic gene network in hepatocellular carcinoma. *Mol. Syst. Biol.* 6, 402.

[181] Zhu, L., Yu, T., Qi, X., Yang, B., Shi, L., Luo, H., He, X., Huang, K., Xu, W., (2016). *miR-122* plays an important role in ochratoxin A-induced hepatocyte apoptosis *in vitro* and *in vivo*. *Toxicology Research*.

[182] Gramantieri, L., Ferracin, M., Fornari, F., Veronese, A., Sabbioni, S., Liu, C.-G., Calin, G.A., Giovannini, C., Ferrazzi, E., Grazi, G.L., (2007). Cyclin G1 is a target of *miR-122*a, a microRNA frequently down-regulated in human hepatocellular carcinoma. *Cancer Res.* 67, 6092-6099.

[183] Fornari, F., Gramantieri, L., Giovannini, C., Veronese, A., Ferracin, M., Sabbioni, S., Calin, G.A., Grazi, G.L., Croce, C.M., Tavolari, S., (2009). *MiR-122*/cyclin G1 interaction modulates p53 activity and affects doxorubicin sensitivity of human hepatocarcinoma cells. *Cancer Res.* 69, 5761-5767.

[184] Zhang, X., Boesch-Saadatmandi, C., Lou, Y., Wolffram, S., Huebbe, P., Rimbach, G., (2009). Ochratoxin A induces apoptosis in neuronal cells. *Genes & nutrition*. 4, 41-48.

[185] Shen, X.L., Zhang, B., Liang, R., Cheng, W.-H., Xu, W., Luo, Y., Zhao, C., Huang, K., (2014). Central role of Nix in the autophagic response to ochratoxin A. *Food Chem. Toxicol.* 69, 202-209.

[186] Deveraux, Q.L., Takahashi, R., Salvesen, G.S., Reed, J.C., (1997). X-linked IAP is a direct inhibitor of cell-death proteases. *Nature*. 388, 300-304.

[187] Deveraux, Q.L., Roy, N., Stennicke, H.R., Van Arsdale, T., Zhou, Q., Srinivasula, S.M., Alnemri, E.S., Salvesen, G.S., Reed, J.C., (1998). IAPs block apoptotic events induced by caspase - 8 and cytochrome c by direct inhibition of distinct caspases. *The EMBO journal*. 17, 2215-2223.

[188] Bruce Alberts, A.J., Martin Raff, Keith Roberts, Peter Walter. (2002) *Molecular Biology of the Cell, 4th edition*. New York: Garland Science New York， America.

[189] Budin, I., Devaraj, N.K., (2012). Membrane assembly driven by a biomimetic coupling reaction. *J Am Chem Soc*. 134, 751-753.

[190] Sergent, O., Pereira, M., Belhomme, C., Chevanne, M., Huc, L., Lagadic-Gossmann, D., (2005). Role for membrane fluidity in ethanol-induced oxidative stress of primary rat hepatocytes. *J. Pharmacol. Exp. Ther.* 313, 104-111.

[191] Gekle, M., Oberleithner, H., Silbernagl, S., (1993). Ochratoxin A impairs "postproximal" nephron function *in vivo* and blocks plasma membrane anion conductance in Madin-Darby canine kidney cells *in vitro*. *Pflugers Archiv : European journal of physiology*. 425, 401-408.

principle to OTA, the primary target of OTA should be inhibition of RNA and protein synthesis [7] (they then proved that the inhibition of RNA synthesis is a secondary regulatory effect). This holds true for at least *Streptococcus faecalis*, *Bacillus subtilis*, and *Bacillus stearothermophilus*, as there is not a functional respiratory chain in *Streptococcus faecalis,* and the uncoupling behavior caused by OTA in *Bacillus subtilis* only occurred at higher concentrations. Thus, in bacteria, the possibility of a specific action of OTA on respiration should be very low [7].

(2) Induction of Free Radicals and Lipid Peroxidation

In the 1970s, many research studies determined that OTA could severely inhibit coupled respiration when applied at low concentrations to rat liver mitochondria [9] and could act as a competitive type of inhibitor for mitochondrial transport carrier proteins [10]. However, there are no mitochondria in bacteria, and in some anaerobic or facultative anaerobic bacteria, there is no functional respiratory chain. However, these bacteria were still inhibited by OTA at concentrations lower than 10 μg/ml. At high doses (12 μg/ml and a lyt-mutant), respiration of *Bacillus subtilis* was increased despite a complete block of growth and decreased intracellular ATP level. It thus acts in a similar manner as the classical uncoupler of the respiratory chain, dinitrophenol. However, at lower doses, which still inhibit growth of *Bacillus subtilis*, the uncoupling effect of OTA was negligible [5].

In 1990, Omar et al. reported that OTA could form a strong complex with ferric ion, which lead to stimulation of lipid peroxidation *in vitro* in a reconstituted system including NADPH, EDTA, Fe^{3+}, the flavoprotein NADPH-cytochrome P450 reductase and phospholipid vesicles [11]. Malaveille et al. [12] reported that, at concentrations of 1-4 mM (400-1600 μg/ml), OTA induced SOS DNA repair in the absence of an exogenous metabolic activation system in *Escherichia coli* strain PQ37. They also reported that the genotoxicity response of *Escherichia coli* strain PQ37, caused by OTA at 2-4 mM (800-1600 μg/ml), could be completely quenched by a water-soluble form of vitamin E at concentrations of 1-2.5 mM.

Hoehler et al. [13] then provided direct evidence of lipid peroxidation and radicals caused by OTA in bacteria. The molecular mechanism of action of OTA at a concentration of 1 mg/ml was studied using *Bacillus brevis*. The results of electron paramagnetic resonance (EPR) spectroscopy showed that the addition of OTA enhanced the generation of free radicals. The EPR signals could be further enhanced by the addition of Ca^{2+} and eliminated by vitamin E. The spin adduct hyperfine splitting constants indicated the presence of α-hydroxy-ethyl radicals resulting from generated hydroxyl radicals, which further suggested the induction of free radical production by OTA by enhancing the permeability of the cellular membrane to Ca^{2+} in this model system.

Both Malaveile et al. and Hoehler et al. studied OTA as well as its analogues. Hoehler et al. [13] suggested that the absence from OTA of either the chlorine atom (OTB) or the phenylalanine side chain (OTα) completely prevented their uptake. OTα generated only a limited amount of the hydroxyl radical, and OTB generated a weak amount of the hydroxyl radical. Malaveile et al. [12] found that OTB, OTα, 5-chlorosalicylic acid and citrinin were inactive towards the induction of SOS DNA repair, but all except OTα were cytotoxic. Thus, the presence of chlorine at the C-5 position of OTA molecules seems to be the determining factor for genotoxicity in these substances.

The concentration of OTA in these experiments proved that the induction of free radicals by OTA must be given attention. In the research by Malaveile et al. and Hoehler et al. the concentration of OTA was approximately 1 mg/ml; however, in the series of studies conducted by Röschenthaler and colleagues to research the inhibition of growth and protein synthesis in gram-positive bacteria, lower than 10 μg/ml was typically used, which is a 100-fold difference in concentration.

(3) Genotoxicity and Induction of Mutations

Whether OTA is a carcinogen to humans is still under debate, and OTA is currently classified as a Group 2B carcinogen. Some of the evidence in animals or mammalian cells to prove that OTA can cause genotoxicity includes dietary feeding of OTA in female mice that induces renal adenomas and hepatocellular carcinomas [14], that OTA induces single-strand breaks of DNA in the kidney, liver, and spleen cells of mice [15], and that OTA also induces DNA adducts in mouse kidneys through a treatment-related pathway [16]. There have been many debates on the genotoxicity of OTA and its actual role in carcinogenicity over several years. Several research groups have concluded that OTA is genotoxic, whereas other groups have indicated that it is unlikely that OTA has an effect through direct genotoxic mechanisms; they think that the carcinogenicity caused by OTA is due to an indirect mechanism, such as the induction of oxidative stress [17].

In bacteria, determination of the genotoxicity of OTA is more puzzling. In 1978, Wehner et al. tested the mutagenicity of 17 mycotoxins produced by various Aspergillus and Penicillum species, including OTA by *Salmonella typhimurium* reverse mutation test (Ames) using *Salmonella typhimurium* TA98, TA100, TA1535, and TA 1537. At concentrations of 0.5, 5.0, 50, and 500 μg/plate, OTA showed no evidence of mutagenic activity [18]. The research of Bendele et al. confirmed this result; their Ames test using *Salmonella typhimurium* TA98, TA100, TA1535, TA1537, and TA1538 also showed that, at concentrations of 50-600 μg/plate of OTA or in a gradient plate assay at concentrations of 0.1-1000 μg/ml of OTA, no mutagenic effect was observed either with or without metabolic activation [14].

However, positive results were obtained in several strains of *Salmonella typhimurium* after a 2 h incubation in the cell-free section from the medium of 24 h incubations of primary rat hepatocytes with 100 μM OTA (40 μg/ml) [19]. Föllmann and Lucas [20] performed a similar research study using much lower concentrations of OTA. After obtaining converted OTA metabolites by incubating primary cultured rat hepatocytes with OTA at concentrations of 0.016 to 0.8 μM, the resulting culture medium containing OTA metabolites was then tested in mutagenicity assays. The results showed no mutagenic effect in Ames tests.

Obrecht-pflumio et al. [21] found an interesting phenomenon. No mutagenicity was induced by OTA in the classical Ames microtest using rat liver microsomal fractions as metabolic activators, but reverse mutations were obtained using mouse kidney microsomal fractions as metabolic activators with three strains, TA1525, TA1538, and TA98, in the presence of either NADP or arachidonic acid as cofactors. Furthermore, both base pair substitutions and frame shift mutations could be caused by OTA after metabolic activation.

Aside from the Ames microtest, Malaveille et al. found that, at concentrations of 1-4 mM (400-1600 μg/ml), OTA could induce SOS DNA repair in the absence of an exogenous metabolic activation system in *Escherichia coli* strain PQ37, and the presence of chlorine at C-5 appears to be one determinant of genotoxicity in OTA and its analogues [12]. They suggested a potential relationship between OTA-derived free radicals and the genotoxicity of OTA.

(4) Induction of Autolysis

During research of protein synthesis inhibition caused by OTA, Röschenthaler and colleagues found that, in *Streptococcus faecalis*, OTA inhibited protein synthesis; this strain was not killed by OTA even at very high concentrations. However, in *Bacillus subtilis*, OTA at concentrations higher than 12 μg/ml induced autolysis in growing cells, and when the concentration of OTA was above 20 μg/ml, the cultures could become sterile. They regarded this phenomenon as quite a unique property, as protein synthesis inhibitors are usually used to prevent lysis.

1.1.3. Bacterial Detoxification of OTA

Many kinds of microorganisms, including bacteria, yeasts, and molds, isolated from a variety of environments and materials have been found to be able to degrade or adsorb OTA. In this section, we will introduce the effects and mechanisms of bacterial degradation or adsorption of OTA; because very few yeasts or molds are present in the guts of human, they are thus not the focus of this chapter. The control and degradation of OTA by yeasts and molds will be introduced in Chapter 2.

(1) Bacterial Species That Can Degrade or Adsorb OTA

In 1983, Wegst and Lingens described the degradation of OTA by the Gram-negative soil bacterium *Phenylobacterium immobile*. This species consumes phenylalanine during their growth and is thus able to utilize the phenylalanine moiety of OTA. This bacterial species could not grow in culture medium with OTA (0.2 g/l) supplied as the only source of carbon and energy. However, this species could grow normally in the medium when both L-phenylalanine (1 g/l) and OTA (0.02 g/l) were supplied. After pre-growth on phenylalanine and then suspending the bacterium in a mineral salts medium with OTA, four metabolites were isolated from the resting culture medium, including OTα. The toxicity of the degradation product was much lower than that of OTA; thus, this transformation of OTA by *Phenylobacterium immobile* could be regarded as a detoxification procedure [22].

In 1994, Hwang and Draughon found that *Acinetobacter calcoaceticus* could degrade OTA. In an ethanol minimal salts medium with the addition 10 μg/ml of OTA, the OTA was removed by *Acinetobacter calcoaceticus* at a rate of 0.1005 and 0.0636 μg/ml/h at 25 and 30°C, respectively. OTA was significantly degraded by this bacterium during and after the log phase of cell growth. The degradation product was found to be OTα [23]

In 1996, Škrinjar et al. found that the content of OTA in a milk sample decreased after fermentation by yogurt bacteria and bifidobacteria. OTA was added to the milk before fermentation at a rate of 0.05-1.5 μg/ml. After fermentation by *Streptococcus salivarius* subsp. *thermophilus*, *Lactobacillus delbrueckii* subsp. *bulgaricus*, and *Bifidobacterium bifidum*, no residues of OTA were found in samples containing 0.05 and 0.1 μg/ml of OTA. In the samples containing 0.5 μg/ml of OTA, only the yogurt bacteria sample containing both *Streptococcus salivarius* subsp. *thermophilus* and *Lactobacillus delbrueckii* subsp. *bulgaricus* could decrease the OTA to no remaining residue. Thus, the yogurt bacteria sample was regarded to be the most effective [24]. However, the authors did not describe the degradation product of OTA by these bacteria.

In 2000, Böhm et al. studied the OTA detoxification effects of several strains of *Lactobacillus acidophilus*, *Lactobacillus bulgaricus*, *Lactobacillus helveticus*, *Bacillus licheniformis*, and *Bacillus subtilis* in MRS broth at 37°C. All of these bacterial strains were found to be able to biodegrade OTA at an initial concentration of 50 ppb (0.05 μg/ml) with the strain *Lactobacillus bulgaricus* 259/2 showing the highest detoxification ratio of 94% after incubation with OTA for 48 h [25]. In the same year, Piotrowska and Zakowska studied the OTA detoxification effects of three other species of the genus *Lactobacillus*, i.e., *Lactobacillus plantarum*, *Lactobacillus brevis*, and *Lactobacillus sanfrancisco,* in MRS broth at 30 °C. Among these three species, *Lactobacillus plantarum* showed the highest OTA degradation ratio of 56% after incubation with OTA for 40 h at the initial concentration of 300 μg/kg (approximately 0.3 μg/ml) [26]. The degradation product was not detected in these research studies either.

In 2006, Schatzmayr et al. screened and isolated several bacterial strains for deactivating OTA from various environments, including one strain (99.8% similarity of 500 bp 16S rDNA sequence with *Lactobacillus vitulinus*) from rumen fluid, four strains (95.7% similarity with *Eubacterium ramulus*, 99.2% similarity with *Streptococcus pleomorphus*, and 99.8% and 99.3% similarity with *Eubacterium callanderi*) and ten strains from soil/water samples [27].

In 2007, Fuchs et al. investigated the detoxification of OTA by a highly effective strain screened from thirty different lactic acid bacterial strains, that is, *Lactobacillus acidophilus* VM 20 causing a decrease of OTA by ≥95% after incubation for 4 hours in MRS broth at 37 °C at the initial OTA concentration of 0.5 or 1.0 μg/ml. The authors did not describe the degradation product of OTA by this strain. However, they found that the viability of the cells played an important role, as only a moderate reduction of OTA (11%) was observed in heat-inactivated cells [28].

In 2007, Petchkongkaew et al. isolated a strain of *Bacillus* from Thai-fermented soybeans (Thua-nao). This *Bacillus spp.*, identified as *Bacillus licheniformis* by ITS sequencing, was able to inhibit the growth of *Aspergillus* strains and to remove both AFB1 (74% decrease) and OTA (92.5% decrease, the initial concentration of OTA was 5 μg/ml) after being incubated at 37 °C for one week. The OTA decrease was due to degradation to OTα [29].

In 2011, Rodriguez et al. reported the degradation of OTA by several *Brevibacterium* spp. strains, which possess a well-known capacity to transform aromatic compounds. *Brevibacterium casei*, *Brevibacterium linens*, *Brevibacterium iodinum*, and *Brevibacterium epidermidis* showed 100% degradation of OTA at the initial concentration of 1 μg/ml. The degradation product in the cell-free supernatants was detected and identified as OTα by high-performance liquid chromatography mass spectrometry (HPLC-MS) [30].

In 2012, Upadhaya et al. obtained one anaerobic bacterium from swine intestinal microbiota. Screening results indicated that this anaerobic bacterial isolate detoxified more than 75% of OTA in M98-5 media supplemented with 100 ppb OTA (0.1 μg/ml) when incubated for 12 h at 39 °C. Furthermore, 1.0 ppm OTA (1 μg/ml) was degraded completely after a 24 h incubation. The results of 16S rDNA sequencing showed that the bacterial strain had 97% sequence similarity with *Eubacterium biforme*. Cell-free extracts after cell lysis were found to have the highest ability to degrade OTA followed by whole cells, and autoclaved cells, medium and PBS showed no ability to degrade OTA. The authors did not detect the degradation product of OTA [31].

In 2014, our research group isolated a strain of *Bacillus* spp. from fresh elk feces, which could inhibit the growth of the OTA-producing species *Aspergillus ochraceus* and *Aspergillus carbonarius*. This strain showed 99% similarity to *Bacillus subtilis* according to 16S rDNA sequencing. At 6 μg/ml of OTA, both viable and autoclaved (121 °C, 20 min) cells of this strain decreased OTA by more than 60%. Furthermore, OTA could also be decreased by the cell-free supernatant of this strain. The cell-free supernatant decreased 97.6% of OTA after 24 h of incubation at 30 °C, and no degradation products were detected by high-performance liquid chromatography. In 3 g samples of contaminated maize, 47.1% of OTA was degraded by 50 ml of inoculant of overnight cultures of this strain [32].

In 2015, Chang et al. isolated a strain of *Bacillus amyloliquefaciens* identified by 16S rDNA sequencing (similarity not shown) from grain depot-stored maize. The HPLC analysis showed that the strain has the capacity to degrade OTA by 98.5% with an initial concentration of 1 μg/ml during the first 24 h, and there was almost no OTA remaining after 72 h. The authors suggested that the OTA level was reduced due to degradation rather than adsorption by the cells because during the first 24 h the number of bacteria was small. Subsequently, a carboxypeptidase enzyme related to the enzymatic conversion of OTA was cloned from this strain and expressed by *Escherichia coli*. This heterogeneously expressed enzyme could dramatically degrade OTA [33].

In summary, strains of (or related to) *Phenylobacterium immobile*, *Acinetobacter calcoaceticus*, *Streptococcus pleomorphus*, *Streptococcus salivarius* subsp. *thermophilus*, *Lactobacillus delbrueckii* subsp. *bulgaricus*, *Lactobacillus plantarum*, *Lactobacillus acidophilus*, *Lactobacillus brevis*, *Lactobacillus sanfrancisco*, *Lactobacillus vitulinus*, *Bifidobacterium bifidum*, *Eubacterium ramulus*, *Eubacterium callanderi*, *Eubacterium biforme*, *Brevibacterium casei*, *Brevibacterium linens*, *Brevibacterium iodinum*, *Brevibacterium epidermidis*, *Bacillus licheniformis*, *Bacillus subtilis*, and *Bacillus amyloliquefaciens* have been isolated with the ability to detoxify OTA. Among these bacteria, *Phenylobacterium immobile* and *Acinetobacter calcoaceticus* are gram-negative species, whereas the others are all gram-positive. The *Streptococcus*, *Lactobacillus*, *Bifidobacterium, Eubacterium*, and *Bacillus* genera are often found in the gut microbiota of animals.

(2) OTA Detoxification Mechanisms of Bacteria

Degradation and adsorption are two main mechanisms for OTA detoxification by bacteria. In the OTA degradation mechanism by *Phenylobacterium*, *Acinetobacter*, *Brevibacterium,* and some cases of *Bacillus*, the final degradation product was OTα. In most research studies on probiotic lactic acid bacteria, no degradation product was detected; thus, though there is still a debate, the detoxification of OTA by lactic acid bacteria is usually thought to be by adsorption.

Two pathways have been reported to achieve the transformation of OTA to OTα in bacteria. The first pathway is direct cleavage of the peptide bond in OTA, forming L-phenylalanine and OTα. This reaction could be catalyzed by Lipase, Protease A, Pancreatin, Carboxypeptidase, and Prolyve PAC. In bacteria, Carboxypeptidase catalyzes most of this reaction according to the research to date. Stander et al. [34] conducted a kinetic study into the hydrolysis of OTA and its analogues by Carboxypeptidase A. They found that the k_{cat}/K_m of Carboxypeptidase A for hydrolysis of OTA is 47 min^{-1} μM^{-1}. The second pathway has to date only been reported in *Phenylobacterium immobile*. There is first a series of reactions on the phenylalanine moiety of OTA to open the benzene ring, and then a hydrolytic cleavage occurs, forming aspartic acid and OTα [22]. The enzymes that catalyze these reactions have not yet been found.

In both of the above pathways, OTα was ultimately formed. However, in most studies of OTA detoxification by *Lactobacillus* species, the formation of OTα, or any other residues of OTA, was not reported. Moreover, the elimination of OTA could be partially reversible in lactic acid bacteria [35]. These results can be explained by desorption of OTA in the bacterial cell structures. Thus, the removal of OTA by lactic acid bacteria is not through degradation but rather by cell binding [36, 37]. However, evidence against this opinion appeared in a study by Fuchs et al. [28]. They found that the viability of the lactic acid bacterial cells plays an important role as heat-inactivated cells showed only a moderate reduction of OTA (11%) and PAT (16%) [28].

1.2. *In Vivo* Research on the Interactions between OTA and Gut Microbiota

1.2.1. Enterotoxicity of OTA for Animals

Before discussing the enterotoxicity of OTA and the interaction of OTA with gut microbiota, we will briefly summarize the fate of OTA in the gastrointestinal track, namely, absorption, biotransformation, and excretion.

Rumen animals are much less sensitive to OTA than non-rumen animals. In rumen animals, following oral administration, OTA can be degraded extensively in the rumen by ruminal microbes, with protozoa being a central group, with disappearance half-lives of 0.6-3.8 h [38]. However, most likely due to some of the OTA bypassing the rumen, an estimated 2-6.5% (maximum 10%) of the OTA dosage was still detected systemically [38]. Thus, 4-8% of the OTA dosage could be detected in urine, and 1-2% of the OTA was excreted in feces. The metabolic product was OTα with 70-80% of OTα excreted by urine and 10-20% by feces [38, 39].

Differing from rumen animals in which most of the OTA was degraded to OTα before absorption, non-rumen animals also have microorganisms with the ability to degrade OTA, but most of them are in the colon. Generally, OTA is absorbed from the stomach and the proximal jejunum in its monoanion (OTA^-) and non-ionized (OTA^0) forms by passive diffusion [40]. Moreover, some *in vitro* research studies using Caco-2 cells in a bicameral system have shown that the organic anion carrier (OAT) is not involved in OTA transport across the cells [41], suggesting that the absorption mechanism is diffusion in the gastrointestinal tract [40]. The percentage of OTA absorbed differs between animals with 66% in pigs [41], 56% in rats and rabbits and 40% in chickens [42]. Evidence for enterohepatic circulation of OTA was discovered using a [^{3}H]OTA radioactivity profile in mice. Between 28 and 68% of conjugated OTA together with OTA cleavage products were found in bile, indicating that the biliary system was one method of excretion for OTA and its metabolites in mice [43].

(1) OTA Causes Intestinal Inflammation

The intestinal epithelium represents the principal site of OTA exposure. It was reported that OTA at the high concentration could damage the intestinal mucosa in rats [44,45], swine [46], dogs [46], and chickens [47] with rapid inflammatory effects associated with desquamation, necrosis, and acute diarrhea. Using differentiated Caco-2 cells as a model and interleukin 8 (IL-8) as an indicator of intestinal inflammation, Maresca et al. [48] investigated whether the intestinal inflammation in human intestinal epithelial cells induced by OTA resulted from direct and/or indirect pro-inflammatory activity. They found that OTA could not directly increase IL-8 secretion but could potentiate the effect of IL-1β on IL-8 secretion and increase the transepithelial passage of commensal bacteria, both of which indirectly stimulate IL-8 secretion [48].

(2) OTA Decreases Epithelial Barrier Function

Because of the phenomenon observed in acute oral toxicity, Maresca et al. [49] suggested that alterations of the intestinal barrier function might contribute to the toxicity of OTA and facilitate its absorption. In their research, two human epithelial intestinal cell lines, HT-29-D4 and Caco 2, were assayed. In both cell lines, the main effects of the OTA were a dramatic decrease of the transepithelial resistance and inhibition of cellular growth. These data suggest that OTA may potentiate its own absorption through paracellular pathways because transepithelial resistance reflects the organization of tight junctions over the cell monolayer [49]. Furthermore, OTA also alters the absorption functions of the intestinal epithelium according to this *in vitro* test [49].

Then, McLaughlin et al. [50] gave more direct and detailed experimental evidence on the effect of OTA decreasing epithelial barrier function. They revealed that the reduction of the barrier properties of the Caco-2 cell line was caused by OTA. Using a Caco-2 monolayer model grown on Transwell filters, OTA was shown to reduce the transepithelial electrical resistance of this model system by approximately 40% over 24 h. Concurrently, the permeability of the monolayer was increased with respect for 4- and 10-kDa FITC dextrans but not for 20- or 40-kDa dextrans [50]. The decreased barrier properties are concomitant with the disappearance of claudins 3 and 4, but not claudin 1, from Caco-2 cell membranes, according to the results of immunoblotting and immunofluorescence [50]. Moreover, in 2007, a study by the same research group showed that OTA is able to remove claudins 3 and 4 from the detergent insoluble membrane microdomains associated with tight junctions in Caco-2 cells [51].

(3) Potential Risk Factors of Chronic Intestinal Inflammatory Diseases

Chronic intestinal inflammatory diseases, including primarily Crohn's diseases and ulcerative colitis, are incurable diseases that still have unclear pathological mechanisms. By reviewing evidence based on various cellular and animal studies, Maresca and Fantini showed that several mycotoxins induce intestinal alterations that are similar to those observed at the onset and during the progression of inflammatory bowel disease [52]. It seems highly plausible that food-associated exposure to certain mycotoxins could lead to the induction and persistence of human chronic intestinal inflammatory diseases in genetically predisposed patients, and this "mycotoxin theory" expands the existing list of previously postulated causes of Crohn's and celiac diseases [52]. Although OTA was in the list of mycotoxins analyzed in this review article, the author proposed that OTA is probably not a risk with respect to intestinal effects, as the doses of mycotoxin required for such effects are far greater (at least 400) than their provisional maximum tolerable daily intake (PMTDI) or their higher exposure doses [52].

1.2.2. Effect of OTA on Gut Microbiota

There was plenty of ancillary evidence to indicate that OTA could influence the gut microbiota of animals *in vivo*, and many reviews that deeply researched this field declared it to be important [52,53]; however, methodological limitations prevented the acquisition of direct evidence until recently. In 2014, our research group conducted the first *in vivo* research on the relationship between OTA and gut microbiota using high-throughput sequencing-based metagenomic approaches [54].

(1) Brief Introduction of the Development of Methodology to Study Gut Microbiota

Culture-based methods are traditionally the most widely applied in the field of microbiology. However, when they were used in studies of gut microbiota, many restrictions became apparent. First, most of the gut bacteria are anaerobes or facultative anaerobes, which makes it hard to isolate and culture them. Some of them still do not have a known suitable culture medium. Second, the structure of the community, i.e., the proportion of each bacterial strain present in the gut, is important for assessing the status of the gut microbiota. Unfortunately, during culture, the proportion changes as the conditions of the culture are not evenly suitable for the growth of each strain. Third, there are too many bacterial strains in the gut microbiota to culture, which makes the workload too large.

Metagenomics is the study of genetic material recovered directly from environmental samples without the process of culture. Early metagenomics using molecular methods such as Temperature Gradient Gel Electrophoresis (TGGE), Denaturing Gradient Gel Electrophoresis (DGGE), and Restriction Fragment Length Polymorphism (RFLP), among others, have overcome the restrictions of culture-based methods and have been popular since the 1990s and early 2000s for studying the field of gut microbiota. Unfortunately, the sensitivities of these methods are unsatisfactory; thus, low abundance strains in the gut microbiota, which may also play important roles for the host, may not be detected.

After 2005, the development of high-throughput sequencing and its application in metagenomics have greatly advanced studies in gut microbiota. High-throughput sequencing-based metagenomics mainly include two strategies, 16S rRNA sequencing and shotgun sequencing, with 16S rRNA sequencing used for the detection of microbiota species composition and shotgun sequencing used when the functional genes of microbiota need to be studied. High-throughput sequencing-based metagenomics could detect strains with an abundance in the community of lower than 0.00001%, which means almost all of the bacterial strains could be discovered in these studies.

(2) Brief Introduction of Experimental Designs of Metagenomics

a) DNA Extraction

DNA extraction is one of the most important procedures for metagenomics. Some extraction methods may result in bias of the results, as the extraction efficiency is different for strains of different taxonomies. We compared several DNA extraction methods and found that the method of differential centrifugation followed by saturated phenol-chloroform is most effective.

b) Selection of Sequencing Platform

Currently, Roche 454 GS FLX, Illumina Hiseq and Miseq, and Life Technologies ion torrent are the main high-throughput sequencing platforms available. In the area of amplicon sequencing, Hiseq and Miseq sequencing platforms have been shown to be more widely applicable according to the published literature. The length of the target sequence and the number of reads are two important bases for choosing the sequencing platform. The Roche 454 GS FLX sequencing platform provides a longer length of each sequence but a lower number of reads. Illumina Hiseq PE250 results in less than 500 nt length of sequencing but provides 10 to 100 times the number of reads as that of the GS FLX. Furthermore, sequencing using Hiseq and Miseq are much cheaper and faster than using GS FLX.

c) Selection of Read Number

The read number of sequencing, i.e., the sequencing depth, is an important factor in the design of microbiota sequencing research. Lundin et al. used three data sets of freshwater and sediment bacteria with each data set containing 6-15 communities of 15000-20000 16S rDNA sequences, where diversity was explored using 454 pyrosequencing [55]. The subsamples of these data were conducted repeatedly to 10 different depths. Diversity estimates varied with sequencing depth, yet trends in diversity among samples were less sensitive. In their results, 1000 denoised sequences per sample were found to explain to 90% the trends in β diversity (Bray-Curtis index) among samples observed for 15000-20000 sequences. Similarly, 5000 denoised sequences were sufficient to describe trends in α diversity (Shannon index) with the same accuracy [55]. For Hiseq or Miseq sequencing, due to the shot length of each read, a higher read number is necessary to acquire this information.

(3) Effect of OTA on the Structure of Gut Microbiota

In 2014, both 16S rRNA sequencing and shotgun sequencing methods were used to study the interaction of OTA and gut microbiota of F344 rats *in vivo* [54]. According to the results of 16S rRNA sequencing, the treatment of OTA at both 70 μg/kg body weight and 210 μg/kg body weight could significantly alter the structure of gut microbiota, which was revealed by non-metric multidimensional scaling (NMDS) plot analysis and analysis of similarities (ANOSIM). The α diversity (within-subject variation) of gut microbiota was assessed using the Simpson index. The results showed that the 210 μg/kg OTA had a slight decreasing effect on the α diversity of gut microbiota ($P < 0.1$). The β diversity (between-subject variation) of gut microbiota was assessed using the Bray-Curtis index. After the treatment of OTA at both concentrations, the β diversity increased compared with that of the control group ($P < 0.05$). The decrease in α diversity indicated that OTA could inhibit the growth of gut microbiota *in vivo*, as OTA has been shown to inhibit the growth of some bacteria in culture medium, whereas the increase in β diversity indicated that the ultimate status of the gut microbiota following OTA exposure was not identical among individuals [54].

Specific changes in the microbial community associated with OTA treatment were also revealed in this study. At the family level, an increase of Lactobacillaceae and a decrease of Bacteroidaceae in relative abundance were observed. At the genus level, *Bacteroides*, *Dorea*, *Escherichia*, *Oribacterium*, *Ruminococcus*, and *Syntrophococcus* were the most significantly decreased genera after OTA treatment ($P < 0.01$), whereas an increase in the relative abundance of *Lactobacillus* was particularly significant and interesting ($P < 0.01$). The changes of the species composition indicated that different bacterial strains had different sensitivities to OTA. As introduced in the previous section, the *in vitro* inhibition assay of OTA showed that gram-positive bacteria are more sensitive to OTA than gram-negative bacteria. However, this *in vivo* study showed an increased proportion of gram-positive bacteria in the gut microbiota after OTA treatment. The most important reason for this phenomenon was due to an increase in the abundance of *Lactobacillus*, a gram-positive genus [54]. *Lactobacillus* species seem to be resistant to OTA both *in vitro* and *in vivo*. According to *in vitro* studies, *Lactobacillus* could detoxify OTA, most likely by adsorption [37]. In *in vivo* conditions of gut microbiota after the OTA treatment, a decrease in the abundance of other sensitive bacterial strains means the resistant strains could acquire more nutritional resources and ecological niches. Thus, it is not difficult to understand the increase in *Lactobacillus* after the treatment with OTA.

(4) Effect of OTA on the Function of Gut Microbiota

Shotgun sequencing was used for functional analysis of gut microbiota. Clusters of orthologous groups (COGs) are the basic units for functional analysis in shotgun sequencing metagenomics. After OTA treatment, changes in functional COGs of gut microbiota included genes belonging to *signal transduction*, *carbohydrate transport*, *transposase*, the *amino acid transport system*, and the *mismatch repair* pathway according to the pathway centric analysis, and according to regression analysis, some COGs belonging to *amino acid transport and metabolism*, *replication, recombination and repair*, *carbohydrate transport* and *metabolism* showed OTA dose-dependent changes.

The signal transduction pathway was one of the most significant COG pathways affected by OTA [54]. In eukaryotes, the effects of OTA on signal transduction pathways, such as MAPKs and calcium homeostasis, have been thought to be related to the carcinogenicity of OTA [56]. Thus, it seems plausible to predict that, in prokaryotes, the effects of OTA on bacterial signal transduction may also be one of the potential ways by which the growth of some bacterial strains was inhibited [54]. However, there should be more studies to confirm this prediction, as the differences between the bacterial and animal signal transduction systems are quite notable.

COGs related to amino acid and oligopeptide transporters also increased significantly, and for bacteria, this could be a potential protective mechanism [54]. According to the introduction of the previous section, OTA could inhibit the activity of phenylalanyl-tRNA synthetase, and the increased intake of phenylalanine due to the increased expression of amino acid transporters may reverse the toxicity of OTA. Supplementation with phenylalanine has been proven to alleviate the effects of OTA in bacteria, animal cell lines, and animals [39].

Additionally, carbohydrate transporters were also affected by OTA in the metagenome of gut microbiota. Some of the *ABC-type carbohydrate transport system*-related COGs decreased, whereas other types of carbohydrate transport systems, such as *putative glucose uptake permease* and *glycerol uptake facilitator and related permeases,* increased [54]. Furthermore, the COG pathway *replication, recombination and repair* is also affected, which is reasonable as genotoxicity is one of the potentially toxic effects of OTA. Particularly, there were five *transposase and inactivated derivatives*-related COGs that significantly changed, indicating that transposases may be one of the neglected factors in the DNA toxicity of OTA [54].

1.2.3. Role of Gut Bacteria in the Detoxification of OTA

In previous *in vitro* studies, many bacterial species with OTA degradation or adsorption abilities have been identified using culture-based methods with some of them present in the gut. However, whether these species play a role *in vivo* has yet to be confirmed. According to the sequencing-based metagenomics study [54], among the known OTA detoxification genera introduced in section 6.1.1.3, only the relative abundance of *Lactobacillus* increased; thus, the other genera of bacteria appeared not to play a role in OTA detoxification, as their relative abundances were quite low and did not increase when the gut microbiota were exposed to OTA. Furthermore, none of the enzymes with a known ability to hydrolyze OTA to OTα, such as proteases, peptidases, esterases, and lipases, were found to be upregulated according to the COGs analysis [54]. Thus, the authors suggested that absorption, rather than hydrolysis, was more likely to be the main way by which *Lactobacillus* detoxifies OTA *in vivo* [54].

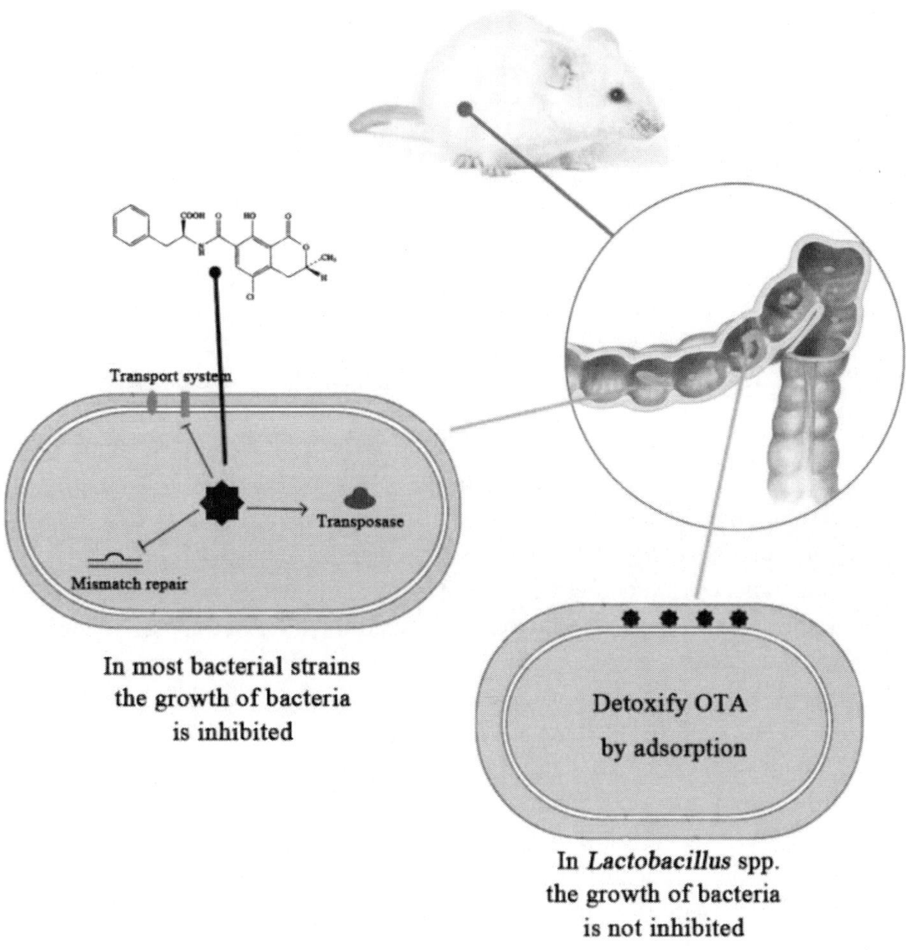

Figure 1. OTA-induced the change of the intestinal flora.

2. Neurotoxicity

2.1. Research Status

OTA induces many types of toxicity, and neurotoxicity is one that cannot be neglected. To date, researchers have performed numerous studies on various aspects of neurotoxicity to tremendous effect.

Belmadani et al. [57] fed young adult rats OTA for 1 to 6 weeks. They discovered that the OTA concentration in the brain was related to time. OTA induced variation in free amino acids (Phe and Tyr) in young adult rat brains, significantly reducing Tyr and significantly increasing Phe expression.

Sava et al. [58] gave a dose of 3.5 mg/kg of OTA, which is approximately one-tenth of the reported LD50. Six hours later, DNA repair was inhibited. After 72 h, oxidative stress was apparent in six areas of the brain. Lipid peroxidation and DNA oxidative damage increased significantly. SOD continued to rise as well. At 24 h, oxidative stress reached its peak in the midbrain, caudate putamen, and the hippocampus.

Belmadani et al. [59] also found that OTA could induce down-regulation of Glu, Gln, and Asp and disrupt neurotransmitters. The ventral mesencephalon and the cerebellum were considered to be selectively sensitive to cytotoxicity of OTA. Compared to the striatum and hippocampus, the ventral mesencephalon and the cerebellum were enriched with more OTA.

In 2011, Doi and Uetsuka [60] found that OTA could result in the acute consummation of striatal dopamine and its metabolites, and even worse, neuronal cell apoptosis can be observed in the substantia nigra, striatum and hippocampus. Neurotoxicity is more pronounced in the ventral mesencephalon, hippocampus (HP), and striatum than in the cerebellum (CB).

Madsen et al. [60] explored the effects of OTA in pigs. They administered OTA to bacon pigs of 8-week-old weaners. No effect on the health of the pigs was observed, but it caused a decrease in the consumption of feed, inhibition of growth, increased water consumption and polyuria. Administration of 0 to 200 μg/kg of OTA had little impact on the daily gain and feed conversion efficiency. Even after a few weeks, replacing with normal feed resulted in the resumption of normal behavior. Naturally contaminated feed did not appear to have a direct influence on carcass quality.

Aside from the *in vivo* experiments, experiments *in vitro* were also useful, meaningful and convenient. Zhang et al. [61] treated SH-SY5Y neuronal cells with OTA. They found that a low concentration of OTA could cause neurotoxicity in nerve cells and that sensitivity to OTA between liver and nerve cells was similar; however, nerve cells were more sensitive than renal cells. OTA, through neurotoxicity, induced apoptosis.

Yoon et al. [62] investigated the potential harm in terms of the effect of exposure to environmental OTA on neuronal cell proteome profiles and viability. They used mouse hippocampal HT22 and human neuroblastoma SH-SY5Y cells. They concluded that OTA treatment caused oxidative stress in both cell lines. However, it is interesting that after treatment with OTA several proteins were up-regulated only in HT22 cells, which contain Atp5b protein, nucleophosmin 1, eukaryotic translation elongation factor 1 delta isoform, ornithine aminotransferase, prohibitin, and peroxiredoxin 6. OTA can only induce apoptosis in HT22 but not SH-SY5Y cells.

Zurich et al. [63] treated brain cell cultures with 10 and 20 nM of OTA. The 10 nM OTA treatment did not change the total protein content or LDH activity until the 9th day. The 20 nM OTA treatment did not change the total protein content or LDH activity at 24 or 48 h, but at 9 days both were significantly reduced. In dealing with immature cells, glutamine synthetase had no significant difference after being treated with 10 nM of OTA, but its mRNA level increased significantly after 24 and 48 h of treatment with 20 nM of OTA. However, glutamine synthetase activity only significantly increased after 48 h. After 48 h of 20 nM OTA treatment in mature cells, glutamine synthetase mRNA levels and glutamine synthetase activity were significantly decreased. After 20 nM of OTA processing, the collagen fiber acidic protein mRNA level was significantly reduced, but this effect was transient and had a greater influence on immature cells. Treatments with 10 and 20 nM of OTA resulted in a significant reduction of the GFAP mRNA level. In the 10 nM group, the level was restored after 48 h, which was not seen in 20 nM OTA-treated group. OTA raises the mRNA level of vimentin. It is interesting that it is dose-related in immature cells, whereas it is concentration-related in mature cells. OTA can also change the astrocytic cytoskeleton and shift the rate of GFAP to vimentin conversion. mRNA levels of the brain inflammation markers PPARγ, iNOS, and HO-1 were elevated by OTA treatment.

In another cell line, neural stem/progenitor cells (NSCs), Sava et al. [64] found that OTA can make the proliferation and differentiation activity of NSCs decrease in a dose- and time-related manner. It was also linked to depression and memory decay. Proliferating cells were more sensitive to OTA than differentiated cells. After 72 h of OTA treatment, the viability of proliferating cells dropped to 41%, and that of differentiated cells dropped to 62%. OTA also led to a significant reduction of the number of living cells and made morphologic changes, cell body contraction, and nerve fiber loss. DNA synthesis was also markedly inhibited. OTA elevated the level of SOD in mitochondria and cytoplasm.

2.2. Pathological Changes in Different Tissues

2.2.1. Different Conditions Used for Various Studies

To date, there have been a number of reports on the neurotoxicity of OTA; Table 1 summarizes the various conditions utilized in these studies.

2.2.2. Brain

Miki et al. [65] found that OTA-treated mice showed a significant 25% deficit in brain weight, and the cortical thickness of the OTA group was decreased in the somatosensory and visual cortex compared to those of the control. Furthermore, there was a significant increase in the numerical density of neurons in the somatosensory cortex. Szczech and Hood [66] found that, in the neuroepithelium of the developing cerebral cortex, many cells were damaged, and even worse, necrosis was prominent around the ventricles. Tamaru et al. [67] injected mice intraperitoneally with 3 mg/kg of OTA in 0.1 M $NaHCO_3$ or with 0.1 M $NaHCO_3$ alone. From the 3rd day after OTA treatment, weight gains in the body and brain of fetuses were suppressed. On the 3rd and 6th day after injection, body weights were reduced to 85 and 80% compared to those of the control groups, and fetal brain weights were reduced to 70% of that of controls.

Table 1. Summary of experimental conditions of studies carried out on neurotoxicity of ochratoxin A

Reference	Species,strain,sex, age	Dose	Route of administratin	Exposure time	Material collection
Belmadani, A.et al. (1998) [57]	Male albino Wistar rats aged 6-8 weeks	289 μg/kg/48h	feeding	1, 2, 3,4, 5 or 6 weeks	taking blood from the retroorbital sinus decapitated
Sava, V., et al.(2006) [58]	Male Swiss ICR mice	0-6 mg/kg of b.w.	injection	6,24, and 72 h	euthanatized with CO_2
Belmadani, A.,et al. (1998) [59]	Male albino Wistar rats aged 6 to 8 weeks	289 μg/kg /24 h	Oral administration	8 days	decapitation
Madsen, A., H.P. Mortensen, and B. Hald (1982) [60]	8-week-old pigs weighed between 14and 18 kg	1400 μg/kg	feeding	6 weeks	slaughtered
Zhang, X., et al.(2009) [61]	SH-SY5Y neuroblastoma cells	0.1, 0.25, 0.5, 1.0 and 2.5 μmol/L	incubation	24 h or 48 h	Centrifugal, collecting cells
Zurich, M.-G., et al. (2005) [63]	cell cultures of fetal rat telencephalon	0, 10, 20 nM	incubation	24 h, 48 h and 9 days	Centrifugal, collecting cells
Sava, V., et al.(2007) [64]	NSC cells (from HP of adult brains of 2-to 3-month- old C57BL/6J mice)	0.01 to 100 μg/ml	incubation	0, 24, 48, and 72 h	Centrifugal, collecting cells
Miki, T., et al.(1994) [65]	ICR mice	3 mg/kg	intraperitoneal injection	6 weeks	perfused intracardially with physiological saline and a fixative containing 2% paraformaldehyde and 2.5% glutaraldehyde in 0.1 M phosphate buffer
Szczech, G.M. and R.D. Hood (1981) [66]	CD-1 mice and ICR mice gestated 8-10 days or 15-17 days	2 to 5 mg/kg/day	gavage or by ip injection	1 or 3 days	kill with CO_2 and all fetuses immediately removed from the uterus
Tamaru, M., Y. Hirata, and T. Matsutani (1988) [67]	ICR mice gestated 11 days	3 mg/kg	injection intraperitoneally	2, 3, 4, 5, 6 days	Pregnant and fetal mice were sacrificed by ether anesthesia. Whole fetal brains were immediately removed, weighed and stored at – 80^0C
Fukui, Y., et al.(1992) [68]	ICR mice gestated 10 days	3 mg /kg	injected i.p.	1 day	Deeply anesthetized and perfused via the heart with fixative containing 2% paraformaldehyde 2.5% glutaraldehyde in 0.1M phosphate buffer
Bruinink, A. and C. Sidler (1997) [69]	cells of brain and neural retina a of Tetra SL chickens	0.002 to 20 μM	incubation	24 h	Centrifugal, collecting cells
Monnet-Tschudi, F., et al. (1996) [70]	cell cultures of foetal rat telencephalon	0.4 nM to 50 microM	incubation	10 days	Centrifugal, collecting cells
Gekle, M., H. Oberleithner, and S. Silbernagl (1993) [71]	male Wistar rats	1.25 μmol/(kg day)		6 days	

On the 2nd day of OTA treatment, DNA levels were decreased by 18% compared to that of the control group. Furthermore, on the 6th day, it remained suppressed. Concurrently, the protein concentration was reduced by 10% and the RNA concentration was reduced 10 to 15%. However, at the 6th day after injection, the protein and RNA levels recovered to normal values. On the 2nd, 4th and 5th days after injection, the RNA/DNA ratio of the OTA-treated groups were prominent elevated by 109-134% that of the controls. Three to four days after OTA treatment, acidic deoxyribonuclease, beta galactose glucoside enzyme, beta glucuronic acid enzyme had peak activities of 1.3-, 1.7-, and 1.7-times that of the control group, respectively. At the 6th day, they were back to normal again. The body weights of the mice three months after birth seemed to be slightly lower than those of the controls. OTA treatment on day 11 of gestation led to a prominent reduction in brain size, which is also called microencephaly. Different regions of the brain had different reductions in weight. The cerebral hemispheres were reduced to 79%, and the remainder of the brain was reduced to 91%. In the OTA-treated mice, the concentrations of dopamine (DA), noradrenaline (NA) and 5-hydroxytryptamine (5-HT) were all increased 15% in the cerebral hemispheres. Contrarily, the total content of monoamines was decreased by 15% ($P < 0.05$).

2.2.3 Neurons

Fukui et al. [68] found that after 3 mg/kg of OTA treatment cortical thickness and the synapse both decreased by 50%. Neurons and glial cells were normal. Synapse neuron number density increased by 39%, but it did not change neurons in the shorter synaptic processing group, meaning that the nuclei of the cerebral cortex neurons remained small, that the average distance between the basal dendritic body was reduced, and that the proportion of neurons was reduced. The dendrites of neurons in the OTA-treated mice may be shorter than those in the control mice. Bruinink and Sidler [69] found that OTA specifically affects neurite formation.

2.3. Research on Mechanisms

OTA exposure may lead to complex mechanisms of action, including oxidative stress, mitochondrial damage, inhibition of protein synthesis, and hippocampal nerve damage, leading to depression and cognitive impairment. Significant progress has been made in elucidating the mechanisms of OTA neurotoxicity, which will be discussed below.

Studies have found that it may be partially related to cell apoptosis. In primary rat cortical neurons and human neuroblastoma SHSY5Y cells, Doi and Uetsuka [72] concluded that caspase-9 and caspase-3 were activated and that caspase inhibitors were effective to a certain extent to offset OTA-induced neurotoxicity. OTA also activated caspase phosphorylation and p53 in various neural cells to induce oxidative stress and increase cell apoptosis. Zhang et al. [61] generated similar results, and Sava et al. [58] determined the effects of OTA on the oxidation of DNA repair.

Another piece of the explanation regards the inhibition of protein synthesis. Monnet-Tschudi et al. measured the intracellular oxidation of 2',7'-dichlorofluorescin and found that reactive oxygen species formation was increased significantly upon a high-concentration OTA treatment (10 to 50 μM). It can be concluded that OTA is toxic to brain cells.

Nanomolar concentrations of OTA inhibit protein synthesis, but ROS is not activated. Sava et al. [58] found that OTA competitively inhibited phenylalanine protein synthesis in the phenylalanine ammonia acylation reaction of tRNA and phenylalanine hydroxylase activity, which affects dopamine in the striatum, dopamine and catecholamine enzyme synthesis. Tamaru et al. [67] found that OTA may use liver damage to affect glycolysis and reduce the concentration of glucose that can be used in a fetus. OTA also inhibits mitochondrial respiration. In their study, they also observed that OTA was toxic to a fetus because it leads to microencephaly. OTA treatment at 11 days of pregnancy in pregnant mice caused significant intrauterine fetal growth retardation and developmental impairment of the fetal brain. Moreover, OTA suppressed total protein expression, and DNA content increased significantly. When OTA was presented at 8 or 9 days of pregnancy, fetal growth was retarded.

What cannot be ignored is the impact of OTA on the central nervous. Gekle et al. [71] studied this field in depth. Their experiments showed that OTA (from 0.1 to 1 mM) does not significantly affect sodium current in an intact myelinated axon. A 1 μM OTA treatment on a pronase treated fiber leads to the following effects: 1) decrease in the amplitude of the tail current with an unappreciable effect on the time constant of its fast-decaying component, 2) reduction of potassium channels by 70%, and 3) decrease in the leakage conductance. OTA reduced the potassium concentration gradient in MDCK (Madin-Darby canine kidney) cells, which may be due to the membrane or cell pH being out of the steady state. Pronase failed to interrupt the conversion of OTA to phenylalanine and OTα. The results of the experiment indicate that OTA exerts a relevant effect on the fast K^+ channels of demyelinated nerve fibers. OTA may stop the excitation by fast recovery current. OTA can lead to abnormal changes in the nerve membrane of K^+, which may lead to abnormal sensations, pain and other symptoms. OTA will affect the inward rectifier and at the same time affect the proliferation of Schwann cells. OTA can lead to neuroblastoma cells and a decrease in nerve fiber inner rectifier inactivation time constants. OTA can damage the potassium buffer, which has a long-term impact on cells, such as the formation of myelin.

3. Immunotoxicity

3.1. Research Status

Ochratoxin A (OTA) is a kind of immunosuppression fungal compound that is produced by Aspergillus and Penicillium fungi in a variety of climates and geographical areas. Almost all types of food can be contaminated. Previous risk assessments focused on other aspects of the compound, including genotoxicity and carcinogenicity, and ignored the immunotoxicity data. However, experimental studies have indicated that OTA modulates an immune response even at levels far below the toxicity threshold [73]. Klinkert et al. [74] found that OTA was able to inhibit chemotactic activity of murine peritoneal macrophages *in vitro*. Chang and Hamilton [75] fed chicks with 4 mg kg^{-1} of OTA, which impaired the motricity and phagocytosis of neutrophils.

Boorman et al. [76] administered 0, 20, 40 or 80 mg/kg of ochratoxin A to 6- to 7-week-old female B6C3Fr mice for 8 days. After 24 h from the last feed, histopathology, bone marrow, and macrophage parameters were assayed. A change in the structure of bone marrow progenitor cells and bone marrow cells were found in the 20 and 40 mg/kg dose groups. Their study indicated a possible mechanism of OTA myelotoxicity by decreasing the bone marrow progenitors of macrophage granulocytes.

Ochratoxin A has also been considered to make animals more susceptible to some diseases. Huff and Ruff [77] mixed ochratoxin A into feed at dosage levels of 0, 0.5, 1.0, 2.0, 4.0, and 8.0 μg/g and fed birds from one day of age. One group contained a combination of OTA and coccidiosis. Decreased body weights were observed with an increase in feed conversions after treatment with OTA. Ochratoxin A and coccidia can interact to cause effects on different growth and physiological parameters. OTA was indicated to increase susceptibility to coccidiosis. Stoev et al. [78] used pigs as the model organism and came to the same conclusions.

Elissalde et al. [79] split young male broiler chicks into two groups. One was treated with only with *Salmonella typhimurium* only and the other was treated with *S. typhimurium* and 3.0 mg/kg of OTA. *S. typhimurium* alone had little effect on chicks except for a 13.2% increase in mortality and a decrease in body weight compared to those of the controls. However, the combination of *S. typhimurium* and ochratoxin A can cause a significant synergistic increase in mortality. It was not more likely that OTA caused carcass contamination, although it would depress immunosuppression of the chicks. Eleven-day-old chickens were administered *Salmonella typhimurium* orally for two days and then given 3 mg/kg of OTA orally, as described by Fukata et al. [80]. The results, an increasing number of Salmonella typhimurium, meant that OTA affected the susceptibility of chickens to Salmonella colonization.

Kumar et al. [81] found that the presence of OTA in poultry feed increased the mortality and the severity of E. coli infections. They performed an experiment to study the effects of OTA in Escherichia coli-challenged broiler chickens using 184 chicks divided into two evenly sized groups. One was regarded as the control group and fed a mash diet, whereas the other was treated with mash diet including 2 ppm of OTA. On the 14th day, the two groups were further divided into another two groups. One was inoculated with E. coli O78, whereas the other group was not. After 14 days of inoculation, each group had 4 chicks taken on days 1, 2, 3, 5, 7, 10, 14, and 21. They discovered that the E. coli infection alone caused dullness, depression, huddling, and diarrhea. The mortality was 14.3% in chicks. However, in the OTA group infected with E. coli, the mortality increased to 35.7%. Additionally, body weight and feed intake decreased. The effects were more pronounced in chicks fed OTA and infected with E. coli. Decreased levels of total protein, albumin, globulins, calcium, and phosphorus and increased serum levels of aspartate aminotransferase, alanine aminotransferase, uric acid, and creatinine were observed in the OTA-fed group. Escherichia coli infection did not cause significant alteration to any of the serum biochemical parameters. It can be concluded that OTA exacerbates the effects of E. Coli in chicks.

Stoev et al. [82] found that ochratoxin A ingestion through contaminated feed increased the susceptibility of pigs to natural infection by Serpulina hyodysenteriae, Salmonella cholerasuis, and Campylobacter coli.

In vitro, OTA is also considered to impair the immune function of cells. Haubeck et al. [83] administered 0.005 pg/kg of body weight of ochratoxin A to sheep red blood cells in BALB/c mice, and OTA was found to suppress the immunogenicity. However, this effect is through the prevention of phenylalanine. Lea et al. [84] incubated peripheral blood human T lymphocytes with 2-fold dilutions of OTA in culture medium ranging in concentration from 12.5 to 100 μM for 60 min. The OTA treatment almost completely prevented the multiplication of cells, and the OTA inhibition of T-cell function can explain the suppressive effect on B-cell function.

Dortant et al. [85] investigated the immunotoxicity of OTA in young adult rats that were 12 weeks old and female rats that were 27-30 months old. The rats were treated with 0, 0.07, 0.34 or 1.68 mg/kg of OTA for 4 weeks. The results implied that the two groups did not differ in their OTA toxicity. The kidney and brain were the primary target organs, not the immune system. As expected, old rats were more sensitive to OTA than young rats. Table 2 summarizes the different effects of OTA on the immune response.

3.2. Pathological Changes in Different Tissues

3.2.1. Different Study Conditions

OTA immunosuppressive activity is an essential characteristic for reducing the size of immune organs, for example, thymus, lymph nodes, spleen, depression of antibody response, change in the number and function of immune cells, and an adjustment in the production of cytokines.

3.2.2. Lymph

In previous studies, it was revealed that OTA could cause lymphoid cell depletion and degenerative changes in the immune organs in the germinal center of chickens, including thymus, lymphoid follicles, spleen, bursa of Fabricius and Peyer's patches of the intestinal mucosa. Kanisawa et al. [90] performed experiments on adult male Wistar rats to study the acute oral toxicity of ochratoxin A. The oral dosage was estimated to be 28 mg/kg. They found that necrosis occurred in cells of the germinal centers of the spleen and lymph nodes, which indicated that lymphatic tissues were sensitive to OTA.

Kitchen et al. [92] administered ochratoxin A (0.1 and 0.2 mg/kg) to beagle dogs for 14 days. After administration of OTA, necrosis can be observed in lymphoid tissues in the tonsil, spleen, thymus, peripheral lymph nodes and lymph nodules of the ileum, colon and rectum of the dogs. Szczech et al. [96] also showed necrosis in mesentery lymph nodes in dogs. Others have confirmed reductions in the size of the bursa of Fabricius. Kozaczyński [97] fed 1 mg/kg of OTA to 1-day-old male chicks for 20 days and found that the size of the bursa of Fabricius was reduced with advanced atrophy of bursal follicles, which led to a marked reduction in the size of both bursal plicae as well as the whole organ.

However, there are some paradoxical conclusions by other researchers. Thuvander et al. [98] determined that the sizes of the lymphoid organs were not altered by OTA in female BALB/c mice or Sprague Dawley rats. Alvarez et al. used 12-week-old Wistar male rats to study the immunotoxic and genotoxic effects of OTA. Rats were gavaged with 50, 150 and 450 mg/kg of OTA for four weeks. The body weight of the 450 mg/kg OTA-treated rats was prominently decreased. The activity of T lymphocytes was decreased in the lowest group, whereas it did not change in the other two groups. Only a few necrotic-like follicular cells were found between lymphoid follicles in the thymus. Such contradictory effects may be due to different times and doses.

Table 2. Summary of experimental conditions of studies carried out on immunotoxic activities of ochratoxin

Reference	Species,strain,sex, age	Dose	Route of administratin	Exposure time	Material collection
Boorman et al. (1984) [76]	6 to 8 weeks B6C3Fi female mice	20,40, or 80 mg/kg body weight.	intraperitoneal injection	8 days	killed with CO_2 for necropsy
CHANG, C.-F. et al. (1980) [86]	One-day-old male turkey poults	0, 1, 2, 4, and 8 μg/g	feeding	1,2,3 weeks	
Dortant, P. et al. (2001) [85]	12 weeks old female Wag/ mlb rats	0, 0.07,0.34,1.68 mg/kg OTA/kg body weight	gavage	26-28 days	Anesthesia to death
Verma, J. et al. (2004) [87]	One-day-old Marek's broilers	1,2,4 mg/kg	feeding	49 days	
Raju,M.,Devegowda, G. (2002) [88]	960 day-old broilers	0 and 2 ppm	feeding	5 weeks	cervical dislocation
Campbell, M. et al. (1982) [89]	Male chicks	0 and 2 μg/g	feeding	1,2,3 weeks	
Kanisawa, M. et al. (1977) [90]	Adult male Wistar rats	15, 20, 25, 30, 35, 40, 45 and 50 mg/kg	oral	1 week	
Haubeck, H.-D.et al. (1981) [83]	8-to 12-week-old BALB/c mice	0.005, 0.05, 0.5, 5, 50 μg/kg	intraperitoneal injection	5 days	
Alvarez, L. et al. (2004) [91]	Wistar male rats (aged 12 weeks)	0, 50, 150 or 450 mg OTA/kg b.w.	gavage	28 days	sacrificed by CO_2 inhalation in a metacrylate chamber
Lea, T. et al. (1989) [84]	T lymphocytes	50 μM	incubation	60 min	Centrifugal, collecting cells
Kitchen, D. et al. (1977) [92]	10-week-old male Beagle dogs	0.1,0.2 mg/kg	oral	14 days	
Huff, W., Ruff, M. (1981) [77]	Broiler Chickens	0, 0.5, 1.0, 2.0, 4.0, and 8.0 μg/g	feeding	14 days	killed by cervical dislocation
Elissalde, M. et al. (1993) [79]	male broiler chicks	0,3 mg/kg	feeding	1,7,14,21 days	
Singh,G et al. (1990) [93]	broiler chicks	0.5 and 2.0 ppm	feeding	21 days	
Stoev, S. et al. (1999) [94]	broiler chicks	0,130,305,790 μg/kg	feeding	10 months	
Stoev, S. et al. (1999) [78]	pigs	0.2 and 0.066 mg /kg b. w.	feeding	21 days	slaughtered
Stoev, S.D. et al. (2002) [95]	2-weeks-old chicks	1 and 5 mg/kg	feeding	6 and 10 weeks	slaughtered

3.2.3. Thymus

Boorman et al. [76] treated female B6C3F1 mice as described above. In the high group, the weight of the mean thymus was decreased to 33% that of the controls. This reduction was observed 24 h after the last dose. In the inner cortex of the kidney, this change was slight, which may suggest why the thymus was more sensitive compared to the common target organs of OTA. The reduction in thymus size was also seen in Fischer rats and in chickens.

Boorman [99] used F344/N rats and gave OTA by gavage. They treated male and female rats with 0, 1, 4, or 16 mg/kg of ochratoxin A in corn oil by gavage 5 days per week for 16 days and 13 weeks, 9 months, 15 months, or 2 years. Within 6 days, all rats that received 16 mg/kg of ochratoxin A died, and the rats receiving the 4 mg/kg treatment lost weight. Compound-related lesions in rats included thymic atrophy, bone marrow hyperplasia, necrosis and hyperplasia of the forestomach epithelium, renal tubular cell degenerative and regenerative changes (nephropathy), and adrenal gland hemorrhage.

Aleksandrov and Dzhurov [100] assessed the effect of ochratoxin A on birds through an experiment carried out with 7-day-old broiler birds. They offered a standard feed mixture containing 4 mg/kg of pure ochratoxin A to the birds over the course of 21 days. There were no severe clinical symptoms or cases of death observed during the experimental period. However, some adverse symptoms were observed, such as the growth and development arrest of the birds, an increase in the relative weight of the kidneys, vascular disturbances and dystrophy more strongly manifested in the kidneys and to a weaker extent seen in the bursa of Fabricius, the thymus, and other organs. Singh et al. [93] added 0.5 and 2.0 ppm of OTA to the feed of broiler chicks. Both the 0.5 and 2.0 ppm OTA groups showed a significant decrease of the thymus, bursa of Fabricius and spleen compared to those of the controls.

In three weeks of age turkey poults, Chang et al. [86] mixed 0, 1, 2, 4, and 8 mg/g of OTA into their diet, which resulted in a decreased growth rate, enlarged proventriculus and gizzard, and regressed thymus in all turkeys of the 4 and 8 mg/g group. However, in this experiment, the sizes of the liver, kidneys, pancreas, spleen, and bursa of Fabricius were unaffected. It was strange that the thymuses of turkeys treated with OTA were degenerated but that the bursa of Fabricius was unchanged. As the thymus is the primary determinant of cell-mediated immunity, these results suggesting that OTA can inhibit the thymus suggest that OTA may damage turkeys' cellular immunity. In this study, a similar conclusion is obtained.

3.2.4. Spleen

In the spleen, a significant size reduction was observed in chickens and in old but not young rats after forced feeding for 4 weeks at 1.68 mg/kg of body weight.

Stoev et al. [94] did several experiments regarding effect of OTA on the spleen of chicks. In one of them, they grouped broiler chicks into five experimental groups with 20 birds in each: control group, diet without OTA; group 1, 130 µg/kg of OTA in the diet; group 3, 305 µg/kg of OTA; group 3, 790 µg/kg of OTA of OTA with treatment lasting for 42 or 70 days. In all four treatment groups, the relative weights of the thymus, bursa of Fabricius and spleen were significantly decreased compared to those of the control group. The strongest decrease of spleen at 42 days was established in group 2, and the strongest decrease of spleen at 70 days was established in group 3.

In another experiment, Stoev et al. [101] treated Bulgarian chickens with OTA and obtained similar results in the lymphoid organs (thymus, bursa of Fabricius, and spleen), which were obviously diminished in size.

In conclusion, all of these reports showed that OTA will change the size of the lymph, thymus, spleen, and bursa of Fabricius, and it can be concluded that the immune system is a target organ for OTA, albeit not more sensitively than the brain or kidney.

3.3. Research on Mechanisms

3.3.1. Ochratoxin and Cellular Immune Response

Johannisson et al. [102] found that exposure of dams at lower doses of OTA of 10 to 50 mg of OTA/kg of body weight during pregnancy can lead to an increase in the proliferative response of splenocytes and lymphocytes to mitogens in pups. Furthermore, at a higher dose of 500 mg/kg of body weight, decreased proliferation of splenic and thymic lymphocytes can be observed in response to mitogens in the pups. After OTA exposure, the cell number declined due to necrosis and apoptosis [103].

Then, Ul-Hassan et al. [104] determined that *in vitro* exposure of purified lymphocyte populations and subpopulations to OTA can change the ability of mitogens in the cells. Størmer and Lea [105] studied the effect of OTA on both early and late events during activation of human T lymphocytes. Grissmer et al. [106] showed that releasing Ca^{2+} from inside cells and Ca^{2+} influx across the plasma membrane leads to the mitogenic stimulation of human T cells and then, as a result, an increase in Ca^{2+}. They also observed that the absolute levels of Ca^{2+}, the duration of the Ca^{2+} response, and the PKC activity are not affected by OTA. They considered that other mechanisms than just a general inhibition of protein synthesis are operating. In one study [98], the pups exposed to a high dose of OTA had more immature, double-positive (CD^{4+} CD^{8+}) cells and fewer mature CD^{4+} cells. However, in another study, they obtained the opposite conclusion. In the spleens of the offspring, 250 mg/kg of OTA induced a decrease in the percentages of splenic CD^{4+} and CD^{8+} cells. Johannisson et al. [102] found that a low concentration of OTA induced an increase in the expression of markers CD69 and CD25 expressed on the cell surface of human lymphocytes after a mitogenic stimulation, whereas a higher concentration induced an inhibition of the expression of CD69, CD25 and CD71.

Lea et al. [84] concluded that a decrease in blastogenesis was linked to the impairment of IL-2 production and of IL-2 receptor expression of activated T lymphocytes. Then, Prior and Sisodia [107] found an interesting phenomenon that the suppression of blastogenesis seems to depend on the type of toxin administration. There was a significant depression of blast transformation in mice treated intraperitoneally with OTA and stimulated with concanavalin A, whereas oral administration of OTA did not depress blast transformation.

Furthermore, OTA is responsible for decreasing the cell viability of other immune cells. In mice, OTA suppressed marrow granulocyte macrophage progenitors [108] and induced lymphocytopenia, neutrophilia and eosinophilia. In neutrophils of healthy or esophageal and breast cancer patients, exposure to 50 mg/ml of OTA induced a reduction in their viability [109].

Short-term exposure of offspring to OTA induced an immunostimulatory effect in the first phase. For example, the proliferative responses of lymphocytes to polyclonal activation will be stimulated when suckling pups are given OTA via the milk. Conversely, Boorman et al. [76] reported that in B6C3F mice treated with 20 to 80 mg/kg of OTA, phagocytosis was stimulated by peritoneal macrophages. The authors considered that direct action of the substance on the cells or from its myelotoxic action could result in the negative effect on phagocytes and lead to a shift in the maturity of the cells.

3.3.2. Ochratoxin and Inflammation

At the beginning of the acute inflammatory process, vascular permeability increases, and neutrophils (PMN) infiltrate the tissue. Several mediators, such as ROS, contribute to the inflammatory response [110]. Ferrante et al. [111] found that a single oral administration of 10 mg/kg of OTA in mice induced oxidative damage and PMN infiltration in parenchymal organs. MPO is a marker of neutrophil infiltration. It has been considered to reflect the severity of inflammation in the tissues, and OTA induces a significant increase in MPO in the duodenum of the small intestine of mice and rats [45].

Although OTA causes intestinal inflammation *in vivo*, the mechanisms responsible for this effect are poorly understood. One of the mechanisms is a direct stimulatory effect on the production of pro-inflammatory cytokines by intestinal epithelial cells. In fact, many experiments have proven this conclusion [112], suggesting a possibly similar direct effect of this mycotoxin on cytokine production by gut epithelial cells.

NF-kB is always considered to be a typical proinflammatory signaling pathway. This is because proinflammatory cytokines, such as IL-1 and TNF-a, can activate NF-kB. NF-kB also plays an important role in the expression of other proinflammatory genes, including cytokines, chemokines, and adhesion molecules [113]. Hong et al. [114] found a significant increase in NF-kB expression after OTA exposure in rat embryonic midbrain cells, and Ramyaa and Padma [115] found the same in a human hepatoma cell line. Furthermore, after OTA treatment, NF-kB nuclear localization occurs, indicating that nuclear localization of NF-kB and its expression can occur simultaneously. In rat liver, Weidenbach et al. [116] found that OTA induced the release of TNF-a from blood-free perfusions. Additionally, Al-Anati et al. [117] acquired similar *in vitro* results. They showed that only macrophages and macrophage-derived cells respond to OTA and are considered as sources for TNF-a release after exposure to the toxin. Then, they showed mediated TNF-a release from the liver, whereas lipoxygenase and cytochrome P-450 have the opposite effect. It was suggested that the NF-kB transcription factor pathway may be related to the release of TNF-a induced by OTA.

The TNF-a receptor TNFRI and its transcriptional inhibitors, such as ActD, are necessary for TNF-a to induce apoptosis. OTA was able to induce apoptosis in different tissues, including liver, but Essid et al. [118] showed that TNF-a/TNFRI is not involved in OTA-mediated apoptosis in hepatocytes. Additionally, it was shown that in both healthy human or esophageal and breast cancer patients, OTA induces a reduction in the number of the TNF-a, G-CSF and CX3-CR1 receptors in neutrophils and lymphocytes [109]. A few studies have shown a decrease in TNF-a synthesis induced by OTA. For example, Mechoud et al. [119] found that, in human peripheral blood mononuclear cells after 24 h of exposure to 5 and 7.5 mg/mL of OTA, TNF-a production was completely inhibited.

OTA can also affect the synthesis of other inflammatory cytokines. One of the important mediators of the inflammatory response is IL-1 beta, which is related to various cellular activities, including cell proliferation, differentiation, and apoptosis [120]. In neutrophils of healthy human or esophageal and breast cancer patients, IL-1 was significantly increased in the supernatants. Interleukin 6 (IL-6), as another pro-inflammatory cytokine, was also reported to be induced from blood-free perfused rat liver by OTA. Nasal epithelial cell cultures also induced a significant increase of IL-6 and IL-8 upon OTA treatment [121].

3.3.3. Ochratoxin and Nitrosative Stress

Nitric oxide (NO) is a signaling molecule that plays a key role in the pathogenesis of inflammation. In pathological conditions, NO may result in the radical production of nitrosative stress. NO is considered to be a proinflammatory mediator and induces inflammation when it is over-produced [122].

Nitric oxide synthases (NOSs) are a kind of enzyme that can catalyze L-arginine into becoming NO. Nitric oxide is regulated and controlled by the isoenzymes eNOS (endothelial NOS) and nNOS (neuronal NOS). Inducible NO synthase (iNOS) is produced after activation by different factors, such as cytokines, to generate NO.

There are only a few studies that refer to the effects of OTA on NO synthesis. Divya et al. [123] found that nitric oxide was low in broiler chickens immunized with S. gallinarum from activated macrophages. Additionally, others have found a decrease in NO synthesis in the presence of NO with a starting concentration of 10 ng/mL of OTA. Thankfully, there are more studies regarding the effect of OTA on the expression of NOSs expression than on NO synthesis. Researchers have found that OTA effects iNOS expression in a dose-dependent manner. Thus, in J774A.1, OTA at lower concentrations (3 mM) will increase iNOS expression in macrophages, whereas when the concentration of OTA is up to 10 mM, iNOS expression is reduced [111].

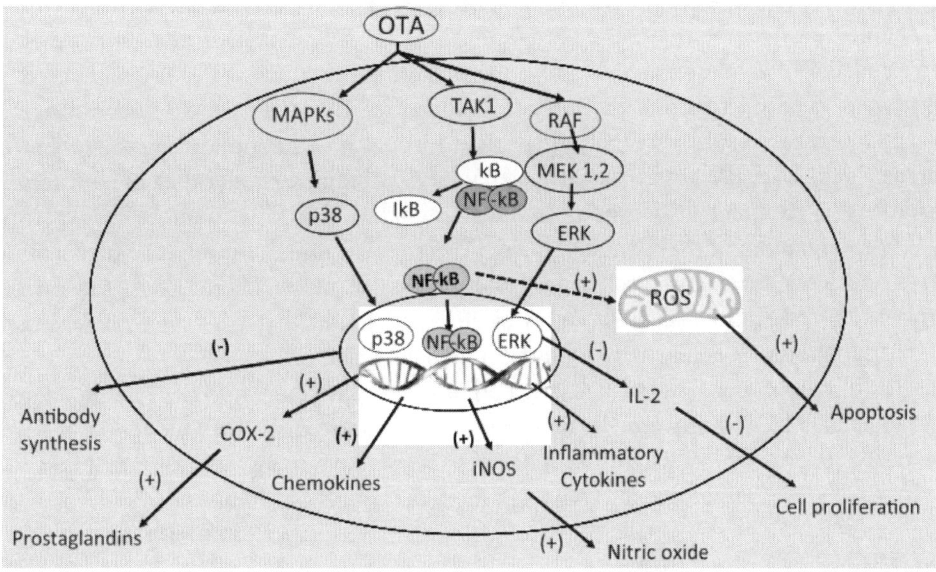

Figure 2. Schematic representation of the pathways affected after exposure to OTA [125].

OTA was able to induce iNOS in both kidney and liver, whereas OTA was able to induce eNOS only in kidney, resulting in increased nitrite levels [17]. OTA was found to stimulate the formation of NO through an NF-kB-dependent induction of iNOS. Increased expression of the iNOS gene was associated with the induction of the expression of genes involved in oxidative stress, for example, heme-oxygenase 1 (HO-1) is a very well-known mediator of oxidative stress [124]. The related pathways are showed in Figure 2.

4. REPRODUCTION TOXICITY

4.1. Research Status

4.1.1. OTA and the Placenta

The human placenta is very important for a fetus as the main interface between the mother and fetus. The placenta provides a pathway for oxygen and nutrition from a mother to a fetus. It also produces hormones to support pregnancy and functions to discharge various metabolites and carbon dioxide. However, it potentially concurrently exchanges some potential xenobiotics, including OTA [126].

Research on the effects of OTA on fetuses has been performed. Possible reasons for toxicity may be transporters, immature metabolism, and low elimination capacity of the fetus. The transfer of OTA is limited by its high-affinity binding to plasma proteins. It has been estimated that approximately 99% of circulating OTA is bound to plasma proteins and that only a very small amount of free OTA is present. Because of the fact that only an unbound molecule is capable of passing through the placental barrier, OTA transfer may be not concentration dependent in human perfused placenta [127].

Studies treating mice with OTA show that OTA in the progeny (the fetus and the offspring) through transplacental contamination can lead to DNA adducts, which reflect a direct genotoxic effect. Petkova-Bocharova et al. [128] found that, in the kidneys of male progeny and the liver of female progeny, there are high amounts of DNA adducts if their mothers were exposed to OTA during late pregnancy. These data are in accordance with previously described results that the target organ for males is the kidney and the target organ for females is the liver. Petkova-Bocharova et al. [129] obtained similar data in hamsters.

After OTA treatment, it can be detected in embryos in a large number of experiments [130]. At the beginning of pregnancy in mice, OTA quickly passes through the placenta. OTA arrives at the fetus after crossing the maternal placental barrier. There, OTA will affect the development of other organs [131]. Woo et al. [132] found that OTA is able to cross the placenta; furthermore, it is during early rather than late gestation.

In the tissue of deceased kittens, specific DNA adducts of OTA can be detected. Wangikar et al. [133] discovered that, during the organogenesis period, if OTA is given continuously, microscopic changes of vital organs of developing fetuses are likely to be observed, as are histopathologic alterations in the fetal organs of rabbits.

4.1.2. OTA and its Impact on Sperm

An important part of the person's overall health is male reproductive health and well-being. In general, the reproductive health of men has been ignored. Reproductive contraception and infertility problems have long been viewed as only relating to women. [134].

Table 3. Summary of experimental conditions of studies carried out on reproduction toxicity of ochratoxin A

Reference	Species,strain,sex, age	Dose	Route of administratin	Exposure time	Material collection
Solti, L. et al. (1999) [139]	boars	20 and 40 μg	feeding	5 weeks	Semen samples were collected two times a week with gloved hands
Ballinger, M.B., T.D. Phillips, and L.F. Kubena, (1986) [147]	Sprague-Dawley rats	2.5 mg/kg	subcutaneous injection	1, 2, 4, 6, 12, 24, and 48 h	euthanasia
Appelgren, L.-E.and R. Arora (1983) [131]	Female mice (C57bl) gestated 9, 10 and 17 days	150--170 mg/kg	injected intravenously	3-4 h	All the injected animals were subjected to the modified Ullberg whole body autoradiography method originally published in 1954. Some sections of pregnant uteri were photographed under UV-light to show the blue fluorescence of ochratoxin
Fenske, M. and J. Fink-Gremmels (1990) [148]	Interstitial cells of Mongolian gerbils	496, 2481, 49618 nM	incubation	4 h	Centrifugal, collecting cells
More,J.andR.Camguilhem (1978)[149]	Male Sprague-Dawley rats	4, 4.2, 4.6, 5 mg/kg	Injection	10 days	euthanized
Verma,R. and D. Chakraborty (2008) [135]	Wistar rats	50, 100 μg/ 0.2 mL olive oil/day	feeding	45 days	cervical dislocation
Biró, K., et al.(2003) [138]	boars	20 μg/day	feeding	6 weeks	Semen was collected twice a week in accordance with daily routine work
Woo, C.S.J. et al.(2012) [127]	Microsomal and cytosolic fractions of human placenta	1 μM	incubation	3 h	Centrifugal, collecting cells
Bouaziz, C., et al.(2010) [150]	HeLa, HCT116, HCT116-Bax-KO, HCT116-Bak-KO	15 μM	incubation	48 h	Centrifugal, collecting cells
Abdel-Wahhab, M., S. Nada, and M.Arbid (1999) [151]	Sprague-Dawley female rats gestated 6-15 days	1 mg/kg b.w	orally	To day 20 of pregnancy	

Table 3. (Continued)

Reference	Species,strain,sex, age	Dose	Route of administratin	Exposure time	Material collection
Gupta, M., et al.(1980) [152]	Albino female rats	5 mg/kg	Injection	twice weekly for 15 day	Immediately after sacrifice, the ovaries and uterus were dissected out from each animal and made free from adherent fat tissues.
Woo, C.S.J.,et al. (2013) [132]	Human placental JEG-3 choriocarcinoma cell line	0,10,50,100 nM; 1,10 μM	incubation	24, 48, 72 and 96 h.	Centrifugal, collecting cells
Castegnaro, M., et al. (1998) [153]	Male and female DA and Lewis rats	0.4 mg OTA per kg body weight	intragastric intubation	3 times a week for their whole lifespan	CO_2 asphyxiation
Gharbi, A., et al. (1993) [154]	Male Wistar rats	289 μg/kg	gastric intubation	every 48 h for 2, 4, 6 and 8 weeks	Euthanasia by decapitation
Chakraborty, D. and R. Verma (2009) [137]	inbred Swiss strain male albino mice	50 μg/0.2 mL olive oil/ animal/day	orally	45 days	cervical dislocation

Verma and Chakraborty [135] found that, in many countries over the last few decades, male reproductive health has deteriorated with the passing of time. Toxins in the environment have always been considered to be one of the reasons why. As toxins are deemed to damage the male reproductive system, male fertility can be affected by impairment of sperm production and/or damage of testis.

Several studies have shown that OTA treatment in mice can significantly decrease sperm count and motility and increase abnormalities in sperm morphology [136]. Chakraborty and Verma [137] gave mice 50 and 100 μg/0.2 mL of OTA and came to the conclusion that OTA can alter sperm count, viability, and motility and fertility rate in a dose-dependent manner. The same phenomenon also appears in pigs [138]. Both 20 and 40 μg of OTA reduces the fertility of pigs, and OTA was found in the seminal liquid of the pigs. The decrease of boar semen sperm production and quality may be caused by an unstable spermatozoal membrane caused by OTA [139].

For the testis of many animals, OTA is toxic. One region of high incidence is the Balkans, where the food is contaminated by OTA. In these areas, there is a high chance of developing testicular cancer. For example, in Vas, Hungary, the incidence of testicular cancer increased fivefold from 1960 to 1985. A similar increase was observed in Slovenia [140]. Recently, some experiments on rodents have shown that carcinogenesis can be induced in testis if exposed to OTA. Ueta et al. [141] fed pregnant mice 2 mg/kg of OTA intraperitoneally. On the 9th day, the expression of the gene Dmrt-I was decreased markedly. Doublesex and mab-3-related transcription factor 1 play an important role in the development of the mammalian testis, and it is also known as a tumor suppressor gene in the testis. If this gene is altered, testicular tumors can emerge more easily in the testes of mice [142] and humans [143]. Mantle and Nolan [144] administered 100 μg/d of OTA to 24 male rats for 35 weeks and found testis tumors in 25% of the rats. Research of Schwartz et al. [145] confirmed the above finding by the phenomenon that DMRT1 expression in mice would decreased after OTA treatment during the pregnancy period, especially in male offspring. DNA adducts are widely considered to be markers of an increased risk of cancer. Jennings-Gee et al. [146] discovered that when adult mice were treated with OTA, DNA adducts occurred in the testis, whereas when adult mice were treated without OTA, DNA adducts did not occur in the testis. Thus, OTA is considered a carcinogenic toxin for males.

Therefore, important molecular evidence supports the hypothesis that OTA may have a causal relationship to testicular tumor in mice and male germ cells. Future epidemiologic and molecular studies in humans should focus on whether OTA is a causative agent in testicular cancer. Table 3 summarizes the different effects of OTA on the reproduction.

4.2. Research on Mechanisms

4.2.1. OTA and Endocrine Disruption

The functions of all living organisms are controlled by hormones. The consequences of endocrine disruption are reproductive barriers (interference with the reproductive cycle to stop ovulation, infertility, loss of libido, interference with spermatogenesis, neoplasmic damage, altered behavior, reduce offspring making it to birth), osteoporosis, bone marrow fibrosis, adenoma and bone deformation [155]. Zinedine et al. [156] found that OTA may disturb the stability of the endocrine system by affecting enzymes in the process of synthesizing steroids or hormones.

In the progress of biosynthesis of all types of steroid hormones, the enzymatic complex 3β-hydroxysteroid dehydrogenase (3β-HSD) is necessary. It is located in the mitochondria and endoplasmic reticulum. The first step of the biosynthesis of androgens and estrogens is catalyzed by 3β-HSD [157]. Gupta et al. [152] chose Albino female rats of 150 to 200 g of body weight. They divided the rats into four groups: group I, vehicle control; group II, 1 mg/kg of ochratoxin A treatment; group III, 3 mg/kg of ochratoxin A treatment; and group IV, 5 mg/kg of ochratoxin A treatment. Each group contained 12 rats. OTA treatment was twice weekly for 2 weeks. Administration of 5 mg/kg of ochratoxin A caused a decrease of the 3β-HSD activities in the three compartments of the ovary and reduces ovarian steroidogenesis. These changes were not found in the case of ochratoxin A-treated rats of groups II and III. The researchers also used male Wistar rats. They distributed the rats into four groups: group I rats were fed OTA at 4 mg/kg; group II rats were fed endosulfan at 5 mg/kg; group III rats were fed OTA at 4 mg/kg and endosulfan at 5 mg/kg; and group IV rats were fed a standard diet. After 30 days, all of the rats were sacrificed. They found that triiodothyronine, testosterone, insulin, and cortisol were decreased and that prolactin and thyroxin were increased. Fenske and Fink-Gremmels [148] found that OTA inhibits testosterone secretion *in vitro*. In rats, OTA inhibits both cortisol and testosterone secretions [158]. In another rat study, Nagar [158] found that 1) OTA will make rat serum T3 iodine (thyroid original glycine) levels decrease while at the same time making the T4 iodine (thyroid original glycine) levels increase; 2) OTA will make serum testosterone levels decrease significantly; 3) OTA can improve blood sugar while at the same time reduce serum insulin levels; and 4) OTA decreases cortisol levels.

4.2.2. OTA and Molecular Mechanisms of Teratogenicity

Wilk-Zasadna and Minta [159] found that OTA induces cytotoxicity of all brain cells of both neurons and astrocytes. They used computer imaging to reveal that the ability of OTA teratogenic function was associated with their effects on the aggregation of neuronal cells and the formation of neurospheres. Katagiri et al. [160] treated Pdn/Pdn embryos intraperitoneally with 2 mg/kg of OTA. They discovered in Pdn/+ females on day 7 and/or 7.5 of gestation that the incidence of NTD in Pdn/Pdn increased from 51.6%. Gli3 can induce polydactyly/arhin encephaly in the (Pdn/Pdn) mice. The Fgf8 and Emx2 genes can lead to down-regulation of the Gli3 gene in the Pdn/Pdn mice. The Fgf8 gene is a growth factor important for the development of the telencephalon. Emx2 plays an important role in the morphogenesis of the telencephalon, diencephalon, and mesencephalon [161], and OTA treatment will cause these genes to be overexpressed and decrease Gli3 gene expression, which was the result seen in the Pdn/Pdn mice.

In recent years, Ueta et al. [141] has studied the role of gender on the incidence of NTDs induced by OTA in Pdn/Pdn mouse. In their research, it has been suggested that many genes are linked to the manifestation of NTDs in the male OTA-treated Pdn/Pdn mice, such as Gli3, Wntb7b, Wntb8b, Fez1, Barx1, Lim1, Dmrt1, Igf1, Fog2, Dax1, and Sox9. Barx1 and Sox9 are both gender-dependent genes, and they are reduced after OTA treatment. Moreover, they found in these mice that OTA exposure brings about the down-regulation of DMRT1 expression. Petkova-Bocharova et al. [128] showed that a 0.5 or 2 mg/kg OTA dose given by gastric intubation to the mother on the 17th or 18th day of gestation induced OTA-DNA adducts in the kidneys and livers of the fetuses and in newborns several months after birth, furthermore, tumors arose in some of the latter. They also obtained similar results in hamsters.

4.2.3. OTA and MicroRNAs

A number of studies have given attention to the impact of OTA on male reproduction. However, studies on a possible mechanism are fewer. Meanwhile, studies on the role of miRNA on the effect are even fewer. Therefore, it is important to investigate the OTA treatment on males and the function of miRNA. Chen et al. [162] performed a study that aimed to characterize the impact of OTA on GC-2 cells and elucidate the role of miRNA in OTA-induced apoptosis in GC-2 cells. This cell line is a spermatocyte-like cell line and was originally created by Hofmann et al. [163] to study the process of germ cell differentiation and apoptosis.

In this study, GC-2 cells were first with gradient OTA concentrations of 0, 1, 5, 10, 20, 30, 50, and 100 μM. After 24 h, Cell Counting Kit-8 (CCK8) was used to determine cell viability. At the same time, cell apoptosis was detected by in situ DNA TUNEL labeling and Annexin V/Propidium Iodide staining. From these results, it can be concluded that the OTA treatment resulted in a significant decrease in cell viability of GC-2 cells from 30 μM and that the decrease was dose-dependent. Additionally, apoptotic GC-2 cells increased along with the increasing concentration of OTA. When the concentration reached 20 μM, there was significant difference between the experiment group and the control group.

MicroRNAs (miRNAs) are short noncoding single-stranded RNA molecules. miRNAs regulate various cellular processes, including spermatogenesis. In recent studies, several miRNAs were considered to have an important role in the regulation of the spermatogenesis process, including *miR-34c* [164], *miR-146* [165], *miR-15a* [166], *miR-122* [167] and others. To study the relationship between miRNAs and the effect of OTA on GC-2 cells, the expression of a series of genes was detected after OTA treatment. It was observed that Dicer was significantly increased in the 20 μM OTA-treated GC-2 cells, whereas Drosha did not change. Moreover, in the OTA-treated GC-2 cells, there was no obvious change in the expression of *miR-449, miR-15a, miR-34c or miR-129.* However, the expression levels of miR-146 and *miR-122* obviously increased, especially the expression of *miR-122*, which was significantly up-regulated in the OTA-treated GC-2 cells. This suggests that *miR-122* may play an important role in OTA toxicity of GC-2 cells. To test this hypothesis, further experiments were conducted. *MiR-122* inhibitor was transferred to the GC-2 cells to downregulate *miR-122* expression. Then, OTA was added for 24 h. Using real-time PCR, the expression of *miR-122* was found to decrease by approximately 50%, and in situ DNA TUNEL labeling results showed that cell apoptosis was also decreased. These data indicate that *miR-122* partially mediates OTA-induced cell apoptosis.

In previous studies, several genes have been proven to be direct targets of *miR-122*, including Bcl-w [168] and Cyclin G1 [169]. With the purpose of determining the collection between *miR-122* regulation and OTA toxicity in GC-2 cells, the expression of *Bcl-w, CCNG1, caspase-3*, and *p53* was detected by real-time PCR. As a result, the expression levels of *Bcl-w* and *cyclin G1* were both reduced and *caspase-3* and *p53* were both increased. As *caspase-3* and p53 are downstream of *Bcl-w* and *CCNG1*, respectively, these results indicate that *miR-122* may induce the down-regulation of *Bcl-w* and suppress the expression of

CCNG1. Then, *caspase-3* could be activated by *Bcl-w*, and *p53* could undergo negative feedback regulation by *CCNG1*. Thus, cell apoptosis occurred after OTA treatment.

REFERENCES

[1] Shotwell, O. L., Hesseltine, C., Goulden, M. L., (1969). Note on the natural occurrence of ochratoxin A. *Journal of the Association of Official Analytical Chemists*. 52, 81-83.

[2] Broce, D., Grodner, R., RL, K., Bonner, F., (1970). Ochratoxins-A-and-B confirmation by microbiological assay using bacillus-cereus-mycoides. *Journal of the Association of Official Analytical Chemists*. 53, 616-&.

[3] Heller, K., Schulz, C., Löser, R., Röschenthaler, R., (1975). The inhibition of bacterial growth by ochratoxin A. *Can. J. Microbiol.* 21, 972-979.

[4] Ali - Vehmas, T., Rizzo, A., Westermarck, T., Atroshi, F., (1998). Measurement of Antibacterial Activities of T - 2 Toxin, Deoxynivalenol, Ochratoxin A, Aflatoxin B1 and Fumonisin B1 Using Microtitration Tray - based Turbidimetric Techniques. *Journal of Veterinary Medicine Series A*. 45, 453-458.

[5] Singer, U., Röschenthaler, R., (1978). Induction of autolysis in Bacillus subtilis by ochratoxin A. *Can. J. Microbiol.* 24, 563-568.

[6] Konrad, I., Röschenthaler, R., (1977). Inhibition of phenylalanine tRNA synthetase from Bacillus subtilis by ochratoxin A. *FEBS Lett.* 83, 341-347.

[7] Heller, K., Röschenthaler, R., (1978). Inhibition of protein synthesis in Streptococcus faecalis by ochratoxin A. *Can. J. Microbiol.* 24, 466-472.

[8] Bunge, I., Dirheimer, G., Röschenthaler, R., (1978). *In vivo in vitro*inhibition of protein synthesis in Bacillus stearothermophilus by ochratoxin A. *Biochem. Biophys. Res. Commun.* 83, 398-405.

[9] Moore, J.H., Truelove, B., (1970). Ochratoxin A: Inhibition of mitochondrial respiration. *Science*. 168, 1102-1103.

[10] Meisner, H., Chan, S., (1974). Ochratoxin A, an inhibitor of mitochondrial transport systems. *Biochemistry*. 13, 2795-2800.

[11] Omar, R.F., Hasinoff, B.B., Mejilla, F., Rahimtula, A.D., (1990). Mechanism of ochratoxin A stimulated lipid peroxidation. *Biochem. Pharmacol.* 40, 1183-1191.

[12] Malaveille, C., Brun, G., Bartsch, H., (1994). Structure-activity studies in E. coli strains on ochratoxin A (OTA) and its analogues implicate a genotoxic free radical and a cytotoxic thiol derivative as reactive metabolites. *Mutation Research/Fundamental and Molecular Mechanisms of Mutagenesis*. 307, 141-147.

[13] Hoehler, D., Marquardt, R.R., McIntosh, A.R., Xiao, H., (1996). Free radical generation as induced by ochratoxin A and its analogs in bacteria (Bacillus brevis). *J. Biol. Chem.* 271, 27388-27394.

[14] Bendele, A.M., Carlton, W.W., Krogh, P., Lillehoj, E., (1985). Ochratoxin A carcinogenesis in the (C57BL/6J× C3H) F1 mouse. *J. Natl. Cancer Inst.* 75, 733-742.

[15] Creppy, E., Kane, A., Dirheimer, G., Lafarge-Frayssinet, C., Mousset, S., Frayssinet, C., (1985). Genotoxicity of ochratoxin A in mice: DNA single-strand break evaluation in spleen, liver and kidney. *Toxicol. Lett.* 28, 29-35.

[16] Pfohl-Leszkowicz, A., Grosse, Y., Kane, A., Creppy, E.E., Dirheimer, G., (1993). Differential DNA adduct formation and disappearance in three mouse tissues after treatment with the mycotoxin ochratoxin A. *Mutation Research/Fundamental and Molecular Mechanisms of Mutagenesis*. 289, 265-273.

[17] Sorrenti, V., Di Giacomo, C., Acquaviva, R., Barbagallo, I., Bognanno, M., Galvano, F., (2013). Toxicity of ochratoxin a and its modulation by antioxidants: A review. *Toxins (Basel)*. 5, 1742-1766.

[18] Wehner, F., Thiel, P., Van Rensburg, S., Demasius, I.P., (1978). Mutagenicity to Salmonella typhimurium of some Aspergillus and Penicillium mycotoxins. *Mutation Research/Genetic Toxicology*. 58, 193-203.

[19] Hennig, A., Fink-Gremmels, J., Leistner, L., (1990). Mutagenicity and effects of ochratoxin A on the frequency of sister chromatid exchange after metabolic activation. *IARC Sci. Publ.*, 255-260.

[20] Föllmann, W., Lucas, S., (2003). Effects of the mycotoxin ochratoxin A in a bacterial and a mammalian *in vitro* mutagenicity test system. *Arch. Toxicol.* 77, 298-304.

[21] Obrecht-Pflumio, S., Chassat, T., Dirheimer, G., Marzin, D., (1999). Genotoxicity of ochratoxin A by Salmonella mutagenicity test after bioactivation by mouse kidney microsomes. *Mutation Research/Genetic Toxicology and Environmental Mutagenesis*. 446, 95-102.

[22] Wegst, W., Lingens, F., (1983). Bacterial degradation of ochratoxin A. *FEMS Microbiol. Lett.* 17, 341-344.

[23] Hwanga, C.-A., Draughon, F.A., (1994). Degradation of ochratoxin A by Acinetobacter calcoaceticus. *Journal of Food Protection®*. 57, 410-414.

[24] Škrinjar, M., Rašić, J.L., Stojičić, V., (1996). Lowering of ochratoxin A level in milk by yoghurt bacteria and bifidobacteria. *Folia Microbiol. (Praha)*. 41, 26-28.

[25] Böhm, J., Grajewski, J., Asperger, H., Cecon, B., Rabus, B., Razzazi, E., (2000). Study on biodegradation of some A-and B-trichothecenes and ochratoxin A by use of probiotic microorganisms. *Mycotoxin research*. 16, 70-74.

[26] Piotrowska, M., Zakowska, Z., (2005). The elimination of ochratoxin A by lactic acid bacteria strains. *Polish J. Microbiol.* 54, 279-286.

[27] Schatzmayr, G., Zehner, F., Täubel, M., Schatzmayr, D., Klimitsch, A., Loibner, A.P., Binder, E.M., (2006). Microbiologicals for deactivating mycotoxins. *Mol. Nutr. Food Res.* 50, 543-551.

[28] Fuchs, S., Sontag, G., Stidl, R., Ehrlich, V., Kundi, M., Knasmüller, S., (2008). Detoxification of patulin and ochratoxin A, two abundant mycotoxins, by lactic acid bacteria. *Food Chem. Toxicol.* 46, 1398-1407.

[29] Petchkongkaew, A., Taillandier, P., Gasaluck, P., Lebrihi, A., (2008). Isolation of Bacillus spp. from Thai fermented soybean (Thua - nao): screening for aflatoxin B1 and ochratoxin A detoxification. *J. Appl. Microbiol.* 104, 1495-1502.

[30] Rodriguez, H., Reveron, I., Doria, F., Costantini, A., De Las Rivas, B., Munoz, R., Garcia-Moruno, E., (2011). Degradation of ochratoxin a by Brevibacterium species. *J. Agric. Food Chem.* 59, 10755-10760.

[31] Upadhaya, S.D., Song, J.Y., Park, M.A., Seo, J.K., Yang, L., Lee, C.H., Cho, K.J., Ha, J.K., (2012). Isolation, screening and identification of swine gut microbiota with

ochratoxin A biodegradation ability. *Asian-Australasian journal of animal sciences*. 25, 114.

[32] Shi, L., Liang, Z., Li, J., Hao, J., Xu, Y., Huang, K., Tian, J., He, X., Xu, W., (2014). Ochratoxin A biocontrol and biodegradation by Bacillus subtilis CW 14. *J. Sci. Food Agric.* 94, 1879-1885.

[33] Chang, X., Wu, Z., Wu, S., Dai, Y., Sun, C., (2015). Degradation of ochratoxin A by Bacillus amyloliquefaciens ASAG1. *Food Additives & Contaminants: Part A*. 32, 564-571.

[34] Stander, M.A., Steyn, P.S., van der Westhuizen, F.H., Payne, B.E., (2001). A kinetic study into the hydrolysis of the ochratoxins and analogues by carboxypeptidase A. *Chem. Res. Toxicol.* 14, 302-304.

[35] Piotrowska, M., Żakowska, Z., (2000). The biodegradation of ochratoxin A in food products by lactic acid bacteria and baker's yeast. *Prog. Biotechnol.* 17, 307-310.

[36] Del Prete, V., Rodriguez, H., Carrascosa, A.V., de las Rivas, B., Garcia-Moruno, E., Munoz, R., (2007). *In vitro* removal of ochratoxin A by wine lactic acid bacteria. *Journal of Food Protection®*. 70, 2155-2160.

[37] Wu, Q., Dohnal, V., Huang, L., Kuca, K., Wang, X., Chen, G., Yuan, Z., (2011). Metabolic pathways of ochratoxin A. *Curr. Drug Metab.* 12, 1-10.

[38] Mobashar, M., Hummel, J., Blank, R., Südekum, K.-H., (2010). Ochratoxin A in ruminants–a review on its degradation by gut microbes and effects on animals. *Toxins (Basel)*. 2, 809-839.

[39] Zanic-Grubisic, T., Zrinski, R., Cepelak, I., Petrik, J., Radic, B., and Pepeljnjak, S. (2000). Studies of ochratoxin A-induced inhibition of phenylalanine hydroxylase and its reversal by phenylalanine. *Toxicol. Appl. Pharmacol.* 167, 132–139.

[40] Ringot, D., Chango, A., Schneider, Y.-J., Larondelle, Y., (2006). Toxicokinetics and toxicodynamics of ochratoxin A, an update. *Chem. Biol. Interact.* 159, 18-46.

[41] Berger, V., Gabriel, A.-F., Sergent, T., Trouet, A., Larondelle, Y., Schneider, Y.-J., (2003). Interaction of ochratoxin A with human intestinal Caco-2 cells: possible implication of a multidrug resistance-associated protein (MRP2). *Toxicol. Lett.* 140, 465-476.

[42] Galtier, P., Alvinerie, M., Charpenteau, J., (1981). The pharmacokinetic profiles of ochratoxin A in pigs, rabbits and chickens. *Food Cosmet. Toxicol.* 19, 735-738.

[43] Roth, A., Chakor, K., EkuéCreepy, E., Kane, A., Roschenthaler, R., Dirheimer, G., (1988). Evidence for an enterohepatic circulation of ochratoxin A in mice. *Toxicology*. 48, 293-308.

[44] Purchase, I., Theron, J., (1968). The acute toxicity of ochratoxin A to rats. *Food Cosmet. Toxicol.* 6, 479-483.

[45] Kanisawa, M., Suzuki, S., Moroi, K., (1989). The mode of action of ochratoxin A in acute enteritis in rats. *Journal of environmental pathology, toxicology and oncology: official organ of the International Society for Environmental Toxicology and Cancer*. 10, 56-63.

[46] Szczech, G., Carlton, W., Tuite, J., Caldwell, R., (1973). Ochratoxin A toxicosis in swine. *Veterinary Pathology Online*. 10, 347-364.

[47] Kubena, L., Phillips, T., Witzel, D., Heidelbaugh, N., (1984). Toxicity of ochratoxin A and penicillic acid to chicks. *Bull. Environ. Contam. Toxicol.* 32, 711-716.

[48] Maresca, M., Yahi, N., Younès-Sakr, L., Boyron, M., Caporiccio, B., Fantini, J., (2008). Both direct and indirect effects account for the pro-inflammatory activity of enteropathogenic mycotoxins on the human intestinal epithelium: Stimulation of interleukin-8 secretion, potentiation of interleukin-1β effect and increase in the transepithelial passage of commensal bacteria. *Toxicol. Appl. Pharmacol.* 228, 84-92.

[49] Maresca, M., Mahfoud, R., Pfohl-Leszkowicz, A., Fantini, J., (2001). The mycotoxin ochratoxin A alters intestinal barrier and absorption functions but has no effect on chloride secretion. *Toxicol. Appl. Pharmacol.* 176, 54-63.

[50] McLaughlin, J., Padfield, P.J., Burt, J.P., O'Neill, C.A., (2004). Ochratoxin A increases permeability through tight junctions by removal of specific claudin isoforms. *American Journal of Physiology-Cell Physiology*. 287, C1412-C1417.

[51] Lambert, D., Padfield, P.J., McLaughlin, J., Cannell, S., O'Neill, C.A., (2007). Ochratoxin A displaces claudins from detergent resistant membrane microdomains. *Biochem. Biophys. Res. Commun.* 358, 632-636.

[52] Maresca, M., Fantini, J., (2010). Some food-associated mycotoxins as potential risk factors in humans predisposed to chronic intestinal inflammatory diseases. *Toxicon*. 56, 282-294.

[53] Grenier, B., Applegate, T.J., (2013). Modulation of intestinal functions following mycotoxin ingestion: Meta-analysis of published experiments in animals. *Toxins (Basel)*. 5, 396-430.

[54] Guo, M., Huang, K., Chen, S., Qi, X., He, X., Cheng, W.-H., Luo, Y., Xia, K., Xu, W., (2014). Combination of metagenomics and culture-based methods to study the interaction between ochratoxin a and gut microbiota. *Toxicol. Sci.* 141, 314-323.

[55] Sun, Z., Mukherjee, B., Brook, R.D., Gatts, G.A., Yang, F., Sun, Q., Brook, J.R., Fan, Z., Rajagopalan, S., (2013). Air-Pollution and Cardiometabolic Diseases (AIRCMD): A prospective study investigating the impact of air pollution exposure and propensity for type II diabetes. *Science Of the Total Environment*. 448, 72-78.

[56] Marin-Kuan, M., Cavin, C., Delatour, T., Schilter, B., (2008). Ochratoxin A carcinogenicity involves a complex network of epigenetic mechanisms. *Toxicon*. 52, 195-202.

[57] Belmadani, A., Tramu, G., Betbeder, A., Creppy, E., (1998). Subchronic effects of ochratoxin A on young adult rat brain and partial prevention by aspartame, a sweetener. *Human & experimental toxicology*. 17, 380-386.

[58] Sava, V., Reunova, O., Velasquez, A., Harbison, R., Sanchez-Ramos, J., (2006). Acute neurotoxic effects of the fungal metabolite ochratoxin-A. *Neurotoxicology*. 27, 82-92.

[59] Belmadani, A., Tramu, G., Betbeder, A., Steyn, P., Creppy, E., (1998). Regional selectivity to ochratoxin A, distribution and cytotoxicity in rat brain. *Archives of toxicology*. 72, 656-662.

[60] Madsen, A., Mortensen, H.P., Hald, B., (1982). Feeding experiments with ochratoxin A contaminated barley for bacon pigs: I. Influence on pig performance and residues. *Acta Agriculturae Scandinavica*. 32, 225-239.

[61] Zhang, X., Boesch-Saadatmandi, C., Lou, Y., Wolffram, S., Huebbe, P., Rimbach, G., (2009). Ochratoxin A induces apoptosis in neuronal cells. *Genes & nutrition*. 4, 41-48.

[62] Yoon, S., Cong, W.-T., Bang, Y., Lee, S.N., Yoon, C.S., Kwack, S.J., Kang, T.S., Lee, K.Y., Choi, J.-K., Choi, H.J., (2009). Proteome response to ochratoxin A-induced apoptotic cell death in mouse hippocampal HT22 cells. *Neurotoxicology*. 30, 666-676.

[63] Zurich, M.-G., Lengacher, S., Braissant, O., Monnet-Tschudi, F., Pellerin, L., Honegger, P., (2005). Unusual astrocyte reactivity caused by the food mycotoxin ochratoxin A in aggregating rat brain cell cultures. *Neuroscience*. 134, 771-782.

[64] Sava, V., Velasquez, A., Song, S., Sanchez-Ramos, J., (2007). Adult hippocampal neural stem/progenitor cells *in vitro* are vulnerable to the mycotoxin ochratoxin-A. *Toxicological Sciences*. 98, 187-197.

[65] Miki, T., Fukui, Y., Uemura, N., Takeuchi, Y., (1994). Regional difference in the neurotoxicity of ochratoxin A on the developing cerebral cortex in mice. *Developmental brain research*. 82, 259-264.

[66] Szczech, G.M., Hood, R.D., (1981). Brain necrosis in mouse fetuses transplacentally exposed to the mycotoxin ochratoxin A. *Toxicology and applied pharmacology*. 57, 127-137.

[67] Tamaru, M., Hirata, Y., Matsutani, T., (1988). Neurochemical effects of prenatal treatment with ochratoxin A on fetal and adult mouse brain. *Neurochemical research*. 13, 1139-1147.

[68] Fukui, Y., Hayasaka, S., Itoh, M., Takeuchi, Y., (1992). Development of neurons and synapses in ochratoxin A-induced microcephalic mice: a quantitative assessment of somatosensory cortex. *Neurotoxicology and teratology*. 14, 191-196.

[69] Bruinink, A., Sidler, C., (1997). The neurotoxic effects of ochratoxin-A are reduced by protein binding but are not affected byl-phenylalanine. *Toxicology and applied pharmacology*. 146, 173-179.

[70] Monnet-Tschudi, F., Sorg, O., Honegger, P., Zurich, M., Huggett, A., Schilter, B., (1996). Effects of the naturally occurring food mycotoxin ochratoxin A on brain cells in culture. *Neurotoxicology*. 18, 831-839.

[71] Gekle, M., Oberleithner, H., Silbernagl, S., (1993). Ochratoxin A impairs "postproximal" nephron function *in vivo* and blocks plasma membrane anion conductance in Madin-Darby canine kidney cells *in vitro*. *Pflügers Archiv*. 425, 401-408.

[72] Doi, K., Uetsuka, K., (2011). Mechanisms of mycotoxin-induced neurotoxicity through oxidative stress-associated pathways. *International journal of molecular sciences*. 12, 5213-5237.

[73] Müller, G., Kielstein, P., Köhler, H., Berndt, A., Rosner, H., (1995). Studies of the influence of ochratoxin A on immune and defence reactions in the mouse model. *Mycoses*. 38, 85-91.

[74] Klinkert, W., Lorkowski, G., Creppy, E., Dirheimer, G., Röschenthaler, R., (1981). Inhibition of macrophage migration by ochratoxin A and citrinin, and prevention by phenylalanine of the ochratoxin A-induced inhibition. *Toxicol. Eur. Res.* 3, 185.

[75] Chang, C.-F., Hamilton, P.B., (1979). Impaired phagocytosis by heterophils from chickens during aflatoxicosis. *Toxicol. Appl. Pharmacol.* 48, 459-466.

[76] Boorman, G., Hong, H., Dieter, M., Hayes, H., Pohland, A., Stack, M., Luster, M., (1984). Myelotoxicity and macrophage alteration in mice exposed to ochratoxin A. *Toxicology and applied pharmacology*. 72, 304-312.

[77] Huff, W., Ruff, M., (1982). Eimeria acervulina and Eimeria tenella infections in ochratoxin A-compromised broiler chickens. *Poultry science*. 61, 685-692.

[78] Stoev, S., Goundasheva, D., Mirtcheva, T., Mantle, P., (2000). Susceptibility to secondary bacterial infections in growing pigs as an early response in ochratoxicosis. *Experimental and toxicologic pathology*. 52, 287-296.

[79] Elissalde, M., Ziprin, R., Huff, W., Kubena, L., Harvey, R., (1994). Effect of ochratoxin A on Salmonella-challenged broiler chicks. *Poultry science*. 73, 1241-1248.

[80] Fukata, T., Sasai, K., Baba, E., Arakawa, A., (1996). Effect of ochratoxin A on Salmonella typhimurium-challenged layer chickens. *Avian Dis.*, 924-926.

[81] Kumar, A., Jindal, N., Shukla, C., Pal, Y., Ledoux, D., Rottinghaus, G., (2003). Effect of ochratoxin A on Escherichia coli-challenged broiler chicks. *Avian Dis.* 47, 415-424.

[82] Stoev, S., Koynarsky, V., Mantle, P., (2002). Clinicomorphological studies in chicks fed ochratoxin A while simultaneously developing coccidiosis. *Vet. Res. Commun.* 26, 189-204.

[83] Haubeck, H.-D., Lorkowski, G., Kölsch, E., Röschenthaler, R., (1981). Immunosuppression by ochratoxin A and its prevention by phenylalanine. *Applied and Environmental Microbiology*. 41, 1040-1042.

[84] Lea, T., Steien, K., Størmer, F.C., (1989). Mechanism of ochratoxin A-induced immunosuppression. *Mycopathologia*. 107, 153-159.

[85] Dortant, P., Peters-Volleberg, G., Van Loveren, H., Marquardt, R., Speijers, G., (2001). Age-related differences in the toxicity of ochratoxin A in female rats. *Food and Chemical Toxicology*. 39, 55-65.

[86] Chang. C.-F., Doerr, J.A., Hamilton, P.B., (1981). Experimental ochratoxicosis in turkey poults. *Poultry science*. 60, 114-119.

[87] Verma, J., Johri, T., Swain, B., Ameena, S., (2004). Effect of graded levels of aflatoxin, ochratoxin and their combinations on the performance and immune response of broilers. *British poultry science*. 45, 512-518.

[88] Raju, M., Devegowda, G., (2002). Esterified-glucomannan in broiler chicken diets-contaminated with aflatoxin, ochratoxin and T-2 toxin: Evaluation of its binding ability (*in vitro*) and efficacy as immunomodulator. *Asian Australasian Journal of Animal Sciences*. 15, 1051-1056.

[89] Campbell, M., May, J., Huff, W., Doerr, J., (1983). Evaluation of immunity of young broiler chickens during simultaneous aflatoxicosis and ochratoxicosis. *Poultry Science*. 62, 2138-2144.

[90] Kanisawa, M., Suzuki, S., Kozuka, Y., Yamazaki, M., (1977). Histopathological studies on the toxicity of ochratoxin A in rats. I. Acute oral toxicity. *Toxicology and applied pharmacology*. 42, 55-64.

[91] Alvarez, L., Gil, A., Ezpeleta, O., Garcıa-Jalón, J., De Cerain, A.L., (2004). Immunotoxic effects of ochratoxin A in Wistar rats after oral administration. *Food and chemical toxicology*. 42, 825-834.

[92] Kitchen, D., Carlton, W., Tuite, J., (1977). Ochratoxin A and Citrinin Induced Nephrosis in Beagle Dogs II. Pathology. *Veterinary Pathology Online*. 14, 261-272.

[93] Singh, G., Chauhan, H., Jha, G., Singh, K., (1990). Immunosuppression due to chronic ochratoxicosis in broiler chicks. *Journal of comparative pathology*. 103, 399-410.

[94] Stoev, S., Anguelov, G., Ivanov, I., Pavlov, D., (2000). Influence of ochratoxin A and an extract of artichoke on the vaccinal immunity and health in broiler chicks. *Experimental and toxicologic pathology*. 52, 43-55.

[95] Stoev, S.D., Djuvinov, D., Mirtcheva, T., Pavlov, D., Mantle, P., (2002). Studies on some feed additives giving partial protection against ochratoxin A toxicity in chicks. *Toxicology letters*. 135, 33-50.

[96] Szczech, G., Carlton, W., Tuite, J., (1973). Ochratoxicosis in beagle dogs. 1. Clinical and clinicopathological features. *Veterinary Pathology Online*. 10, 135-154.

[97] Kozaczyński, W., (1993). Experimental ochratoxicosis A in chickens. Histopathological and histochemical study. *Archivum veterinarium Polonicum/Polish Academy of Sciences, Committee of Veterinary Sciences*. 34, 205-219.

[98] Thuvander, A., Breitholtz-Emanuelsson, A., Olsen, M., (1995). Effects of ochratoxin A on the mouse immune system after subchronic exposure. *Food Chem. Toxicol.* 33, 1005-1011.

[99] Boorman, G.A. (1989), *NIH publication*. Department of Health and Human Services, Vol. 89.

[100] Aleksandrov, M., Dzhurov, A., (1986). [Effect of ochratoxin on the health status of broilers]. *Veterinarno-meditsinski nauki*. 24, 38-43.

[101] Stoev, S.D., Daskalov, H., Radic, B., Domijan, A., Peraica, M., (2002). Spontaneous mycotoxic nephropathy in Bulgarian chickens with unclarified mycotoxin aetiology. *Vet. Res.* 33, 83-94.

[102] Johannisson, A., Thuvander, A., Gadhasson, I., (1995). Activation markers and cell proliferation as indicators of toxicity: a flow cytometric approach. *Cell Biol. Toxicol.* 11, 355-366.

[103] Sauvant, C., Holzinger, H., Gekle, M., (2005). The nephrotoxin ochratoxin A induces key parameters of chronic interstitial nephropathy in renal proximal tubular cells. *Cell. Physiol. Biochem.* 15, 125-134.

[104] Ul-Hassan, Z., Zargham Khan, M., Khan, A., Javed, I., (2012). Immunological status of the progeny of breeder hens kept on ochratoxin A (OTA)-and aflatoxin B1 (AFB1)-contaminated feeds. *J. Immunotoxicol.* 9, 381-391.

[105] Størmer, F.C., Lea, T., (1995). Effects of ochratoxin A upon early and late events in human T-cell proliferation. *Toxicology*. 95, 45-50.

[106] Grissmer, S., Lewis, R.S., Cahalan, M.D., (1992). Ca (2+)-activated K+ channels in human leukemic T cells. *The Journal of general physiology*. 99, 63-84.

[107] Prior, M., Sisodia, C., (1982). The effects of ochratoxin A on the immune response of Swiss mice. *Can. J. Comp. Med.* 46, 91.

[108] Hong, H.H.L., Jameson, C.W., Boorman, G.A., (1988). Residual hematropoietic effect in mice exposed to ochratoxin a prior to irradiation. *Toxicology*. 53, 57-67.

[109] Odhav, B., Adam, J.K., Bhoola, K.D., (2008). Modulating effects of fumonisin B1 and ochratoxin A on leukocytes and messenger cytokines of the human immune system. *Int. Immunopharmacol.* 8, 799-809.

[110] Hedenborg, M., Klockars, M., (1987). Production of reactive oxygen metabolites induced by asbestos fibres in human polymorphonuclear leucocytes. *J. Clin. Pathol.* 40, 1189-1193.

[111] Perrone, M.G., Gualtieri, M., Consonni, V., Ferrero, L., Sangiorgi, G., Longhin, E., Ballabio, D., Bolzacchini, E., Camatini, M., (2013). Particle size, chemical composition, seasons of the year and urban, rural or remote site origins as determinants of biological effects of particulate matter on pulmonary cells. *Environmental Pollution*. 176, 215-227.

[112] AL - ANATI, L., Petzinger, E., (2006). Immunotoxic activity of ochratoxin A. *J. Vet. Pharmacol. Ther.* 29, 79-90.

[113] Lawrence, T., (2009). The nuclear factor NF-κB pathway in inflammation. *Cold Spring Harb. Perspect. Biol.* 1, a001651.

[114] Hong, J.T., Lee, M.K., Park, K.S., Jung, K.M., Lee, R.D., Jung, H.K., Park, K.L., Yang, K.H., Chung, S.Y., (2002). Inhibitory effect of peroxisome proliferator-activated receptor gamma agonist on ochratoxin A-induced cytotoxicity and activation of transcription factors in cultured rat embryonic midbrain cells. *Journal of Toxicology and Environmental Health Part A.* 65, 407-418.

[115] Ramyaa, P., Padma, V.V., (2014). Quercetin modulates OTA-induced oxidative stress and redox signalling in HepG2 cells —up regulation of Nrf2 expression and down regulation of NF-κB and COX-2. *Biochimica et Biophysica Acta (BBA)-General Subjects.* 1840, 681-692.

[116] Weidenbach, A., Schuh, K., Failing, K., Petzinger, E., (2000). Ochratoxin A induced TNFα release from the isolated and blood-free perfused rat liver. *Mycotoxin research.* 16, 189-193.

[117] Al-Anati, L., Katz, N., Petzinger, E., (2005). Interference of arachidonic acid and its metabolites with TNF-α release by ochratoxin A from rat liver. *Toxicology.* 208, 335-346.

[118] Essid, E., Dernawi, Y., Petzinger, E., (2012). Apoptosis induction by OTA and TNF-α in cultured primary rat hepatocytes and prevention by silibinin. *Toxins (Basel).* 4, 1139-1156.

[119] Mechoud, M.A., Juarez, G.E., de Valdez, G.F., Rodriguez, A.V., (2012). Lactobacillus reuteri CRL 1098 and Lactobacillus acidophilus CRL 1014 differently reduce *in vitro* immunotoxic effect induced by Ochratoxin A. *Food Chem. Toxicol.* 50, 4310-4315.

[120] Dinarello, C.A., (2009). Immunological and inflammatory functions of the interleukin-1 family. *Annu. Rev. Immunol.* 27, 519-550.

[121] Cremer, B., Soja, A., Sauer, J.-A., Damm, M., (2012). Pro-inflammatory effects of ochratoxin A on nasal epithelial cells. *Eur. Arch. Otorhinolaryngol.* 269, 1155-1161.

[122] Oldreive, C., Rice-Evans, C., (2001). The mechanisms for nitration and nitrotyrosine formation *in vitro* and *in vivo*: impact of diet. *Free Radic. Res.* 35, 215-231.

[123] Divya, A., Suresh, K., Naresh, J., (2006). Ochratoxin A feeding to broiler chickens causes reduction in nitric oxide production in peripheral blood mononuclear cells following immunization with S. gallinarum vaccinal preparation. *J Immunol Immunopathol.* 8, 125.

[124] Cavin, C., Delatour, T., Marin-Kuan, M., Fenaille, F., Holzhäuser, D., Guignard, G., Bezençon, C., Piguet, D., Parisod, V., Richoz-Payot, J., (2009). Ochratoxin A-mediated DNA and protein damage: roles of nitrosative and oxidative stresses. *Toxicol. Sci.*, kfp090.

[125] Marin, D.E., Taranu, I., (2014). Ochratoxin A and its effects on immunity. *Toxin Rev.* 34, 11-20.

[126] Myllynen, P., Vähäkangas, K., (2013). Placental transfer and metabolism: An overview of the experimental models utilizing human placental tissue. *Toxicol. In vitro.* 27, 507-512.

[127] Woo, C.S.J., Partanen, H., Myllynen, P., Vähäkangas, K., El-Nezami, H., (2012). Fate of the teratogenic and carcinogenic ochratoxin A in human perfused placenta. *Toxicol. Lett.* 208, 92-99.

[128] Petkova-Bocharova, T., Stoichev, I., Chernozemsky, I., Castegnaro, M., Pfohl-Leszkowicz, A., (1998). Formation of DNA adducts in tissues of mouse progeny through transplacental contamination and/or lactation after administration of a single dose of ochratoxin A to the pregnant mother. *Environ. Mol. Mutagen.* 32, 155-162.

[129] Petkova-Bocharova, T., Pfohl-Leszkowicz, A., Hadjiolov, D., Spirov, K., (1998). Investigation on the genotoxicity of ochratoxin A by measurement of DNA-adducts in different organs of Syrian hamsters after chronic intoxication. *Rev. Med. Vet. (Toulouse).* 149.

[130] Minervini, F., Giannoccaro, A., Nicassio, M., Panzarini, G., Lacalandra, G.M., (2013). First evidence of placental transfer of ochratoxin A in horses. *Toxins (Basel).* 5, 84-92.

[131] Appelgren, L.-E., Arora, R., (1983). Distribution studies of 14 C-labelled aflatoxin B 1 and ochratoxin A in pregnant mice. *Veterinary research communications.* 7, 141-144.

[132] Woo, C.S.J., Wan, M.L.Y., Ahokas, J., El-Nezami, H., (2013). Potential endocrine disrupting effect of ochratoxin A on human placental 3β-hydroxysteroid dehydrogenase/isomerase in JEG-3 cells at levels relevant to human exposure. *Reproductive Toxicology.* 38, 47-52.

[133] Wangikar, P., Dwivedi, P., Sinha, N., Sharma, A., Telang, A., (2005). Teratogenic effects in rabbits of simultaneous exposure to ochratoxin A and aflatoxin B 1 with special reference to microscopic effects. *Toxicology.* 215, 37-47.

[134] Wijesinha, S., Piterman, L., Kirby, C., (2013). The male reproductive system: An overview of common problems. *Aust. Fam. Physician.* 42, 276.

[135] Verma, R., Chakraborty, D., (2008). Emblica Officinalis Aqueous Extract Ameliorates Ochratox-In-Induced Lipid Peroxidation In The Testis Of Mice.

[136] Farag, I., Abdel-Aziz, K., Nada, S., Tawfek, N., Farouk, T., Darwish, H., (2010). Modulation of ochratoxin-induced oxidative stress, genotoxicity and spermatotoxic alterations by Lactobacillus rhamnosus GG in male albino mice. *J. Am. Sci.* 6, 575-587.

[137] Chakraborty, D., Verma, R., (2009). Spermatotoxic effect of ochratoxin and its amelioration by Emblica officinalis aqueous extract. *Acta poloniac pharmaccutica-Drug Res.* 66, 689-695.

[138] Biró, K., Barna-Vetró, I., Pécsi, T., Szabó, E., Winkler, G., Fink-Gremmels, J., Solti, L., (2003). Evaluation of spermatological parameters in ochratoxin A—Challenged boars. *Theriogenology.* 60, 199-207.

[139] Solti, L., Pécsi, T., Barna-Vetró, I., Szász, F., Biró, K., Szabó, E., (1999). Analysis of serum and seminal plasma after feeding ochratoxin A with breeding boars. *Animal reproduction science.* 56, 123-132.

[140] Huyghe, E., Plante, P., Thonneau, P.F., (2007). Testicular cancer variations in time and space in Europe. *Eur. Urol.* 51, 621-628.

[141] Ueta, E., Kodama, M., Sumino, Y., Kurome, M., Ohta, K.i., Katagiri, R.i., Naruse, I., (2010). Gender - dependent differences in the incidence of ochratoxin A - induced neural tube defects in the Pdn/Pdn mouse. *Congenital anomalies.* 50, 29-39.

[142] Krentz, A.D., Murphy, M.W., Kim, S., Cook, M.S., Capel, B., Zhu, R., Matin, A., Sarver, A.L., Parker, K.L., Griswold, M.D., (2009). The DM domain protein DMRT1 is

a dose-sensitive regulator of fetal germ cell proliferation and pluripotency. *Proceedings of the National Academy of Sciences*. 106, 22323-22328.

[143] Turnbull, C., Rapley, E.A., Seal, S., Pernet, D., Renwick, A., Hughes, D., Ricketts, M., Linger, R., Nsengimana, J., Deloukas, P., (2010). Variants near DMRT1, TERT and ATF7IP are associated with testicular germ cell cancer. *Nat. Genet.* 42, 604-607.

[144] Mantle, P.G., Nolan, C.C., (2010). Pathological outcomes in kidney and brain in male Fischer rats given dietary ochratoxin A, commencing at one year of age. *Toxins (Basel)*. 2, 1100-1110.

[145] Schwartz, G.G., Manderville, R.A., Pfohl-Leszkowicz, A., (2010). Response to comments of Peter G. Mantle. *Toxins*. 2, 2337-2339.

[146] Jennings-Gee, J.E., Tozlovanu, M., Manderville, R., Miller, M.S., Pfohl-Leszkowicz, A., Schwartz, G.G., (2010). Ochratoxin A: In utero exposure in mice induces adducts in testicular DNA. *Toxins*. 2, 1428-1444.

[147] Ballinger, M. B., Phillips, T. D., Kubena, L. F., (1986). Assessment of the Distribution and Elimination Of Ochratoxin A In The Pregnant Rat1. *Journal of Food Safety*. 8, 11-24.

[148] Fenske, M., Fink-Gremmels, J., (1990). Effects of fungal metabolites on testosterone secretion *in vitro*. *Archives of toxicology*. 64, 72-75.

[149] More, J., Camguilhem, R., (1979). Effects of low doses of ochratoxin A after intratesticular injection in the rat. *Experientia*. 35, 890-892.

[150] Bouaziz, C., Martel, C., Golli, E.E., Abid - Essefi, S., Brenner, C., Lemaire, C., Bacha, H., (2011). Molecular events involved in ochratoxin A induced mitochondrial pathway of apoptosis, modulation by Bcl - 2 family members. *Environmental toxicology*. 26, 579-590.

[151] Abdel-Wahhab, M., Nada, S., Arbid, M., (1999). Ochratoxicosis: prevention of developmental toxicity by L-methionine in rats. *Journal of Applied Toxicology*. 19, 7-12.

[152] Gupta, M., Bandyopadhyay, S., Mazumdar, S., Paul, B., (1980). Ovarian steroidogenesis in rats following ochratoxin A treatment. *Toxicology and applied pharmacology*. 53, 515-520.

[153] Castegnaro, M., Mohr, U., Pfohl-Leszkowicz, A., Esteve, J., Steinmann, J., Tillmann, T., Michelon, J., Bartsch, H., (1998). Sex- and strain-specific induction of renal tumors by ochratoxin A in rats correlates with DNA adduction. *International Journal of Cancer*. 77, 70-75.

[154] Gharbi, A., Trillon, O., Betbeder, A., Counord, J., Gauret, M., Pfohl-Leszkowicz, A., Dirheimer, G., Creppy, E., (1993). Some effects of ochratoxin A, a mycotoxin contaminating feeds and food, on rat testis. *Toxicology*. 83, 9-18.

[155] Malir, F., Ostry, V., Pfohl-Leszkowicz, A., Novotna, E., (2013). Ochratoxin A: developmental and reproductive toxicity—an overview. *Birth Defects Research Part B: Developmental and Reproductive Toxicology*. 98, 493-502.

[156] Zinedine, A., Soriano, J.M., Molto, J.C., Manes, J., (2007). Review on the toxicity, occurrence, metabolism, detoxification, regulations and intake of zearalenone: an oestrogenic mycotoxin. *Food Chem. Toxicol.* 45, 1-18.

[157] Thomas, J. L., Myers, R.P., Strickler, R.C., (1989). Human placental 3β-hydroxy-5-ene-steroid dehydrogenase and steroid 5→ 4-ene-isomerase: purification from mitochondria

and kinetic profiles, biophysical characterization of the purified mitochondrial and microsomal enzymes. *J. Steroid Biochem.* 33, 209-217.

[158] Nagar, I., (2011). Experimentally Induced Toxicity of Ochratoxin A and Endosulfan in Male Wistar Rats: A Hormonal Disorder 1Shashi Nandar Kumar, 2Avinash Gopal Telang, 2Karan Pal Singh, 1Arun Kumar Iain, 3leasarrat Afroz and 4Raj endra Damu Patil. *Journal ofAnimal and Veterinary Advances.* 10, 1750-1755.

[159] Shah, A.S.V., Langrish, J.P., Nair, H., McAllister, D.A., Hunter, A.L., Donaldson, K., Newby, D.E., Mills, N.L., (2013). Global association of air pollution and heart failure: a systematic review and meta-analysis. *Lancet.* 382, 1039-1048.

[160] Cesaroni, G., Badaloni, C., Gariazzo, C., Stafoggia, M., Sozzi, R., Davoli, M., Forastiere, F., (2013). Long-Term Exposure to Urban Air Pollution and Mortality in a Cohort of More than a Million Adults in Rome. *Environmental Health Perspectives.* 121, 324-331.

[161] Correia, A.W., Pope, C.A., III, Dockery, D.W., Wang, Y., Ezzati, M., Dominici, F., (2013). Effect of Air Pollution Control on Life Expectancy in the United States An Analysis of 545 US Counties for the Period from 2000 to 2007. *Epidemiology.* 24, 23-31.

[162] Chen, R., Deng, L., Yu, X., Wang, X., Zhu, L., Yu, T., Zhang, Y., Zhou, B., Xu, W., Chen, L., (2015). *MiR-122* partly mediates the Ochratoxin A-induced GC-2 cell apoptosis. *Toxicology in vitro.*

[163] Hofmann, M.-C., Hess, R.A., Goldberg, E., Millan, J.L., (1994). Immortalized germ cells undergo meiosis *in vitro. Proceedings of the National Academy of Sciences.* 91, 5533-5537.

[164] Bouhallier, F., Allioli, N., Lavial, F., Chalmel, F., Perrard, M.-H., Durand, P., Samarut, J., Pain, B., Rouault, J.-P., (2010). Role of miR-34c microRNA in the late steps of spermatogenesis. *Rna.* 16, 720-731.

[165] Huszar, J.M., Payne, C.J., (2013). MicroRNA 146 (Mir146) modulates spermatogonial differentiation by retinoic acid in mice. *Biology of reproduction.* 88, 15.

[166] Teng, Y., Wang, Y., Fu, J., Cheng, X., Miao, S., Wang, L., (2011). Cyclin T2: a novel miR-15a target gene involved in early spermatogenesis. *FEBS letters.* 585, 2493-2500.

[167] Liu, T., Huang, Y., Liu, J., Zhao, Y., Jiang, L., Huang, Q., Cheng, W., Guo, L., (2013). MicroRNA-122 influences the development of sperm abnormalities from human induced pluripotent stem cells by regulating TNP2 expression. *Stem cells and development.* 22, 1839-1850.

[168] Lin, C.J.-F., Gong, H.-Y., Tseng, H.-C., Wang, W.-L., Wu, J.-L., (2008). *miR-122* targets an anti-apoptotic gene, Bcl-w, in human hepatocellular carcinoma cell lines. *Biochemical and biophysical research communications.* 375, 315-320.

[169] Gramantieri, L., Ferracin, M., Fornari, F., Veronese, A., Sabbioni, S., Liu, C.-G., Calin, G.A., Giovannini, C., Ferrazzi, E., Grazi, G.L., (2007). Cyclin G1 is a target of *miR-122*a, a microRNA frequently down-regulated in human hepatocellular carcinoma. *Cancer research.* 67, 6092-6099.

Author's Contact Information

Dr. Wentao Xu
Beijing Advanced Innovation Center for Food Nutrition and Human Health,
College of Food Science & Nutritional Engineering,
China Agricultural University,
Beijing, 100083, China
Email: xuwentao@cau.edu.cn

Prof. Dr. Wentao Xu was born in Anhui Province, China on Nov 11th, 1979, obtained Ph. D. in China Agricultural University in 2006. He is currently a principle investigator / doctoral supervisor of the College of Food Science and Nutritional Engineering in China Agricultural University, an Editor in Chief of "Journal of DNA and RNA Research".

INDEX

L

M

N

O

P

Q

R